**Operator Theory**
**Advances and Applications**
**Vol. 66**

**Editor**
**I. Gohberg**

# Differentiable Operators and Nonlinear Equations

Victor Khatskevich
David Shoiykhet

Translated from the Russian by Mircea Martin

Birkhäuser Verlag
Basel · Boston · Berlin

Authors

Victor Khatskevich
Department of Mathematics
University of Haifa
Afula Research Institute
Mount Carmel, Haifa 31905
Israel

David Shoiykhet
Department of Mathematics
International College of Technology
Ort Braude, College Campus
P.O.B. 78
Karmiel 20101
Israel

A CIP catalogue record for this book is available from the Library of Congress, Washington D.C., USA

Deutsche Bibliothek Cataloging-in-Publication Data
**Khatskevich, Victor:**
Differentiable operators and nonlinear equations / Victor Khatskevich ; David Shoiykhet. Transl. from the Russ. by Mircea Martin. - Basel ; Boston ; Berlin : Birkhäuser, 1994
(Operator theory ; Vol. 66)
ISBN 3-7643-2929-7 (Basel ...)
ISBN 0-8176-2929-7 (Boston)
NE: Šoiḥeṭ, Dāwid:; GT


Printed on acid-free paper produced from chlorine-free pulp
Cover design: Heinz Hiltbrunner, Basel
Printed in Germany
ISBN 3-7643-2929-7
ISBN 0-8176-2929-7

9 8 7 6 5 4 3 2 1

# Table of Contents

Introduction

Chapter 0: Preliminaries

1. Sets and relations ........ 1
2. Topological spaces ........ 5
3. Convergence. Directedness ........ 8
4. Metric spaces ........ 13
5. Spaces of mappings ........ 16
6. Linear topological spaces ........ 19
7. Normed spaces ........ 21
8. Linear operators and functionals ........ 24
9. Conjugate space. Conjugate operator ........ 27
10. Weak topology and reflexivity ........ 29
11. Hilbert spaces ........ 31

Chapter I: Differential calculus in normed spaces

1. The derivate and the differential of a nonlinear operator ........ 35
2. Lagrange formula and Lipschitz condition ........ 39
3. Examples of Fréchet differentiable operators ........ 41
4. Lemmas about differentiable operators ........ 43
5. Partial derivatives ........ 45
6. Multilinear operators. Duality. Homogeneous forms ........ 48
7. Higher order derivatives ........ 50
8. Complete continuity of operators and of their derivatives ........ 55

Chapter II: Integration in normed spaces
1. Riemann - Stieltjes integrals of vector-functions ........ 61
2. Pettis integral and the connection with Riemann - Stiltjes integral ..... 65
3. Antiderivatives of vector-functions. Integral representations ........ 66
4. Integrals of operators in Banach spaces ........ 71

Chapter III: Holomorphic (analytic) operators and vector-functions on complex Banach spaces
1. Differentiability in complex and real sense. Cauchy - Riemann conditions ........ 76
2. The $\rho$-topology and holomorphy ........ 80
3. Cauchy integral theorems and their consequences ........ 85
4. Uniqueness theorems and maximum principles
5. Schwartz Lemma and its generalizations ........ 92
6. Uniformly bounded families of $\rho$-holomorphic (holomorphic) operators. Montel property ........ 98

Capter IV: Linear operators
1. The spectrum and the resolvent of a linear operator ........ 103
2. Spectral radius ........ 108
3. Resolvent and spectrum of the adjoint operator ........ 111
4. The spectrum of a completely continuous operator ........ 114
5. Normally solvable operators ........ 117
6. Noether and Fredholm operators ........ 119
7. Projections. Splitable operators ........ 122
8. Invariant subspaces ........ 127

Chapter V: Nonlinear equations with differentiable operators
1. Fixed points. Banach principle ........ 133
2. Non-expansive operators ........ 137
3. Fixed points for differentiable operators ........ 143
4. Some applications of fixed point principle ........ 147
5. Implicite and inverse operators. Connection with fixed points ........ 160

Chapter VI: Nonlinear equations with holomorphic operators
1. s-fixed points for holomorphic operators. A converse of Banach principle ........ 171
2. Criterions for the existence of an s-fixed point and its extension with respect to a parameter ........ 177
3. Regular fixed points. Geometric criterions ........ 182

4. Apriori estimates and the extension of an s-solution to the boundary of the domain . . . 189
5. Local inversion of holomorphic operators and a posteriori error estimates . . . 195
6. Single-valued small solutions in some degenerate cases . . . 200

Chapter VII: Banach manifolds
1. Basic definitions . . . 211
2. Smooth mappings . . . 213
3. Submanifolds . . . 214
4. Complex manifolds and Stein manifolds . . . 218

Chapter VIII: Non-regular solutions of nonlinear equations
1. Ramification of solutions. Statement of the problem . . . 223
2. Equations of ramification . . . 225
3. Equations of ramification for an analytic operator. The problem of the coefficients . . . 231
4. The description of the set of fixed points for an analytic operator . . . 232

Chapter IX: Operators on spaces with indefinite metric
1. Spaces with indefinite metric . . . 239
2. Angle operators . . . 242
3. Plus-operators . . . 244
4. Symmetric properties of a plus-operator and its adjoint . . . 249
5. The problem of invariant semi-definite subspaces . . . 258
6. An application of fixed point principles for holomorphic operators to the invariant semi-definite subspace problem . . . 262

References . . . 267

List of Symbols . . . 277

Subject Index . . . 279

# Introduction

We have considered writing the present book for a long time, since the lack of a sufficiently complete textbook about complex analysis in infinite dimensional spaces was apparent. There are, however, some separate topics on this subject covered in the mathematical literature. For instance, the elementary theory of holomorphic vector-functions and mappings on Banach spaces is presented in the monographs of E. Hille and R. Phillips [1] and L. Schwartz [1], whereas some results on Banach algebras of holomorphic functions and holomorphic operator-functions are discussed in the books of W. Rudin [1] and T. Kato [1].

Apparently, the need to study holomorphic mappings in infinite dimensional spaces arose for the first time in connection with the development of nonlinear analysis. A systematic study of integral equations with an analytic nonlinear part was started at the end of the 19th and the beginning of the 20th centuries by A. Liapunov, E. Schmidt, A. Nekrasov and others. Their research work was directed towards the theory of nonlinear waves and used mainly the undetermined coefficients and the majorant power series methods. The most complete presentation of these methods comes from N. Nazarov.

In the forties and fifties the interest in Liapunov's and Schmidt's analytic methods diminished temporarily due to the appearence of variational calculus methods (M. Golomb, A. Hammerstein and others) and also to the rapid development of the mapping degree theory (J. Leray, J. Schauder, G. Birkhoff, O. Kellog and others). These new methods were particularly attractive since they enabled the study of many classes of nonlinear equations, and therefore they were highly developed. (Important results were obtained by M. Krasnoselski, P. Zabreiko, V. Odinetz, Yu. Borisovich and B. Sadovski.) However, these new techniques retarded the development of specific methods for solving equations with an analytic nonlinear part. That is why in the sixties some mathematicians (P. Rybin, V. Pokornyi, M. Vainberg, V. Trenogyn and others) interested in the theory of integral equations and their applications returned to the Liapunov-Schmidt and Nekrasov-Nazarov analytic methods.

At the same time the theory of functions of one or several complex variables was enriched with more significant and subtle results. Parallel with these achievements, the first results on holomorphic mappings on infinite dimensional spaces appeared in the works of A. Cartan, R. Phillips, L. Nachbin, L.Harris, T. Suffridge, W.Rudin, M. Herve, E. Vesentini, J.-P. Vigue, P. Mazet, K. Goebel, and of many others.

We consider that it is now about the right time "to set a bridge" between nonlinear analysis and the theory of holomorphic mappings on infinite dimensional spaces. Of course, to this end it is necessary to put together results and techniques from the homology theory, sheaf theory, vector fields theory and from a lot of other modern theories in analysis — a task difficult to achieve within the limits of but one book. That is why we decided to start this vast project, by presenting only the theory of differentiable and holomorphic mappings on Banach spaces, as well as some prerequisites from functional analysis and topology.

In all chapters with the exception of Chapter 0 which has the character of a dictionary, we tried to give a complete account of definitions and proofs, and to make this book accesible not only to specialists, but also to students and to those engineers who are currently using the solutions of some specific integral and differentiable equations.

We conclude the work by mentioning the interesting relationship between the theory of holomorphic mappings and the theory of linear operators on spaces with indefinite metrics. More precisely, our last chapter is a brief exposition of the theory of spaces with indefinite metrics and of some relevant applications of the holomorphic mappings theory in this setting.

In closing, we draw our readers to a few technical points. Throughout the book we strove to use a uniform notation for objects of the same type. The most used notations are presented in Chapter 0. At the end of the book we give a list of some standard symbols, and also a subject index. We used the symbols "$\lhd$" and "$\blacktriangleright$" for the beginning and the end of a proof, respectively. The references in the text contain the name(s) of the author(s) followed by a number in brackets, which corresponds to the one in the reference list.

*THE AUTHORS*

# Chapter 0

# Preliminaries

This is an auxiliary chapter: we collect here the basic definitions and theorems which will be used in what follows. Accordingly, some results are mentioned here merely for a relative completeness and orderliness of the presentation, and not only for pure pragmatic reasons.

## §1. Sets and relations

**1.** We will work with sets and with their elements. For us the words "set", "class", "collection", "family" will have the same meaning. Generally, we denote the sets by Gothic capitals, and their elements or — as one also says — their points, by small Latin characters. The fact that an element $x$ belongs to a set $\mathfrak{X}$ is denoted by $x \in \mathfrak{X}$. We write $\mathfrak{X} = \mathfrak{Y}$ when $x \in \mathfrak{X}$ if and only if $x \in \mathfrak{Y}$. The notation $\{x : x \in \mathfrak{X}\}$ means that we consider all elements of the set $\mathfrak{X}$. Any (not necessarily proper) part of a set is called a *subset*. The fact that a set $\mathfrak{A}$ is a subset of the set $\mathfrak{X}$ is denoted by $\mathfrak{A} \subseteq \mathfrak{X}$. In this respect, the *empty set* $\emptyset$ is considered a subset of any set. As usually, the *union* and the *intersection* of the sets $\mathfrak{X}$ and $\mathfrak{Y}$ are denoted by $\mathfrak{X} \cup \mathfrak{Y}$ and $\mathfrak{X} \cap \mathfrak{Y}$, respectively. The *difference* of the sets $\mathfrak{X}$ and $\mathfrak{Y}$ (or the *complement* of $\mathfrak{Y}$ in $\mathfrak{X}$) is denoted by $\mathfrak{X} \backslash \mathfrak{Y}$. Union and intersection are dual operations by de Morgan formulas:

$$\mathfrak{X} \backslash (\mathfrak{A} \cup \mathfrak{B}) = (\mathfrak{X} \backslash \mathfrak{A}) \cap (\mathfrak{X} \backslash \mathfrak{B}), \quad \mathfrak{X} \backslash (\mathfrak{A} \cap \mathfrak{B}) = (\mathfrak{X} \backslash \mathfrak{A}) \cup (\mathfrak{X} \backslash \mathfrak{B}), \tag{1.1}$$

where $\mathfrak{X}$ is a set, and $\mathfrak{A}$ and $\mathfrak{B}$ are subsets of $\mathfrak{X}$.

De Morgan formulas are valid not only for two, but for any family of subsets of the set $\mathfrak{X}$.

The *direct* (or *Cartezian*) *product* $\mathfrak{X} \times \mathfrak{Y}$ of two sets $\mathfrak{X}$ and $\mathfrak{Y}$ is the set of all ordered pairs $(x, y)$, where $x \in \mathfrak{X}$ and $y \in \mathfrak{Y}$.

**2.** Any subset of the direct product $\mathfrak{X} \times \mathfrak{Y}$ is called a *relation* between the sets $\mathfrak{X}$ and $\mathfrak{Y}$. If $\mathcal{R}$ is a relation, then the notations $x\mathcal{R}y$ and $(x, y) \in \mathcal{R}$ are equivalent; in this case we are saying that $x$ is in the relation $\mathcal{R}$ with $y$. The *domain* of a relation $\mathcal{R}$ is the set $D(\mathcal{R})$ of all first coordinates of the pairs in $\mathcal{R}$, and the *range* of $\mathcal{R}$ is the set $R(\mathcal{R})$ of their second coordinates.

The *inverse* of the relation $\mathcal{R}$ is denoted by $\mathcal{R}^{-1}$ and is defined by $\mathcal{R}^{-1} = \{(y, x) : (x, y) \in \mathcal{R}\}$; it follows that $x\mathcal{R}y$ if and only if $y\mathcal{R}^{-1}x$.

The *composition*, or the *product*, of the relations $\mathcal{R}$ and $\mathcal{S}$ is the relation $\mathcal{R} \circ \mathcal{S}$ (denoted also by $\mathcal{RS}$) consisting of all pairs $(x, z)$ such that there exists an $y$ with $(x, y) \in \mathcal{S}$ and $(y, z) \in \mathcal{R}$. Composition is not commutative, since $\mathcal{R} \circ \mathcal{S}$ and $\mathcal{S} \circ \mathcal{R}$ are, in general, different.

The *identity relation* on a set $\mathfrak{X}$ (or the *identity* on $\mathfrak{X}$) is the set of pairs of the form $(x, x)$ with $x \in \mathfrak{X}$. It is denoted by $I_{\mathfrak{X}}$ or $Id_{\mathfrak{X}}$ and it has the property that $\mathcal{R} \circ I_{\mathfrak{X}} = I_{\mathfrak{X}} \circ \mathcal{R} = \mathcal{R}$ for any relation with the domain and the range included in $\mathfrak{X}$. (The subscript $\mathfrak{X}$ could by omitted when no confusions arise.) It is easy to see that $\mathcal{R}^{-1} \circ \mathcal{R} \supseteq I_{D(\mathcal{R})}$, $\mathcal{R} \circ \mathcal{R}^{-1} \supseteq I_{R(\mathcal{R})}$ for any relation $\mathcal{R}$.

The next result is useful when working with relations:

**THEOREM 1.1.** *Let $\mathcal{R}, \mathcal{S}$ and $\mathcal{T}$ be relations. Then:*

(a) $(\mathcal{R}^{-1})^{-1} = \mathcal{R}$ *and* $(\mathcal{R} \circ \mathcal{S})^{-1} = \mathcal{S}^{-1} \circ \mathcal{R}^{-1}$;

(b) $\mathcal{R} \circ (\mathcal{S} \circ \mathcal{T}) = (\mathcal{R} \circ \mathcal{S}) \circ \mathcal{T}$.

Let us consider now some special types of relations, which are very important in all mathematics.

A relation $\mathcal{R}$ is *reflexive* if $x\mathcal{R}x$; in other words $I \subseteq \mathcal{R}$, that is, $\mathcal{R}$ contains the diagonal. A relation $\mathcal{R}$ is called *symmetric* if $x\mathcal{R}y$ always implies $y\mathcal{R}x$, that is, $\mathcal{R} = \mathcal{R}^{-1}$. On the other hand, $\mathcal{R}$ is called *antisymmetric* if $x\mathcal{R}y$ and $y\mathcal{R}x$ do not hold simultaneously, that is, $\mathcal{R} \cap \mathcal{R}^{-1} = \emptyset$. A relation $\mathcal{R}$ is called *transitive* if $x\mathcal{R}y$ and $y\mathcal{R}z$ imply $x\mathcal{R}z$. In terms of composition, $\mathcal{R}$ is transitive if and only if $\mathcal{R} \circ \mathcal{R} \subseteq \mathcal{R}$. In this case $\mathcal{R}^{-1} \circ \mathcal{R}^{-1} = (\mathcal{R} \circ \mathcal{R})^{-1} \subseteq \mathcal{R}^{-1}$, so the inverse of a transitive relation is also transitive. If $\mathcal{R}$ is transitive and reflexive, then $\mathcal{R} \circ \mathcal{R} = \mathcal{R}$.

A transitive relation on a set $\mathfrak{X}$ is called an *ordering* (*partial ordering*, *order relation*) on $\mathfrak{X}$; it is often denoted by $\geqslant$. When $x \geqslant y$ one usually says that $x$ follows $y$ ($y$ precedes $x$), or that $x$ is greater than $y$ ($y$ is smaller than $x$).

Let $\mathfrak{A}$ be a subset of a set $\mathfrak{X}$ ordered by the relation $\geqslant$. An element $x \in \mathfrak{A}$ is called *maximal* (in $\mathfrak{A}$) if for all $y \in \mathfrak{A}$, either $x$ follows $y$, or $x$ does not precede $y$. In the latter case, $x$ is not connected with $y$ by the relation $\mathcal{R}$, or, as one says, $x$ is not comparable with $y$. An element $x \in \mathfrak{X}$ is called an *upper bound* of the set $\mathfrak{A}$ if for any $y \in \mathfrak{A}$, either $x \geqslant y$ or $x = y$.

The set $\mathfrak{A}$ is called *linearly ordered* if all its elements are comparable, that is, for any $x, y \in \mathfrak{A}$ either $x \geqslant y$ or $y \leqslant x$, and, in addition, from $x \geqslant y$ and $y \geqslant x$ it follows that $x = y$.

**LEMMA 1.1** (Zorn). *If any linearly ordered subset of a partially ordered set $\mathfrak{X}$ has an upper bound, then $\mathfrak{X}$ has a maximal element.*

A *direction* in $\mathfrak{X}$ is a reflexive order relation which fulfils the condition:

(a) for any $x, y \in \mathfrak{X}$ there exists $z \in \mathfrak{X}$ such that $z \geqslant x$ and $z \geqslant y$.

A nonempty subset $\mathfrak{D}$ with a given direction $\geqslant$ on it is called a *directed set* and is denoted by $(\mathfrak{D}, \geqslant)$.

An *equivalence relation* (or simply an *equivalence*) on $\mathfrak{X}$ is a reflexive, symmetric and transitive relation.

**3.** We consider now a special class of relations called mappings.

A *mapping* is a single-valued relation. That is, a relation $\mathcal{R}$ between the sets $\mathfrak{X}$ and $\mathfrak{Y}$ is a mapping if $x\mathcal{R}y$ and $x\mathcal{R}z$ imply $y = z$, where $x \in \mathfrak{X}$, $y, z \in \mathfrak{Y}$. We make no distinction between a mapping and its graph. For us the terms correspondence, transformation, mapping, operator have the same meaning. A mapping is more often denoted by the letter $f$. If $f$ is mapping and $x \in D(f)$, then $f(x)$ (or $fx$, or $\langle x, f\rangle$) is the only element $y \in R(f)$ with $xfy$, that is, the element corresponding to $x$ by the relation $f$. The point $f(x)$ is called the *value* of $f$ at the point $x$, or the *image* of $x$ by $f$, while $x$ is a preimage of $y$; one also says that $f$ maps $x$ into $y$. A mapping $f$ is defined *on* $\mathfrak{X}$ if $D(f) = \mathfrak{X}$; when $D(f)$ is only a part of $\mathfrak{X}$, it is said that $f$ is defined *in* $\mathfrak{X}$. If the mapping $f$ is defined in $\mathfrak{X}$ and $R(f) \subseteq \mathfrak{Y}$, then one says, that $f$ acts from $\mathfrak{X}$ into $\mathfrak{Y}$; this is written $f\colon \mathfrak{X} \to \mathfrak{Y}$.

We will use the term *function* for a mapping defined in the set $\mathbb{R}$ of real numbers or in the set $\mathbb{C}$ of complex numbers, and the term *functional* for a mapping which ranges in $\mathbb{R}$ or $\mathbb{C}$.

A mapping $f: \mathfrak{X} \to \mathfrak{Y}$ is called *injective* (or an injection) if it is one-to-one, that is $f(x) = f(y)$ implies $x = y$ for $x, y \in D(f)$. A mapping $f: \mathfrak{X} \to \mathfrak{Y}$ is called *surjective* (or a surjection, or onto) if $R(f) = \mathfrak{Y}$. A mapping defined on $\mathfrak{X}$ which is injective and surjective is called *bijective*, or a bijection. In this case, $f$ is also called an *isomorphism* between $\mathfrak{X}$ and $\mathfrak{Y}$ and we say that $\mathfrak{X}$ and $\mathfrak{Y}$ are isomorphic. A set $\mathfrak{X}$ is called *countable* if either $\mathfrak{X}$ is finite or is isomorphic with the set $\mathbb{N}$ of natural numbers. Notice that any injection is an isomorphism between $D(f)$ and $R(f)$, and that the inverse relation $f^{-1}$ is in this case a mapping from $R(f)$ onto $D(f)$. An isomorphism is often denoted by $\leftrightarrow$.

A mapping is a set, so two mappings $f$ and $g$ coincide if and only if $D(f) = D(g)$ and $f(x) = g(x)$ for all $x \in D(f)$. Consequently, a mapping is defined when its domain and its value at each point of this domain are specified. If $f: \mathfrak{X} \to \mathfrak{Y}$ is a mapping and $\mathfrak{A} \subseteq \mathfrak{X}$, then $f \cap (\mathfrak{A} \times \mathfrak{Y})$ is also a mapping, called the *restriction* of the mapping $f$ to (the set) $\mathfrak{A}$ and denoted by $f \mid \mathfrak{A}$. A mapping $g$ is called an *extension* of the mapping $f$ if $f \subseteq g$. In other words, the mapping $g$ is an extension of the mapping $f$ if and only if $f$ is the restriction of $g$ to a subset of $D(g)$.

Let $f$ be a mapping and $\mathfrak{A} \subseteq D(f)$. We define $f(\mathfrak{A}) = \{y \mid y = f(x) \text{ for some } x \in \mathfrak{A}\}$; the set $f(\mathfrak{A})$ is called the *image* of $\mathfrak{A}$ under $f$. Analogously, if $\mathfrak{B} \subseteq R(f)$, we put $f^{-1}(\mathfrak{B}) = \{x \mid x \in D(f),\ f(x) \in \mathfrak{B}\}$. (Note that $f^{-1}$ is *not* necessarily a mapping.) The set $f^{-1}(\mathfrak{B})$ is called the *preimage* of $\mathfrak{B}$ by $f$. For any two sets $\mathfrak{A}$ and $\mathfrak{B}$ we have $f(\mathfrak{A} \cup \mathfrak{B}) = f(\mathfrak{A}) \cup f(\mathfrak{B})$ and $f(\mathfrak{A} \cap \mathfrak{B}) \subseteq f(\mathfrak{A}) \cap f(\mathfrak{B})$. As far as the preimagine is concerned, we have:

**THEOREM 1.2.** *Let $f: \mathfrak{X} \to \mathfrak{Y}$ be a mapping and $\mathfrak{A}, \mathfrak{B} \subseteq R(f)$. Then:*

(a) $f^{-1}(\mathfrak{A} \setminus \mathfrak{B}) = f^{-1}(\mathfrak{A}) \setminus f^{-1}(\mathfrak{B})$;

(b) $f^{-1}(\mathfrak{A} \cup \mathfrak{B}) = f^{-1}(\mathfrak{A}) \cup f^{-1}(\mathfrak{B})$;

(c) $f^{-1}(\mathfrak{A} \cap \mathfrak{B}) = f^{-1}(\mathfrak{A}) \cap f^{-1}(\mathfrak{B})$.

Rules (b) and (c) can be extended for the union and the intersection of an arbitrary family of sets.

The composition of two mappings is still a mapping. For any mapping $f$ the composition $f^{-1} \circ f$ is an equivalence relation, and $f \circ f^{-1}$ is the identity of $R(f)$.

**4.** We close this paragraph by pointing out how the notion of mapping can be used to define the direct product of an arbitrary family of factors.

Let $\mathfrak{A}$ be an arbitrary set, called in what follows the *indexing set*. Its elements are called *indexes,* and for any $a \in \mathfrak{A}$ let $\mathfrak{X}_a$ be a given set. The *direct product* of the sets $\mathfrak{X}_a$, denoted by $\prod\limits_{a\in\mathfrak{A}} \mathfrak{X}_a$, is defined as the set of all mappings $x$ defined on $\mathfrak{A}$, such that $x(a) \in \mathfrak{X}_a$ for each $a \in \mathfrak{A}$. In this context, the argument is usually written as index; accordingly, $\prod\limits_{a\in\mathfrak{A}} \mathfrak{X}_a = \{x \mid x$ is a mapping defined on $\mathfrak{A}$ such that $x_a \in \mathfrak{X}_a$ for all $a \in \mathfrak{A}\}$. The set $\mathfrak{X}_a$ is called the $a$-th *coordinate set* and the point $x_a$ is called the $a$-th *coordinate* of the point $x$ in the product. The mapping $P_a$, which maps each point $x$ into its $a$-th coordinate $x_a$ is called the *projector*, or the *projection operator*, or the *coordinate projector* on the $a$-th coordinate set.

Suppose that all sets $\mathfrak{X}_a$ are nonempty. Is their direct product nonempty? Zorn lemma implies a positive answer to this question.

**AXIOM OF CHOICE.** *Let $\mathfrak{X}_a$ be a nonempty set for each element $a$ in an indexing set $\mathfrak{A}$. Then there exists a mapping $c$ on $\mathfrak{A}$ such that $c(a) \in \mathfrak{X}_a$ for any $a \in \mathfrak{A}$.*

The case of equal factors is of special interest. Let us assume that each coordinate set $\mathfrak{X}_a$ coincides with a fixed set $\mathfrak{Y}$. Then $\prod\limits_{a\in\mathfrak{A}} \mathfrak{X}_a = \prod\limits_{a\in\mathfrak{A}} \mathfrak{Y} = \{x \mid x : \mathfrak{A} \to \mathfrak{Y}$ is a mapping$\}$. In other words, in this case the direct product is exactly the set of all mappings from $\mathfrak{A}$ into $\mathfrak{Y}$; this is often denoted by $\mathfrak{Y}^{\mathfrak{A}}$. An example is the $n$--dimensional Euclidean space.

## §2. Topological spaces

**1.** Let $\mathfrak{X}$ be a set. A *topology* on $\mathfrak{X}$ is a nonempty family $\tau$ of subsets of the set $\mathfrak{X}$, satisfying three conditions:

a) the intersection of any two elements of $\tau$ is an element of $\tau$;

b) the union of the elements of any subfamily of the family $\tau$ belongs to $\tau$;

c) the set $\mathfrak{X}$ is contained in the union of the elements of the family $\tau$, i.e., $\mathfrak{X} \subseteq \subseteq \bigcup\{\mathfrak{U} \mid \mathfrak{U} \in \tau\}$.

The last condition is obviously equivalent to the equality $\mathfrak{X} = \bigcup\{\mathfrak{U} \mid \mathfrak{U} \in \tau\}$ and therefore (using (b)) $\mathfrak{X} \in \tau$.

The set $\mathfrak{X}$ is called the *space* of the topology $\tau$ and the pair $(\mathfrak{X}, \tau)$ is called a *topological space.* When no confusions arise, we simply write "$\mathfrak{X}$ is a topological space". The elements of the topology $\tau$ are called $\tau$-*open* (or simply *open*) subsets.

The set $\mathbb{R}$ ($\mathbb{C}$) of real (complex) numbers is an example of a topological space, with the usual topology consisting of unions and finite intersections of open intervals (open disks).

The space of a topology is always open. The empty set is also open (as the union of the empty subfamily of the family $\tau$ — see (b)). The topology consisting of $\mathfrak{X}$ and $\emptyset$ only, is called *antidiscrete* or *trivial*; in this case $(\mathfrak{X}, \tau)$ is an antidiscrete (topological) space. Obviously, the antidiscrete topology is the smallest of all possible topologies on $\mathfrak{X}$. The largest topology on $\mathfrak{X}$ is the one containing all subsets of $\mathfrak{X}$. This is called the *discrete* topology and a space endowed with the discrete topology is said to be discrete. In a discrete space every set is open.

Any topology on $\mathfrak{X}$ contains the antidiscrete topology and is contained in the discrete topology. Let $\tau_1$ and $\tau_2$ be two topologies on $\mathfrak{X}$; $\tau_1$ is *weaker* (smaller, rougher) than $\tau_2$, or $\tau_2$ is *stronger* (greater, finer) than $\tau_2$, if $\tau_1 \subseteq \tau_2$. It is possible that, for two given topologies $\tau_1$ and $\tau_2$ on $\mathfrak{X}$, neither $\tau_1$ is greater that $\tau_2$, nor $\tau_2$ is greater that $\tau_1$; in this case $\tau_1$ and $\tau_2$ are said to be *not comparable*.

**2.** A *neighborhood* ($\tau$-neighborhood) of a point $x$ in a topological space $(\mathfrak{X}, \tau)$ is any subset of this space which contains an element $\mathfrak{U}$ of the topology $\tau$ with the property $x \in \mathfrak{U}$. For example, in the case of the space $\mathbb{C}$ of complex numbers with the usual topology, a neighborhood of a point is any subset of $\mathbb{C}$ containing an open disk which contains the considered point.

**THEOREM 2.1.** *A set $\mathfrak{A} \subseteq \mathfrak{X}$ is open if and only if for every $x \in \mathfrak{A}$, $\mathfrak{A}$ is a neighborhood of $x$.*

The family of all neighborhoods of a given point is called the *neighborhood system* of the point. A subfamily $\beta$ of a topology $\tau$ is a *base* for $\tau$ if any element in $\tau$ is the union of some elements of $\beta$. A family $\beta_x$ of neighborhoods of a point $x$ is called a *base of the neighborhood system* of $x$, or a base of the topology in $x$, if any neighborhood of $x$ contains a neighborhood from $\beta_x$. We are saying that the topological space $(\mathfrak{X}, \tau)$ satisfies the *first axiom of countability* if the neighborhood system of any point has a countable base. If the topology $\tau$ has a countable base, then it is said that $\mathfrak{X}$ satisfies the *second axiom of countability*. Spaces satisfying the first, or better the second, axiom of countability have many useful properties; some of these properties will be mentioned below.

**3.** Let $\mathfrak{X}$ be a topological space and let $\mathfrak{A}$ be a subset of it. A point $x \in \mathfrak{X}$ is *adherent* to $\mathfrak{A}$ if every neighborhood of $x$ contains points of $\mathfrak{A}$. A point $x \in \mathfrak{X}$ is an *accumulation point* (*limit point*) of $\mathfrak{A}$ if in every neighborhood of $x$ there exist points of $\mathfrak{A}$ different from $x$. A set $\mathfrak{A} \subseteq \mathfrak{X}$ is *closed* if it contains all its adherent points. In an antidiscrete space $\mathfrak{X}$ only $\mathfrak{X}$ and $\emptyset$ are closed sets, but in a discrete space $\mathfrak{X}$ every set $\{x\}$ (having only one element $x \in \mathfrak{X}$) is closed.

Open and closed sets are connected by:

**THEOREM 2.2.** *A subset $\mathfrak{A}$ of a topological space $(\mathfrak{X}, \tau)$ is open if and only if its complement $\mathfrak{X}\backslash\mathfrak{A}$ is closed.*

The *closure* of a set $\mathfrak{A}$ in the topological space $(\mathfrak{X}, \tau)$ is, by definition, the union of $\mathfrak{A}$ with the set of its limit points, and is denoted by $\overline{\mathfrak{A}}$. It is easy to see that $\overline{\mathfrak{A}}$ is closed for every $\mathfrak{A}$, and that $\mathfrak{A}$ is closed if and only if $\mathfrak{A} = \overline{\mathfrak{A}}$.

**4.** INTERIOR AND BOUNDARY. A point $x$ of a subset $\mathfrak{A}$ in a topological space is called *interior* for $\mathfrak{A}$ if $\mathfrak{A}$ is a neighborhood of $x$. The set of all interior points of $\mathfrak{A}$ is called the *interior* of $\mathfrak{A}$ and is denoted by $\mathfrak{A}^\circ$.

**THEOREM 2.3.** *Let $\mathfrak{A}$ be a subset of the topological space $\mathfrak{X}$. The interior $\mathfrak{A}^\circ$ of the set $\mathfrak{A}$ is an open set, and moreover, $\mathfrak{A}^\circ$ is the largest open set contained in $\mathfrak{A}$. The set $\mathfrak{A}$ is open if and only if $\mathfrak{A} = \mathfrak{A}^\circ$. The set of all points of $\mathfrak{A}$ which are not limit points for $\mathfrak{X}\backslash\mathfrak{A}$ is exactly $\mathfrak{A}^\circ$. The closure $\mathfrak{X}\backslash\mathfrak{A}$ coincides with $\mathfrak{X}\backslash\mathfrak{A}^\circ$.*

The *boundary* of a subset $\mathfrak{A}$ of a topological space $\mathfrak{X}$ is the set of all points $x \in \mathfrak{X}$ which are interior points neither for $\mathfrak{A}$, nor for $\mathfrak{X}\backslash\mathfrak{A}$. Clearly, the boundary of $\mathfrak{A}$ coincides with the boundary of $\mathfrak{X}\backslash\mathfrak{A}$. For example, in an antidiscrete space the boundary of any nonempty subset is the whole $\mathfrak{X}$. In contradistinction, in a discrete space, the boundary of any set is empty. In the set $\mathbb{R}$ of real numbers with the usual topology, the boundary of an interval (open, closed or half-open) consists of the end-points of the interval, while the boundary of the set of rational numbers (as well as the boundary of the set of irrational numbers) is the whole $\mathbb{R}$.

The relations between boundary, closure and interior are given in the next theorem.

**THEOREM 2.4.** *Let $\mathfrak{A}$ be a subset of a topological space $\mathfrak{X}$ and let $b(\mathfrak{A})$ be the boundary of $\mathfrak{A}$. Then:*

$$b(\mathfrak{A}) = \overline{\mathfrak{A}} \cap \overline{(\mathfrak{X}\backslash\mathfrak{A})} = \overline{\mathfrak{A}}\backslash\mathfrak{A}^\circ,$$

$$\mathfrak{X}\backslash b(\mathfrak{A}) = \mathfrak{A}^\circ \cup (\mathfrak{X}\backslash\mathfrak{A})^\circ,$$
$$\overline{\mathfrak{A}} = \mathfrak{A} \cup b(\mathfrak{A}) \text{ and } \mathfrak{A}^\circ = \mathfrak{A}\backslash b(\mathfrak{A}).$$

A set is closed if and only if it contains its boundary; a set is open if and only if it is disjoint of its boundary.

A set $\mathfrak{A}$ is called *dense* (or everywhere dense) in a topological space $\mathfrak{X}$ if $\overline{\mathfrak{A}} = \mathfrak{X}$. The space $\mathfrak{X}$ is *separable* if $\mathfrak{X}$ has a countable dense subset.

A separable topological space satisfies the first axiom of countability, but not necessarily the second one. On the other hand, a topological space satisfying the second axiom of countability is separable.

**5.** INDUCED TOPOLOGY. Let $(\mathfrak{X}, \tau)$ be a topological space and $\mathfrak{Y}$ a subset of $\mathfrak{X}$. The *induced topology* $\alpha$ on $\mathfrak{Y}$ is, by definition, the family of all intersections of the elements of $\tau$ with the set $\mathfrak{Y}$. Thus, $\mathfrak{U} \in \alpha$ if and only if $\mathfrak{U} = \mathfrak{V} \cap \mathfrak{Y}$ for some $\mathfrak{V} \in \tau$. The topological space $(\mathfrak{Y}, \alpha)$ is called a *subspace* of the space $(\mathfrak{X}, \tau)$. Note that if $(\mathfrak{Y}, \alpha)$ is a subspace of $(\mathfrak{X}, \tau)$ and $(\mathfrak{Z}, \beta)$ is a subspace of $(\mathfrak{Y}, \alpha)$ then $(\mathfrak{Z}, \beta)$ is a subspace of $(\mathfrak{X}, \tau)$.

**THEOREM 2.5.** *Let $(\mathfrak{X}, \tau)$ be a topological space, $(\mathfrak{Y}, \alpha)$ a subspace of it, and $\mathfrak{A}$ a set in $\mathfrak{Y}$. Then:*

(a) *the set $\mathfrak{A}$ is $\alpha$-closed if and only if it is the intersection of $\mathfrak{Y}$ with some $\tau$-closed set;*

(b) *a point $y \in \mathfrak{Y}$ is an $\alpha$-limit point for $\mathfrak{A}$ if and only if it is a $\tau$-limit for $\mathfrak{A}$;*

(c) *the $\alpha$-closure of the set $\mathfrak{A}$ is the intersection of the set $\mathfrak{Y}$ with the $\tau$-closure of $\mathfrak{A}$.*

## §3. Convergence. Directedness

**1.** The topology of a space can be also described using the notion of convergence. Let $(\mathfrak{D}, \geqslant)$ be a directed set (see §1) and $\mathfrak{X}$ a topological space. A *directedness* on $\mathfrak{X}$ is a mapping $S: \mathfrak{D} \to \mathfrak{X}$ defined on the whole $\mathfrak{D}$. The value of a directedness $S$ at a point $d \in \mathfrak{D}$ is often denoted by $S_d$, and the directedness itself is denoted by $\{S_d \mid d \in \mathfrak{D}\}$.

A directedness $S$ belongs to a set $\mathfrak{A}$ starting with a moment $m \in \mathfrak{D}$ if $S_d \in \mathfrak{A}$ for all $d \geqslant m$. A directedness $S$ on the space $(\mathfrak{X}, \tau)$ *converges* to a point $s \in \mathfrak{X}$ if for any $\tau-$neighborhood of $s$, there exists $m \in \mathfrak{D}$ such that $S$ belongs to that neighborhood starting with $m$. For example, if $\mathfrak{X}$ is the discrete space, then the directedness $S$

converges to a point $s$ if and only if starting with a certain moment the values of $S$ coincide with $s$. On the contrary, in an antidiscrete space any directedness is convergent to every point of the space. Thus, a directedness may be simultaneously convergent to a lot of different points.

The notions of limit points and closure of a set are characterized in terms of convergence as follows.

**THEOREM 3.1.** *Let $\mathfrak{X}$ be a topological space. Then:*

(a) *A point $s$ is a limit point of a set $\mathfrak{A} \subseteq \mathfrak{X}$ if and only if there is a directedness in $\mathfrak{A}\backslash\{s\}$ which is convergent to $s$.*

(b) *A point $s$ belongs to the closure of a set $\mathfrak{A}$ ($s \in \overline{\mathfrak{A}}$) if and only if there is a directedness in $\mathfrak{A}$ which is convergent to $s$.*

(c) *A set $\mathfrak{A}$ is closed in $\mathfrak{X}$ ($\mathfrak{A} = \overline{\mathfrak{A}}$) if and only if no directedness contained in $\mathfrak{A}$ is convergent to some point of $\mathfrak{X}\backslash\mathfrak{A}$.*

A directedness $\{T_d \mid d \in \mathfrak{D}\}$ is called a *subdirectedness* of a directedness $\{S_e \mid e \in \mathfrak{E}\}$ if there exists a mapping $\mu\colon \mathfrak{D} \to \mathfrak{E}$ such that

a) $T = S \circ \mu$, or equivalently, $T_i = S_{\mu_i}$ for each $i \in \mathfrak{D}$;

b) for each $e \in \mathfrak{E}$ there is a $d \in \mathfrak{D}$ such that if $j \in \mathfrak{D}$ and $j \geqslant d$ then $\mu_j \geqslant e$.

**THEOREM 3.2.** *A point $s$ of a topological space is a limit point of a directedness $S$ if and only if there exists a subdirectedness of $S$ convergent to $s$.*

**2.** For an important class of topological spaces, the convergence can be characterized in terms of sequences.

A *sequence* is a mapping defined on the set of natural numbers, i.e., a directedness for which the directed set is the linearly ordered set $\mathbb{N} = \{1, 2, \ldots\}$. A *subsequence* is a subdirectedness with $\mu$ mapping $\mathbb{N}$ into $\mathbb{N}$ (i.e., $\mathfrak{D} = \mathfrak{E} = \mathbb{N}$). In other words, a subsequence is an infinite part of a sequence. Note that a sequence may have subdirectednesses which are not subsequences.

In a topological space satisfying the first axiom of countability the convergence is completely described in terms of sequences.

**THEOREM 3.3.** *Let $\mathfrak{X}$ be a topological space satisfying the first axiom of countability. Then:*

(a) *A point $s$ is a limit point of a set $\mathfrak{A}$ if and only if there is a sequence in $\mathfrak{A}\backslash\{a\}$ which is convergent to $s$.*

(b) *A set $\mathfrak{A}$ is open if and only if any sequence convergent to a point of $\mathfrak{A}$ is situated in $\mathfrak{A}$ starting with a certain moment.*

(c) *If $s$ is a limit point of a sequence then the sequence contains a subsequence which is convergent to $s$.*

**3.** We return now to the case of general topological spaces. A point $s$ is called an *$\omega$-limit point* of a subset $\mathfrak{A}$ in a topological space $\mathfrak{X}$ if there is a sequence in $\mathfrak{A}\backslash\{s\}$ which is convergent to $s$. A set $\mathfrak{A}$ is *sequentially closed* if it contains all its $\omega$-limit points. Any closed set is sequentially closed, but there exist non-closed, sequentially closed sets.

**4.** Some important properties of topological spaces are expressed in terms of separability. Roughly speaking, separability means the possibility of separating different points or points from sets. For a later use, let us mention the following definition.

A topological space is called a *Hausdorff* space if any two distinct points of the space have two disjoint neighborhoods. Hausdorff separability of a space is characterized as follows.

**THEOREM 3.4.** *A topological space is Hausdorff if and only if no directedness on this space converges to two distinct points.*

**5.** TOPOLOGICAL PRODUCT OF SPACES. On the direct product of a family of topological spaces (see §1) it is possible to define a canonical topology. Let us consider first the case of two topological spaces $(\mathfrak{X}, \alpha)$ and $(\mathfrak{Y}, \beta)$. The family of all direct products of the form $\mathfrak{U} \times \mathfrak{V}$, where $\mathfrak{U}$ is an $\alpha$-open set in $\mathfrak{X}$ and $\mathfrak{V}$ is a $\beta$-open set in $\mathfrak{Y}$ forms a base for a topology $\tau$ on $\mathfrak{X} \times \mathfrak{Y}$; $\tau$ is called the *product topology* on $\mathfrak{X} \times \mathfrak{Y}$. A subset $\mathfrak{W}$ of $\mathfrak{X} \times \mathfrak{Y}$ is open in the product topology if and only if for any element $(x, y) \in \mathfrak{W}$ there exist two open neighborhoods $\mathfrak{U}$ and $\mathfrak{V}$, of $x$ and $y$, respectively, such that $\mathfrak{U} \times \mathfrak{V} \subseteq \mathfrak{W}$. The space $(\mathfrak{X} \times \mathfrak{Y}, \tau)$ is called the *topological product* of the spaces $\mathfrak{X}$ and $\mathfrak{Y}$. It is straightforward to extend this definition to any finite set of coordinate spaces.

Consider now the case of an arbitrary family of topological spaces. Let $\mathfrak{A}$ be an indexing set and, for any $a \in \mathfrak{A}$, let $(\mathfrak{X}_a, \tau_a)$ be a topological space. In §1 we defined the direct product $\prod_{a\in\mathfrak{A}} \mathfrak{X}_a$ of the sets $\mathfrak{X}_a$ as the set of all mappins $x$ on $\mathfrak{A}$ such that $x_a \in \mathfrak{X}_a$, for any $a \in \mathfrak{A}$. Let us define a topology $\tau$ on $\prod_{a\in\mathfrak{A}} \mathfrak{X}_a$ by taking as its base the family of all finite intersections of sets of the form $P_a^{-1}(\mathfrak{U})$, where $\mathfrak{U}$ is an arbitrary $\tau_a$-open set in $\mathfrak{X}_a$ and $P_a$ is the projection onto $\mathfrak{X}_a$. The space $\left(\prod_{a\in\mathfrak{A}} \mathfrak{X}_a, \tau\right)$ is called the topological product of the spaces $\mathfrak{X}_a$.

The next two theorems give two answers to the natural question about some properties of the factors which are inherited by their product.

**THEOREM 3.5.** *The topological product of Hausdorff spaces is a Hausdorff space. Each coordinate projector maps open sets onto open sets.*

**THEOREM 3.6.** *Let $\mathfrak{X}_a$, $a \in \mathfrak{A}$, be the arbitrary family of topological spaces satisfying the second axiom of countability. Their topological product satisfies the second axiom of countability if and only if all $\mathfrak{X}_a$ for $a \in \mathfrak{A}$ but a countable subset of $\mathfrak{A}$, are antidiscrete.*

Other properties of topological products will be discussed below.

**6.** CONTINUOUS MAPPINGS. A mapping $f$ from a topological space $(\mathfrak{X}, \alpha)$ into a topological space $(\mathfrak{Y}, \beta)$ is called *continuous* if the preimage of any $\beta$-open set in $\mathfrak{Y}$ is an $\alpha$-open in $\mathfrak{X}$. It is easily seen that the composition of continuous mappings is continuous. Let $f: \mathfrak{X} \to \mathfrak{Y}$ be a continuous mapping and $\mathfrak{A} \subseteq \mathfrak{X}$. Then the restriction $f \mid \mathfrak{A}$ is also continuous. A mapping $f$ for which $f \mid \mathfrak{A}$ is continuous is said to be continuous on $\mathfrak{A}$.

The continuity of a mapping is characterized as follows.

**THEOREM 3.7.** *Let $\mathfrak{X}$ and $\mathfrak{Y}$ be topological spaces and $f: \mathfrak{X} \to \mathfrak{Y}$ a mapping. The following assertions are equivalent:*

(a) *The mapping $f$ is continuous.*

(b) *The preimage of any closed set is closed.*

(c) *For each point $x \in \mathfrak{X}$, the preimage of any neighborhood of $f(x)$ is a neighborhood of $x$.*

(d) *For each point $x \in \mathfrak{X}$ and any neighborhood $\mathfrak{U}$ of the point $f(x)$, there exists a neighborhood $\mathfrak{V}$ of the point $x$ such that $f(\mathfrak{V}) \subseteq \mathfrak{U}$.*

(e) *For any directedness $S$ in $\mathfrak{X}$, convergent to a point $s$, the composition $f \circ S$ is convergent to $f(s)$.*

(f) $f(\overline{\mathfrak{A}}) \subseteq \overline{f(\mathfrak{A})}$ *for every* $\mathfrak{A} \subseteq \mathfrak{X}$.

(g) $\overline{f^{-1}(\mathfrak{B})} \subseteq f^{-1}(\overline{\mathfrak{B}})$ *for every* $\mathfrak{B} \subseteq \mathfrak{Y}$.

A mapping $f$ from a topological space $\mathfrak{X}$ into a topological space $\mathfrak{Y}$ is called continuous at the point $x \in \mathfrak{X}$ if the preimage of any neighborhood of $f(x)$ is a neighborhood of $x$. It is plain that $f$ is continuous on a set if and only if is continuous at each point of that set.

Let us mention a useful property of continuous mappings with values in a topological product of spaces.

**THEOREM 3.8.** *A mapping $f$ from a topological space into a product space $\prod_{a \in \mathfrak{A}} \mathfrak{X}_a$ is continuous if and only if the composition $P_a \circ f$ is continuous for every $a \in \mathfrak{A}$.*

**7.** A *homeomorphism* (or a *topological isomorphism*) of the topological spaces $(\mathfrak{X}, \alpha)$ and $(\mathfrak{Y}, \beta)$ is an isomorphism $f$ of the sets $\mathfrak{X}$ and $\mathfrak{Y}$ (see §1) such that both $f$ and its inverse $f^{-1}$ are continuous. Two spaces are called *homeomorphic* if there exists a homeomorphism between them. For example, all isomorphic discrete spaces are homeomorphic and all isomorphic antidiscrete spaces are homeomorphic. The identity mapping is a homeomorphism; the inverse mapping of a homeomorphism is an homeomorphism. Plainly, the composition of any two homeomorphisms is a homeomorphism. A property of a topological space, inherited by all spaces homeomorphic with it, is called a *topological invariant.* A property defined only in terms of elements and of the topology of a space is in fact a topological invariant. For example, the first and the second axiom of countability and the Hausdorff separability are topological invariants.

**8.** COMPACTNESS. A family $\gamma$ of subsets of a topological space is called a *cover* of a set $\mathfrak{A}$ if $\mathfrak{A} \subseteq \bigcup\{\mathfrak{U} \mid \mathfrak{U} \in \gamma\}$. A cover $\gamma$ is called *open* if each set $\mathfrak{U} \in \gamma$ is open. A *subcover* of a cover $\gamma$ is a subfamily of $\gamma$ which is a cover itself. A topological space is called *compact* if every open cover of the space contains a finite subcover. A subset $\mathfrak{A}$ is compact if it is compact in the induced topology. A set with compact closure is called *precompact* (or *relatively compact*).

**THEOREM 3.9.** *A topological space $\mathfrak{X}$ is compact if and only if any directedness in $\mathfrak{X}$ has a limit point.*

A topological space $\mathfrak{X}$ is called *sequentially compact* if every sequence in $\mathfrak{X}$

contains a subsequence convergent to a point $s \in \mathfrak{X}$. For a topological space $\mathfrak{X}$ satisfying the first axiom of countability, sequentially compactness is equivalent with the fact that every countable open cover of the space contains a finite subcover.

**9.** COMPACTNESS AND SEPARABILITY. A closed subset of a compact space is compact. The converse is not generally true: any nonempty proper subset of an antidiscrete space is compact but not closed. This situation does not occur in Hausdorff spaces.

**THEOREM 3.10.** *If $\mathfrak{A}$ is a compact set of a Hausdorff space $\mathfrak{X}$ and $x \in \mathfrak{X}\backslash\mathfrak{A}$, then $x$ and $\mathfrak{A}$ have disjoint neighborhoods.*

Consequently, a compact subset of a Hausdorff space is closed.

**THEOREM 3.11.** *For any disjoint compact subsets $\mathfrak{A}$ and $\mathfrak{B}$ in a Hausdorff space there exist disjoint neighborhoods.*

Let us consider now mappings on compact spaces.

**THEOREM 3.12.** *Let $f$ be a continuous mapping from a compact space $\mathfrak{X}$ onto a topological space $\mathfrak{Y}$. Then the space $\mathfrak{Y}$ is compact. Moreover, if $\mathfrak{Y}$ is Hausdorff and $f$ is an isomorphism, then $f$ is a homeomorphism.*

Consequently, a real functional continuous on a compact set $\mathfrak{A}$ is bounded and attains its greatest and its smallest value on $\mathfrak{A}$.

We close this paragraph with a result concerning the product of compact spaces.

**THEOREM 3.13** (A. N. Tihonov). *The topological product of compact topological spaces is compact.*

## §4. Metric spaces

**1.** An important class of topological spaces is the one consisting of those spaces with the topology defined using a distance function (or a metric). A *metric* on a set $\mathfrak{X}$ is a nonnegative real-valued functional $d$ defined on the direct product $\mathfrak{X} \times \mathfrak{X}$ which satisfies (for $x, y, z \in \mathfrak{X}$) the following conditions (the axioms of a metric):

a) $d(x,y) = d(y,x)$;

b) $d(x,y) + d(y,z) \geqslant d(x,z)$ (the triangle inequality);

c) $d(x,y) = 0$ if and only if $x = y$.

The pair $(\mathfrak{X}, d)$, where $\mathfrak{X}$ is a space and $d$ is a metric on $\mathfrak{X}$, is called a *metric space*. For $(x, y) \in \mathfrak{X}$, the number $d(x, y)$ is called the *distance* between $x$ and $y$ (or the $d$-distance). Let $r > 0$. The set $\{y \mid d(x, y) < r\}$ is called the *open r-ball* centered at $x$; the set $\{y \mid d(x, y) \leqslant r\}$ is called the *closed r-ball* centered at $x$. The family of all open balls provides a base for a topology $\tau$ on $\mathfrak{X}$, called the *metric topology*, or the topology induced on $\mathfrak{X}$ by the given metric. According to this definition any closed $r$-ball is closed in $(\mathfrak{X}, \tau)$. For example, let $\mathfrak{X} = \mathbb{C}$ and $d(x, y) = |x - y|$; $d$ is a metric on $\mathbb{C}$ called the usual metric on complex numbers. The topology induced on $\mathbb{C}$ by the usual metric coincides with the usual topology. Let $\mathfrak{X}$ be a set, and set $d(x, y) = 0$ if $x = y$ and $d(x, y) = 1$ if $x \neq y$. Then $d$ is a metric on $\mathfrak{X}$ and the open $r$-ball centered at any point $x$ coincides with $\{x\}$ if $r \leqslant 1$ and with $\mathfrak{X}$ if $r > 1$. It follows that the topology induced by $d$ is discrete. The closed $r$-ball centered at any point $x$ is $\{x\}$ if $r < 1$ and $\mathfrak{X}$ if $r \geqslant 1$. The open 1-ball centered at $x$ is $\{x\}$ and the closed 1-ball centered at $x$ is $\mathfrak{X}$.

**2.** METRIC SPACES ARE HAUSDORFF (see §3). By the distance from a point $x$ to a set $\mathfrak{A}$ we mean the number $\inf\limits_{a \in \mathfrak{A}} d(x, a)$.

**THEOREM 4.1.** *The closure of a set $\mathfrak{A}$ in a metric space $\mathfrak{X}$ is the set of all points of the space at distance zero from $\mathfrak{A}$.*

Consequently, for any $x \in \mathfrak{X}$ the one-point set $\{x\}$ is closed.

**THEOREM 4.2.** *Any metric space is normal, i.e, any two disjoint closed sets have disjoint neighborhoods.*

From Theorems 4.1 and 4.2 it follows:

**THEOREM 4.3.** *Any metric space is Hausdorff.*

**THEOREM 4.4.** *Any metric space satisfies the first axiom of countability. A metric space satisfies the second axiom of countability if and only if the space is separable.*

From Theorems 4.4 and 3.3 it follows that the convergence in a metric space can be described completely in terms of sequences.

**THEOREM 4.5.** *A sequence $\{s_n \mid n \in \mathbb{N}\}$ in a metric space $(\mathfrak{X}, d)$ is convergent to $s$ if and only if the sequence of real numbers $\{d(s_n, s) \mid n \in \mathbb{N}\}$ is convergent to zero.*

**3.** Is the topological product of metric spaces a metric space? In other words, is it possible to define a metric on a product of metric spaces such that the topology induced by this metric coincides with the product topology? Theorems 4.4 and 3.6 imply a negative answer in the case of arbitrary families of factors. However, in the case of countable families the next result holds.

**THEOREM 4.6.** *Let* $\{(\mathfrak{X}_n, d_n) \mid n \in \mathbb{N}\}$ *be a sequence of metric spaces. For any* $x = \{x_n \mid x_n \in \mathfrak{X}_n,\ n \in \mathbb{N}\}$ *and* $y = \{y_n \mid y_n \in \mathfrak{X}_n,\ n \in \mathbb{N}\}$, *let* $d(x,y) = \sum_{i=1}^{\infty} 2^{-n} \min\{1,\ d_n(x_n, y_n)\}$. *Then the functional* $d$ *is a metric on the product of the spaces* $\mathfrak{X}_n$, $n \in \mathbb{N}$, *such that the corresponding metric topology coincides with the product topology.*

**4.** Consider now the following general problem: what are those properties a topological space $(\mathfrak{X}, \tau)$ must satisfy, such that there exists a metric on $\mathfrak{X}$ which induces the topology $\tau$? If this is the case, $\tau$ is called *metrizable.* In order to answer the question we first need some definitions. A family $\gamma$ of subsets of a topological space is called *locally finite* if any point of the space has a neighborhood disjoint from all but a finite set of elements of $\gamma$. The family $\gamma$ is called *discrete* if every point of the space has a neighborhood disjoint from all but at most one element of the family $\gamma$. It is plain that a discrete family is locally finite. Finally, the family $\gamma$ is called *$\sigma$-locally finite* (*$\sigma$-discrete*) if $\gamma$ is the union of a countable set of locally finite (discrete) subfamilies.

**THEOREM 4.7** (Metrizability criterion). *The following three conditions are equivalent for a topological space:*

(a) *The space is metrizable.*

(b) *All one-point sets are closed, the space is regular and has a $\sigma$-locally finite base.*

(c) *All one-point sets are closed, the space is regular and has a $\sigma$-discrete base.*

**5.** COMPLETENESS. Let $(\mathfrak{X}, d)$ be a metric space. A sequence $\{s_n \mid n \in \mathbb{N}\}$ is called *Cauchy* (or *fundamental*) if $d(s_n, s_m) \to 0$ for $n, m \to \infty$, i.e., for any $\varepsilon > 0$ there is $\ell \in \mathbb{N}$ such that $d(x_n, x_m) < \varepsilon$ for all $n, m > \ell$. A metric space is called *complete* if all its Cauchy sequences are convergent.

**THEOREM 4.8.** *In a metric space, any metric convergent sequence is Cauchy. If a Cauchy sequence has a limit point, then it converges to that point.*

**THEOREM 4.9.** *A closed subspace of a complete metric space is complete. A complete subspace of a metric space is closed.*

By Theorem 4.6, a countable topological product of metric spaces is metrizable. We consider now the problem of its completeness.

**THEOREM 4.10.** *The product of a countable set of metric spaces is complete if and only if each coordinate space is complete.*

*A sequence in a product space is fundamental if and only if each of its projections on a coordinate space is fundamental.*

**6.** COMPACTNESS AND MONTEL SPACES. In a metric space the compactness of a set is equivalent with its sequentially compactness. More precisely, we have:

**THEOREM 4.11.** *A subset $\mathfrak{A}$ of a metric space is compact if and only if every sequence in $\mathfrak{A}$ contains a subsequence convergent to a point in $\mathfrak{A}$.*

The next definition allows us to give another compactness criterion for subsets of a metric space. A set $\mathfrak{N}$ in a metric space $(\mathfrak{X}, d)$ is called an *$\varepsilon$-array* for a set $\mathfrak{A} \subseteq \mathfrak{X}$ if for any point $x \in \mathfrak{A}$ there exists a point $y \in \mathfrak{N}$ such that $d(x, y) < \varepsilon$.

**THEOREM 4.12.** (Hausdorff). *A set $\mathfrak{A}$ in a metric space $\mathfrak{X}$ is compact if — and (when $\mathfrak{X}$ is complete) only if — for any $\varepsilon > 0$ there exists a finite $\varepsilon$-array for $\mathfrak{A}$.*

A subset $\mathfrak{A}$ of a topological space $\mathfrak{X}$ is called *bounded* if there exist a point $x \in \mathfrak{X}$ and a number $r > 0$ such that $\mathfrak{A}$ is included in the $r$-ball centered at $x$. From Hausdorff Theorem above it follows that a compact subset of a metric space is bounded. Consequently, Theorem 4.1 implies that any precompact subset of a metric space is bounded. In particular, any convergent sequence is bounded.

A metric space is called a *Montel space* if all its bounded subsets are precompact. A simple example of a Montel space is given by the space $\mathbb{C}$ of complex numbers, with the usual metric. The space $\mathbb{C}^n$, the topological product of $n$ copies of $\mathbb{C}$, is also a Montel space.

## §5. Spaces of mappings

**1.** We consider now spaces of mappings from a set $\mathfrak{X}$ into a fixed topological space $\mathfrak{Y}$ and some special topologies on such spaces. Let $\mathcal{F}$ be a family of mappings from

$\mathfrak{X}$ into $\mathfrak{Y}$. It follows from the definition of the direct product (see §1) that $\mathcal{F}$ is a subset of the space $\mathfrak{Y}^{\mathfrak{X}} = \prod_{x \in \mathfrak{X}} \mathfrak{Y}$. The topology induced on $\mathcal{F}$ by the direct product topology is called the topology of *pointwise convergence* (*coordinatewise convergence, simple convergence*) or, shortly, pointwise topology. A directedness $\{f_d \mid d \in \mathfrak{D}\}$ in $\mathfrak{Y}^{\mathfrak{X}}$ is convergent to $f \in \mathfrak{Y}^{\mathfrak{X}}$ if and only if $\{f_d(x) \mid d \in \mathfrak{D}\}$ converges to $f(x)$ for every $x \in \mathfrak{X}$. If $\mathfrak{Y}$ is Hausdorff or regular, then the space $\mathcal{F}$ with pointwise topology has the same property. Let $\mathcal{F}(x) = \{f(x) \mid f \in \mathcal{F}\}$. Tychonoff Theorem (Theorem 3.13) implies the following.

**THEOREM 5.1.** *A family $\mathcal{F}$ of mappings from a set $\mathfrak{X}$ into a Hausdorff space $\mathfrak{Y}$ is compact in pointwise topology if and only if the next conditions are satisfied:*

(a) *the family $\mathcal{F}$ is closed in $\mathfrak{Y}^{\mathfrak{X}}$ in the direct product topology;*

(b) *for every $x \in \mathfrak{X}$, the set $\mathcal{F}(x)$ is precompact in $\mathfrak{Y}$.*

Condition (b) above is a criterion for the precompactness of the family $\mathcal{F}$ in the pointwise topology.

**2.** The definition of pointwise topology does not depend on any topology existing on $\mathfrak{X}$. In fact, $\mathfrak{X}$ plays in that definition merely the role of an indexing set. More sofisticated topologies on $\mathcal{F}$ can be constructed using topologies on $\mathfrak{X}$. Thus, let $\mathfrak{X}$ be a topological space and $(\mathfrak{Y}, \rho)$ a metric space. A directedness $\{f_d \mid d \in \mathfrak{D}\}$ in $\mathfrak{Y}^{\mathfrak{X}}$ is said to be uniformly convergent on a set $\mathfrak{A} \subseteq \mathfrak{X}$ to a mapping $f \in \mathfrak{Y}^{\mathfrak{X}}$ if for any $\varepsilon > 0$ there exists $\ell \in \mathfrak{D}$ such that for $d \in \mathfrak{D}$ with $d \geqslant \ell$ we have $\rho(f_d(a), f(a)) < \varepsilon$ for all $a \in \mathfrak{A}$. We denote by $C(\mathfrak{X}, \mathfrak{Y})$ the family of all continuous mappings from $\mathfrak{X}$ into $\mathfrak{Y}$. The topology of *compact convergence* (*uniform convergence on compacts*) on $C(\mathfrak{X}, \mathfrak{Y})$ is defined as follows: a directedness $\{f_d \mid d \in \mathfrak{D}\}$ is convergent to $f \in C(\mathfrak{X}, \mathfrak{Y})$ if the directedness is uniformly convergent to $f$ on each compact $\mathfrak{A} \subseteq \mathfrak{X}$.

A family $\mathcal{F} \subseteq C(\mathfrak{X}, \mathfrak{Y})$ is called *equicontinuous* at a point $x$ if for every $\varepsilon > 0$ there exists a neighborhood $\mathfrak{U}$ of $x$ such that $\rho(f(y), f(z)) < \varepsilon$ for all $y, z \in \mathfrak{U}$ and for all $f \in \mathcal{F}$.

**THEOREM 5.2.** *Pointwise topology and topology of compact convergence coincide on any equicontinuous family.*

Theorems 5.1 and 5.2 immediately imply:

**THEOREM 5.3.** *The following two conditions are sufficient for the precompactness* *of a family* $\mathcal{F} \subseteq C(\mathfrak{X}, \mathfrak{Y})$*:*

(a) $\mathcal{F}$ *is equicontinuous;*

(b) *for every* $x \in \mathfrak{X}$, $\mathcal{F}(x)$ *is precompact in* $\mathfrak{Y}$.

The next theorem will be used in Chapter II below.

**THEOREM 5.4.** *Let* $\{F_n \mid n \in \mathbb{N}\} \subseteq C(\mathfrak{X}, \mathfrak{Y})$ *be such that the sequence* $\{F_n x' \mid n \in \mathbb{N}\}$ *is convergent for any* $x' \in \mathfrak{X}'$, *where* $\mathfrak{X}'$ *is a dense subset of* $\mathfrak{X}$. *Then the sequence* $\{F_n \mid n \in \mathbb{N}\}$ *is convergent in the topology of compact convergence to some* $F \in C(\mathfrak{X}, \mathfrak{Y})$.

**3.** Let $\mathfrak{X} = \mathfrak{L}$ be a compact topological space and $f \in C(\mathfrak{L}, \mathfrak{Y})$. The functional $\rho(f(x), \theta)$ is continuous on $\mathfrak{L}$ for every fixed $\theta \in \mathfrak{Y}$. By Theorem 3.12 this functional is bounded: $\rho(f(x), \theta) \leqslant c_f < \infty$. Thus, we can define on $C(\mathfrak{L}, \mathfrak{Y})$ a metric $\tilde{\rho}$ by the formula:

$$\tilde{\rho}(f, g) = \sup_{x \in \mathfrak{L}} \rho\left(f(x), g(x)\right).$$

The metric $\tilde{\rho}$ is called the *uniform metric.*

A family $\mathcal{F} \subseteq C(\mathfrak{L}, \mathfrak{Y})$ is called *uniformly bounded* if it is bounded as a subset of the metric space $(C(\mathfrak{L}, \mathfrak{Y}), \tilde{\rho})$, i.e., there exists $g \in C(\mathfrak{L}, \mathfrak{Y})$ and $r > 0$ such that $\sup_{x \in \mathfrak{L}} \rho\left(f(x), g(x)\right) \leqslant r\ (< \infty)$ for all $f \in \mathcal{F}$.

Let now $\mathfrak{Y} = \mathfrak{M}$ be a Montel space. Then the uniform boundedness of a family $\mathcal{F}$ from $C(\mathfrak{L}, \mathfrak{M})$ is equivalent to its precompactness in the pointwise topology. Moreover, if the family $\mathcal{F}$ is equicontinuous, then, by Theorem 5.2 the pointwise topology coincides on $\mathcal{F}$ with the topology of compact convergence, and the last one clearly coincides with the topology induced by the metric $\tilde{\rho}$. Hence, by Theorem 5.1, we obtain a criterion of precompactness for a family $\mathcal{F}$ in the metric space $(C(\mathfrak{L}, \mathfrak{M}), \tilde{\rho})$ (or, equivalently, in the space $C(\mathfrak{L}, \mathfrak{M})$ with the topology of compact convergence):

**THEOREM 5.5.** *A subset* $\mathcal{F}$ *of the space* $C(\mathfrak{L}, \mathfrak{M})$ *is precompact in the topology of the uniform metric if and only if* $\mathcal{F}$ *is uniformly bounded and equicontinuous.*

Theorems of this type (i.e., about precompactness of families of continuous mappings) are usualy credited to Ascoli, or Arzelà-Ascoli. Arzelà is quoted here because the case $\mathfrak{L} = [0, 1]$ and $\mathfrak{M} = \mathbb{R}$ — with the usual metric — is due to him.

A consequence of Theorem 5.5 is:

*An equicontinuous family of mappings from a compact space* $\mathfrak{L}$ *into a Montel* *metric space $\mathfrak{M}$ is itself a Montel metric space relatively to the uniform metric.*

## §6. Linear topological spaces

**1.** LINEAR SPACES. Let $\mathfrak{X}$ be a set and let $\mathbb{K}$ be the field $\mathbb{R}$ of real numbers or the field $\mathbb{C}$ of complex numbers. The set $\mathfrak{X}$ is called a *linear space* (or a vector space) provided that two operations called addition of elements in $\mathfrak{X}$ and multiplication by scalars are given, satisfying the following axioms ($x, y, z \in \mathfrak{X}$, and $\alpha, \beta \in \mathbb{K}$):

I. Addition

1) commutativity: $x + y = y + x$;

2) associativity: $(x + y) + z = x + (y + z)$;

3) the existence of a unique element $0 \in \mathfrak{X}$ such that $x + 0 = x$;

4) for any $x \in \mathfrak{X}$ there exists a unique element $(-x)$ such that $x + (-x) = 0$.

Instead of $x + (-y)$ we write $x - y$. The element 0 is called the *zero element*, or the zero, of the space $\mathfrak{X}$; the element $-x$ is called the *opposite* of $x$.

II. Multiplication by scalars

1) associativity of multiplication: $\alpha(\beta x) = (\alpha\beta)x$;

2) distributivity: $(\alpha + \beta)x = \alpha x + \beta x$, $\alpha(x + y) = \alpha x + \alpha y$;

3) $1 \cdot x = x$.

A linear space is called real or complex according to the field $\mathbb{R}$ or $\mathbb{C}$ which the scalars are from. For example, the set $\mathbb{R}$ ($\mathbb{C}$) relatively to the usual addition and multiplication is a real (complex) linear space. Here $\mathbb{K}$ is $\mathbb{R}$ ($\mathbb{C}$) itself.

The following are some consequences of the axioms of a linear space:

1. $0 \cdot x = 0$ (the first zero is the real number zero, while the second one is the zero vector);

2. $(-1) \cdot x = -x$;

3. $\alpha \cdot 0 = 0$;

4. if $\alpha x = \beta x$ and $x \neq 0$, then $\alpha = \beta$; if $\alpha x = \alpha y$ and $\alpha \neq 0$, then $x = y$.

Two linear spaces are linearly isomorphic if there exists an isomorphism of the underlying sets (see §1) which preserves the algebraic operations: if $x \leftrightarrow x'$ and $y \leftrightarrow y'$, then $x + y \leftrightarrow x' + y'$ and $\alpha x \leftrightarrow \alpha x'$.

**2.** A set $\mathfrak{E}$ of vectors in a linear space $\mathfrak{X}$ is *linearly independent* if for any finite subset $\{x_1, \ldots, x_n\}$ of $\mathfrak{E}$ the equality $\alpha_1 x_1 + \ldots + \alpha_n x_n = 0$ implies $\alpha_1 = \ldots = \alpha_n = 0$.

Otherwise, the set $\mathfrak{E}$ is called *linearly dependent*. In a linearly dependent set $\mathfrak{E} \neq \{0\}$ there exists an element which is a linear conbination of a finite number of elements in $\mathfrak{E}$, different from it.

A *linear basis*, or a *Hamel basis*, of a linear space $\mathfrak{X}$ is a linearly independent subset $\mathfrak{E}$ with the property that for each $x \in \mathfrak{X}$, there exists $\{x_1, \ldots, x_n\} \subseteq \mathfrak{E}$ such that $x = \alpha_1 x_1 + \ldots + \alpha_n x_n$ for some $\alpha_1, \ldots, \alpha_n \in \mathbb{K}$. Any linear space $\mathfrak{X}$ has a basis, and any two bases are isomorphic. If a basis of $\mathfrak{X}$ contains $n$ elements, the space $\mathfrak{X}$ is called *n-dimensional.* If $\mathfrak{X}$ has no finite basis, then it is called *infinite dimensional.*

A nonempty set $\mathfrak{L}$ of vectors is called a *lineal* if from $\{x_1, \ldots, x_n\} \subseteq \mathfrak{L}$ it follows that $\alpha_1 x_2 + \ldots + \alpha_n x_n \in \mathfrak{L}$ for any $\alpha_1, \ldots, \alpha_n \in \mathbb{K}$. A lineal clearly contains the zero vector.

**3.** DIRECT SUM. Let $\mathfrak{X}$ be a linear space and let $\mathfrak{L}_1, \ldots, \mathfrak{L}_m$ be a finite family of lineals in $\mathfrak{X}$. If each vector $x \in \mathfrak{X}$ has a unique representation of the form

$$x = x_1 + \ldots + x_m, \quad x_i \in \mathfrak{L}_i, \ i = 1, 2, \ldots, m,$$

then we say that $\mathfrak{X}$ is the *direct sum* of the lineals $\mathfrak{L}_1, \ldots, \mathfrak{L}_m$, and we write $\mathfrak{X} = \mathfrak{L}_1 \dot{+} \ldots \dot{+} \mathfrak{L}_m$ or $\mathfrak{X} = \sum_{i=1}^{m} \dot{+} \mathfrak{L}_i$. It is easy to see that in this case $\bigcap_{i=1}^{m} \mathfrak{L}_i = \{0\}$ and that the direct sum $\sum_{i=1}^{m} \dot{+} \mathfrak{L}_i$ is linearly isomorphic with the direct product $\prod_{i=1}^{m} \mathfrak{L}_i$, where the linear operations in the last space are defined coordinatewisely: for $x = (x_1, \ldots, x_m)$ and $y = (y_1, \ldots, y_m)$, put $x + y = (x_1 + y_1, \ldots, x_m + y_m)$ and $\alpha x = (\alpha x_1, \ldots, \alpha x_m)$, where $\alpha \in \mathbb{K}$. Analogously, the direct sum of the linear spaces $\mathfrak{X}_1, \ldots, \mathfrak{X}_m$ is the direct product $\prod_{i=1}^{m} \mathfrak{X}_i$ with the coordinatewise-linear operations given. An example of direct sum is the $n$-dimensional space $\mathbb{C}^n$ of vectors $x = (x_1, \ldots, x_n)$, $x_i \in \mathbb{C}$, $i = 1, \ldots, n$, i.e., $\mathbb{C}^n = \sum_{i=1}^{n} \dot{+} \mathbb{C}$.

**4.** TOPOLOGICAL LINEAR SPACES. Let $\tau$ be a topology on a linear space $\mathfrak{X}$. A natural assumption which enables an effective use of both the algebraic and topological structures, is the continuity of linear operations on $\mathfrak{X}$ with respect to the topology $\tau$. If this is the case, then $\mathfrak{X}$ is called a *topological linear space* (TLS).

Here are some definitions needed in what follows.

For $\lambda \in \mathbb{K}$ and $\mathfrak{U} \subseteq \mathfrak{X}$, $\lambda\mathfrak{U}$ is the set $\{\lambda u \mid u \in \mathfrak{U}\}$. A set $\mathfrak{U} \subseteq \mathfrak{X}$ *absorbs* a set $\mathfrak{M} \subseteq \mathfrak{X}$ if there exists $\lambda_0 \in \mathbb{K}$ such that $\mathfrak{M} \subseteq \lambda\mathfrak{U}$ for all $\lambda$ with $|\lambda| \geqslant |\lambda_0|$. A set $\mathfrak{M} \subseteq \mathfrak{X}$ is called *bounded* if it is absorbed by any neighborhood of 0 in $\mathfrak{X}$.

A mapping $f$ from a topological space $\mathfrak{X}$ into a topological space $\mathfrak{Y}$ is called *proper* if the preimage $f^{-1}(\mathfrak{A})$ of any precompact set $\mathfrak{A} \subseteq \mathfrak{Y}$ is precompact in $\mathfrak{X}$. A mapping $g$ from a linear topological space $\mathfrak{X}$ into a topological space $\mathfrak{Y}$ is called *compact* if the image $f(\mathfrak{A})$ of any bounded set $\mathfrak{A} \subseteq \mathfrak{X}$ is precompact in $\mathfrak{Y}$. A compact continuous mapping is called *completely continuous.*

We return to the problem of metrizable topologies. In TLS case, the metrizability theorem of §2 can be reformulated more accurately as follows.

**THEOREM 6.1.** *A Hausdorff* TLS *is metrizable if and only if it has a countable neighborhood base for zero.*

A TLS is called *locally bounded* if the zero vector has a bounded neighborhood.

**THEOREM 6.2.** *Any locally bounded Hausdorff* TLS *is metrizable.*

## §7. Normed spaces

**1.** In a topological linear space, the linear structure is compatible with the topology. This compatibility is more at hand if the topology is induced by a norm. A *norm* on a linear space $\mathfrak{X}$ is a real-valued functional, usually denoted by $\| \ \|$, which verifies the conditions ($x, y \in \mathfrak{X}$, $\alpha \in \mathbb{K}$):

1) $\|x\| \geqslant 0$; $\|x\| = 0$ if and if $x = 0$;
2) $\|x + y\| \leqslant \|x\| + \|y\|$ (the triangle inequality);
3) $\|\alpha x\| = |\alpha| \cdot \|x\|$.

A linear space $\mathfrak{X}$ with a given norm $\| \ \|$ on it is called a normed space and is denoted by $(\mathfrak{X}, \| \ \|)$. In a normed space, the formula $\rho(x, y) = \|x - y\|$ defines a metric. This metric induces on $\mathfrak{X}$ a topology called the *norm topology.* Two norms $\| \ \|_1$ and $\| \ \|_2$ on $\mathfrak{X}$ are said to be *equivalent* if there exist $c_1, c_2 > 0$ such that $c_1\|x\|_1 \leqslant \|x\|_2 \leqslant c_2\|x\|_1$ for every $x \in \mathfrak{X}$. A normed space which is complete in norm topology is called a *Banach space.*

**EXAMPLES. 7.1.** The $n$-dimensional vector space $\mathbb{C}^n$ is a Banach space relatively to any of the following norms:

$$(1)\quad \|x\| = \sqrt{\sum_{i=1}^{n} \|x_i\|^2}; \quad (2)\quad \|x\| = \max_{1\leqslant i\leqslant n} |x_i|; \quad (3)\quad \|x\| = \sum_{i=1}^{n} |x_i|.$$

The norm (2) is called the *Tchebycheff norm.*

**7.2.** The space $\mathfrak{M}$ of all bounded real or complex sequences $x = (x_1, \ldots, x_n, \ldots)$ with the norm $\|x\| = \sup_i |x_i|$ is an infinite dimensional Banach space.

**7.3.** The space $L_p[a, b]$ of all complex-valued functions $x(t)$ defined on an interval $[a, b] \subseteq \mathbb{R}$ and such that $|x(t)|^p$ $(p \geqslant 1)$ is absolutely integrable on $[a, b]$, with the norm

$$\|x\| = \left( \int_a^b |x(t)|^p \mathrm{d}t \right)^{1/p},$$

is an infinite dimensional Banach space.

**7.4.** Consider now the infinite dimensional linear space of all functions $x(t)$ defined on an interval $[a, b]$, continuous on $[a, b]$, and having continuous derivatives up to an order $k$ inclusively. On this space we define a norm by

$$\|x\| = \max\left\{ \max_{t\in[a,b]} |x(t)|, \max_{t\in[a,b]} |x'(t)|, \ldots, \max_{t\in[a,b]} |x^{(k)}(t)| \right\}.$$

Thus we obtain a Banach space denoted by $C^k[a, b]$.

**2.** A normed space is a metric space, so all properties of metric spaces mentioned above are true for normed spaces. For convenience, we remind some of them.

One-point sets in a normed space are closed, normed spaces are normal, so Hausdorff.

A normed space satisfies the first axiom of countability. The second axiom of countability is satisfied if and only if the normed space is separable.

The norm topology can be described using sequences (a sequence $\{s_n \mid n \in \mathbb{N}\}$ is norm convergent to $s$ if and only if $\|s_n - s\| \to 0$ as $n \to \infty$).

Any normed space $\mathfrak{X}$ is isometrically isomorphic with a dense subset of a Banach space $\mathfrak{B}$. The space $\mathfrak{B}$ is called a *completion* of $\mathfrak{X}$.

A subset $\mathfrak{A}$ of a normed space $\mathfrak{X}$ is bounded if and only if there exists a number $M > 0$ such that $\|a\| \leqslant M$ for all $a \in \mathfrak{A}$.

Let us introduce now a few definitions:

The segment joining the points $x_1$ and $x_2$ is the set of elements of the form $y = (1-t)x_1 + tx_2$, $0 \leqslant t \leqslant 1$. A set $\mathfrak{A} \subseteq \mathfrak{X}$ is called *convex* if $x_1, x_2 \in \mathfrak{A}$ imply that the segment joining $x_1$ and $x_2$ is completely contained in $\mathfrak{A}$. It follows easily that in a normed space any $r$-ball (open or closed) is convex.

A lineal $\mathfrak{L}$, closed in the norm topology, is called a *subspace* of the normed space. Any subspace is obviously convex. Every finite dimensional lineal is a subspace.

**THEOREM 7.1.** *A normed space is a Montel space if and only if it is finite dimensional.*

**3.** TOPOLOGICAL DIRECT SUM. Let $(\mathfrak{X}, \|\ \|)$ be a normed space, and let $\mathfrak{X} = \sum_{i=1}^{n} \dot{+} \mathfrak{L}_i$ be the decomposition of the linear space $\mathfrak{X}$ in a direct sum of lineals $\mathfrak{L}_i$, $i = 1, \ldots, n$. This means that each vector $x \in \mathfrak{X}$ has a unique representation $x = x_1 + \ldots + x_n$, $x_i \in \mathfrak{L}_i$, $i = 1, \ldots, n$. If the coordinate projections $P_i : \mathfrak{X} \to \mathfrak{L}_i$ are continuous in the norm topology, the direct sum $\sum_{i=1}^{n} \dot{+} \mathfrak{L}_i$ is called the *topological direct sum* of the lineals $\mathfrak{L}_i$.

**THEOREM 7.2.** *Let $\mathfrak{X}$ be a Banach space and assume that the lineals $\mathfrak{L}_i$, $i = 1, \ldots, n$ are subspaces. If $\mathfrak{X} = \sum_{i=1}^{n} \dot{+} \mathfrak{L}_i$ then the direct sum is a topological direct sum.*

**4.** We close this section considering the connection between real and complex normed spaces.

Let $\mathfrak{X}$ be a real normed space and let i denote the imaginary unit: $\mathrm{i}^2 = -1$. Set $\mathrm{i}\mathfrak{X} = \{\mathrm{i}x \mid x \in \mathfrak{X}\}$ and form the topological direct sum $\mathfrak{X} \dot{+} \mathrm{i}\mathfrak{X} = \{x + \mathrm{i}y \mid x, y \in \mathfrak{X}\}$. Define a norm in $\mathfrak{X} + \mathrm{i}\mathfrak{X}$ by $\|x + \mathrm{i}y\| = \sqrt{\|x\|^2 + \|y\|^2}$. Thus, $\mathfrak{X} + \mathrm{i}\mathfrak{X}$ is a complex normed space, called the *complexification* of $\mathfrak{X}$. On the other hand, a complex normed space can be represented in many cases as a direct sum of two real normed spaces. Thus, let $\mathfrak{X}$ be a complex normed space with a semilinear involution on it, i.e., a bijection $g : \mathfrak{X} \to \mathfrak{X}$ such that $g(x + y) = g(x) + g(y)$, $g(\alpha x) = \overline{\alpha} g(x)$ and $g(g(x)) = x$ for all $x, y \in \mathfrak{X}$, $\alpha \in \mathbb{C}$. An element $x \in \mathfrak{X}$ with $g(x) = x$ is called real, and an element $v \in \mathfrak{X}$ with $g(v) = -v$ is called pure imaginary. Then any element $x \in \mathfrak{X}$ can be uniquely represented in the form $x = u + \mathrm{i}v$ where $u$ is real and $v$ is pure imaginary, namely $u = \frac{1}{2}(x + g(x))$, $v = \frac{1}{2\mathrm{i}}(x - g(x))$. Therefore $\mathfrak{X} = \mathfrak{Z} \dot{+} \mathrm{i}\mathfrak{Z}$, where

$\mathfrak{Z} = \{u \in \mathfrak{X} \mid g(u) = u\}$. So the complex space $\mathfrak{X}$ is a direct sum of two real linear spaces. If the involution $g$ is continuous, then $\mathfrak{Z}$ is a subspace of $\mathfrak{X}$ and therefore (by Theorem 7.2) the considered direct sum is topological.

For example, consider the complex space $\mathbb{C}^n$ of vectors $x = (x_1, \ldots, x_n)$, $x_i \in \in \mathbb{C}$, $i = 1, \ldots, n$. As an involution $g$ on $\mathbb{C}^n$ take the complex conjugation: $g(x) = = \overline{x}\ (= (\overline{x}_1, \ldots, \overline{x}_n))$. Therefore $\mathbb{C}^n$ may be considered as a $2n$-dimensional real space.

## §8. Linear operators and functionals

**1.** LINEAR OPERATORS. By an established tradition, we will use, as a rule, the term "operator" for a mapping between normed spaces (see §1). Also as a rule, the operators will be denoted by capital Latin characters. The terms "function" and "functional" will have the previously established meanings.

Let $\mathfrak{X}$ and $\mathfrak{Y}$ be normed spaces, both real or complex. An operator $A: \mathfrak{X} \to \mathfrak{Y}$ is called *linear* if its domain $D(A)$ is a lineal of $\mathfrak{X}$ and $A(\alpha x + \beta y) = \alpha A(x) + \beta A(x)$ for any $x, y \in D(A)$, $\alpha, \beta \in \mathbb{K}$. A linear operator $A$ is *bounded* if $\|Ax\| \leqslant M < \infty$ for all $x \in D(A)$ with $\|x\| = 1$. In this case, the (finite) number $\sup\limits_{x \in D(A),\, \|x\|=1} \|Ax\|$ is called the *norm* of the operator $A$ and is denoted by $\|A\|$.

**THEOREM 8.1.** *A linear operator is continuous if and only if it is bounded.*

**EXAMPLES. 8.1.** Consider the space $\mathbb{C}^n$ (Example 7.1) with any of the norm (1)–(3), and a rectangular matrix $(a_{ik})$ of order $n \times m$, $i \in \{1, \ldots, m\}$, $k \in \{1, \ldots \ldots, n\}$, $a_{ik} \in \mathbb{C}$. The equalities

$$y_i = \sum_{k=1}^{n} a_{ik} x_k, \qquad i = 1, \ldots, m$$

define a linear continuous operator $A: \mathbb{C}^n \to \mathbb{C}^m$ ($y = Ax$ for $x = (x_1, \ldots, x_n) \in \in \mathbb{C}^n$, $y = (y_1, \ldots, y_m) \in \mathbb{C}^m$). Note that in a finite dimensional normed space any linear operator is continuous.

**8.2.** On $L_p[a, b]$ (Example 7.3) with $p = 1$, let us consider the integral operator given by

$$y(t) = \int_a^b K(t, s) x(s) \mathrm{d}s,$$

where $K(t,s)$ is a continuous function on the square $a \leqslant t, s \leqslant b$. This is a linear and bounded operator from $L_p[a,b]$ into itself.

**8.3.** Consider the space $C^k[a,b]$ (Example 7.4) and define a differential operator by

$$y(t) = \frac{d^{k+1}}{dt^{k+1}} x(t).$$

This operator is defined on the dense lineal of $C^k[a,b]$ consisting of all functions which have continuous derivatives on $[a,b]$ up to the order $k+1$ inclusively and takes values in $C^0[a,b]$; it is linear but it is not bounded.

**2.** The set $L(\mathfrak{X},\mathfrak{Y})$ of all continuous linear operators defined from the whole space $\mathfrak{X}$ into $\mathfrak{Y}$, where $\mathfrak{X}$ and $\mathfrak{Y}$ are normed spaces, becomes a normed space with the linear operations:

1) $(A+B)x = Ax + Bx$,

2) $(\alpha A)x = \alpha(Ax)$,

(for $A, B \in L(\mathfrak{X},\mathfrak{Y})$, $x \in \mathfrak{X}$, $\alpha \in \mathbb{K}$) and the norm $\|A\| = \sup\limits_{\|x\|=1} \|Ax\|$. In addition, if $\mathfrak{Y}$ is a Banach (i.e., complete) space, then $L(\mathfrak{X},\mathfrak{Y})$ is a Banach space, too. The composition $A \circ B$ of the operators $A \in L(\mathfrak{Y},\mathfrak{Z})$ and $B \in L(\mathfrak{X},\mathfrak{Y})$ is usually called the product of the operators $A$ and $B$, and is also denoted by $AB$. We have $\|AB\| \leqslant \|A\| \cdot \|B\|$, thus $AD \in L(\mathfrak{X},\mathfrak{Z})$.

If $\mathfrak{X} = \mathfrak{Y}$, we write $L(\mathfrak{X})$ instead of $L(\mathfrak{X},\mathfrak{X})$.

The convergence of a sequence of bounded linear operators in the norm of the space $L(\mathfrak{X},\mathfrak{Y})$ is called the *uniform convergence*, and the corresponding topology is called the *uniform operator topology*. We consider also a weaker topology, called the *strong operator topology*. The convergence of a directedness $\{A_d \mid d \in \mathfrak{D}\} \subseteq L(\mathfrak{X},\mathfrak{Y})$ in the strong operator topology is defined as the convergence of any directedness $\{A_d x \mid d \in \mathfrak{D}\} \subseteq \mathfrak{Y}$, for all $x \in \mathfrak{X}$. This topology coincides with the topology induced on $L(\mathfrak{X},\mathfrak{Y})$ by the product topology of $\prod\limits_{x\in\mathfrak{X}} \mathfrak{Y} = \mathfrak{Y}^{\mathfrak{X}}$, where we endow $\mathfrak{Y}$ with the norm topology.

A sequence $\{A_n \mid n \in \mathbb{N}\} \subseteq L(\mathfrak{X},\mathfrak{Y})$ is a Cauchy sequence with respect to the strong operator topology if $\{A_n x \mid n \in \mathbb{N}\}$ is a Cauchy sequence in $\mathfrak{Y}$ for every $x \in \mathfrak{X}$.

**THEOREM 8.2.** *If $\mathfrak{X}$ and $\mathfrak{Y}$ are Banach spaces, then any Cauchy sequence with respect to the strong operator topology in $L(\mathfrak{X},\mathfrak{Y})$ has a limit.*

3. INVERSE OPERATORS. Let $\mathfrak{X}$ and $\mathfrak{Y}$ be normed spaces and $A: \mathfrak{X} \to \mathfrak{Y}$ be a linear operator. If there exists an operator $A^{-1}$ defined on $R(A)$ with values in $D(A)$ such that

$$A^{-1}A = I_{D(A)}, \quad AA^{-1} = I_{R(A)},$$

then $A$ is called *invertible* (see the general definition of the inverse relation in §1); $A^{-1}$ is called the *inverse* of $A$. Clearly $(A^{-1})^{-1} = A$, and $A^{-1}$ is linear.

A linear operator $A$ is invertible if and only if from $x \in D(A)$ and $Ax = 0$ it follows that $x = 0$.

The following three theorems deal with the problem of existence and continuity of the inverse operator. In the first two of them, the continuity of the operator is not assumed.

**THEOREM 8.3.** *Let $A: \mathfrak{X} \to \mathfrak{Y}$ be a linear operator satisfying the condition*

$$\|Ax\| \geqslant c\|x\|$$

*for any $x \in \mathfrak{X}$, where $c$ is a positive constant. Then $A^{-1}$ exists, is bounded, and $\|A^{-1}\| \leqslant c^{-1}$.*

**THEOREM 8.4.** *Let $A: \mathfrak{X} \to \mathfrak{Y}$ be a linear operator such that $A^{-1}$ exists and is continuous; let $B$ be a continuous operator on $D(A)$ such that $\|B\| < \|A^{-1}\|^{-1}$. Then the operator $C = A + B$ has continuous inverse $C^{-1}: R(A) \to D(A)$ and*

$$\|C^{-1} - A^{-1}\| \leqslant \frac{\|B\|}{1 - \|A^{-1}\| \, \|B\|} \|A^{-1}\|^2.$$

**THEOREM 8.5** (Banach). *A continuous linear isomorphism of the Banach spaces $\mathfrak{X}$ and $\mathfrak{Y}$ is a homeomorphism.*

A linear isomorphism between two normed spaces, which is also a homeomorphism is called a linear homeomorphism.

**THEOREM 8.6.** *Let $n \in \mathbb{N}$ be fixed. All $n$-dimensional normed spaces are linearly homeomorphic with $\mathbb{R}^n$, so all of them are linearly homeomorphic.*

*Consequently, all norms on a given finite dimensional space are equivalent.*

4. LINEAR FUNCTIONALS. Let $\mathfrak{X}$ be a normed space, $\mathfrak{L}$ a lineal of $\mathfrak{X}$ and $f$ a linear functional defined on $\mathfrak{L}$. The following theorem shows the possibility of extending the

functional $f$ on the whole space $\mathfrak{X}$ without increasing its norm. For real spaces this theorem was proved by Hahn and Banach; it was extended to the case of complex spaces by G. A. Suchomlynoff.

**THEOREM 8.7** (Hahn-Banach-Suchomlynoff). *Any linear functional $f$, defined on a lineal $\mathfrak{L}$ of a normed space $\mathfrak{X}$, can be extended on $\mathfrak{X}$ with the preservation of the norm, i.e., there exists a linear functional $F$ on $\mathfrak{X}$ such that*

1) $F(x) = f(x)$ *for all* $x \in \mathfrak{L}$;
2) $\|F\| = \|f\|$.

**COROLLARY 8.1.** *Let $0 \neq x_0$ be an arbitrary element of the normed space $\mathfrak{X}$. Then there exists a linear functional $f$ defined on $\mathfrak{X}$ such that*

1) $\|f\| = 1$,
2) $f(x_0) = \|x_0\|$.

**COROLLARY 8.2.** *Let $\mathfrak{X}$ be a normed space, $\mathfrak{L}$ a lineal in $\mathfrak{X}$, and let $x_0$ be a point in $\mathfrak{X}$ at the distance $d > 0$ from $\mathfrak{L}$ ($d = \inf\limits_{x \in \mathfrak{L}} \|x - x_0\|$). Then there exists a linear functional $f$ defined on $\mathfrak{X}$ such that*

1) $f(x) = 0$ *for* $x \in \mathfrak{L}$;
2) $f(x_0) = 1$;
3) $\|f\| = d^{-1}$.

One of the basic results in Banach Space Theory is next theorem.

**THEOREM 8.8** (Banach-Steinhaus principle of uniform boundedness). *Let $\{x_n \mid n \in \in \mathbb{N}\}$ be a sequence of vectors in (a Banach space) $\mathfrak{X}$ such that the numerical sequence $\{\langle x_n, f\rangle \mid n \in \mathbb{N}\}$ is bounded for each continuous linear functional $f$ on $\mathfrak{X}$. Then $\|x_n\| \leqslant M < \infty$.*

## §9. Conjugate space. Conjugate operator

**1.** Let $\mathfrak{X}$ be a normed space. The space of all continuous linear functionals on $\mathfrak{X}$ is called the *conjugate space* of $\mathfrak{X}$ and is denoted by $\mathfrak{X}^*$. Thus, $\mathfrak{X}^* = L(\mathfrak{X}, \mathbb{K})$, where $\mathbb{K}$ is $\mathbb{R}$ or $\mathbb{C}$ depending on the fact that $\mathfrak{X}$ is a real or a complex space. Since $\mathbb{R}$ and $\mathbb{C}$ are complete, $\mathfrak{X}^*$ is a Banach space.

**EXAMPLES. 9.1.** Using Example 8.1 for $m = 1$, it follows that for each linear functional $f$ on $\mathbb{C}^n$ there exist $n$ complex numbers $a_1, \ldots, a_n$ such that $f(x) = \sum_{k=1}^{n} a_k x_k$, for all $x = (x_1, \ldots, x_n) \in \mathbb{C}^n$. Thus, $f$ can be identified with $(a_1, \ldots, a_n) \in \mathbb{C}^n$. According to the definition of the norm on $L(\mathfrak{X}, \mathfrak{Y})$, the norms (2) and (3) from Example 7.1 induce in the conjugate space the norms (3) and (2), respectively; the norm (1) induces the same norm, so $(\mathbb{C}^n, \| \|_1)^* = (\mathbb{C}^n, \| \|_1)$.

**9.2.** For the space $L_p[a, b]$ described in Example 7.3, the conjugate space is $L_q[a, b]$, where $q = \dfrac{p}{p-1}$.

**2.** A normed space is called *strictly convex* if from $x, y \in \mathfrak{X}$ and $\|x + y\| = \|x\| + \|y\|$ it follows that either $x = \lambda y$, or $y = \mu x$, $\lambda, \mu \in \mathbb{K}$. A strictly convex space is called *uniformly convex* if from $x, y \in \mathfrak{X}$, $\|x\| = \|y\| = 1$ and $\|x - y\| \geqslant \varepsilon > 0$ it follows that $\|x + y\| \leqslant 2 - \delta$ for a certain $\delta = \delta(\varepsilon) > 0$.

**THEOREM 9.1.** *Let $\mathfrak{X}$ be a strictly convex normed space and $x, y \in \mathfrak{X}$ such that $\|x\| = \|y\| > 0$ and $\langle x, f\rangle = \langle y, f\rangle$ for a certain $f \in \mathfrak{X}^*$ such that $\|f\| = 1$ and $\langle x, f\rangle = \|x\|$. Then $x = y$.*

**3.** Let $\mathfrak{X}$ be a normed space and $\mathfrak{X}^*$ its conjugate space. Since $\mathfrak{X}^*$ is a normed space, we may construct $\mathfrak{X}^{**} = (\mathfrak{X}^*)^*$, and so on. Let $f \in \mathfrak{X}^*$. The expresion $f(x)$ for a fixed $x \in \mathfrak{X}$ can be considered as a continuous linear functional $F_x$ (in the variable $f$) on the space $\mathfrak{X}^*$: $F_x(f) = f(x)$. (Based on these considerations, we will often use the notation $\langle x, f\rangle$ — see §1 — for the value of $f$ at $x$.) In addition, $\|F_x\| = \|x\|$. Thus, the space $\mathfrak{X}$ is isometrically linearly isomorphic to a subspace of $\mathfrak{X}^{**}$. Up to this isomorphism we can consider that $\mathfrak{X} \subseteq \mathfrak{X}^{**}$. If this isomorphism is onto $\mathfrak{X}^{**}$ the space $\mathfrak{X}$ is called *reflexive*. If $\mathfrak{X}$ is nonreflexive, all spaces in the sequence $\mathfrak{X}, \mathfrak{X}^*, \mathfrak{X}^{**}, \mathfrak{X}^{***}, \ldots$ are different. For instance, the spaces in Examples 7.1 and 7.3 are reflexive, and the spaces in Examples 7.2 and 7.4 are nonreflexive.

**4.** Let $\mathfrak{X}$ and $\mathfrak{Y}$ be normed spaces and $A: \mathfrak{X} \to \mathfrak{Y}$ a linear operator densely defined in $\mathfrak{X}$ ($D(A)$ is dense in $\mathfrak{X}$). For any $y^* \in \mathfrak{Y}^*$, consider the linear functional on $\mathfrak{X}$ given by $\langle Ax, y^*\rangle$, $x \in \mathfrak{X}$. If there exists an element $x^* \in \mathfrak{X}^*$ such that $\langle Ax, y^*\rangle = \langle x, x^*\rangle$ for all $x \in \mathfrak{X}$, then one says that the *conjugate operator* $A^*$ is defined at $y^*$ and $A^* y^* = x^*$. All these $y^*$ form the domain $D(A^*)$ of $A^*$ (the case $D(A^*) = \{0\}$ may occur), so $A^*: \mathfrak{Y}^* \to \mathfrak{X}^*$. If, in particular, $A \in L(\mathfrak{X}, \mathfrak{Y})$, i.e., the operator $A$ is defined on $\mathfrak{X}$ and is continuous, then $A^* \in L(\mathfrak{Y}^*, \mathfrak{X}^*)$ and $\|A^*\| = \|A\|$. For instance, the conjugate of the operator $A$ in Example 8.1, given by the matrix $(a_{ik})$, $i \in \{1, \dots, m\}$, $k \in \{1, \dots, n\}$ is the operator $A^*$ given by the matrix $(\overline{a}_{ki})$ — i.e., the transposed complex-conjugate matrix.

**5.** ORTHOGONALITY. ORTHOGONAL COMPLEMENT. Let us return to the expression $\langle x, f\rangle$ — the value of $f \in \mathfrak{X}^*$ at $x \in \mathfrak{X}$ — and consider it as a semilinear functional in the second variable when the first one is fixed, i.e., $\langle x, \alpha f_1 + \beta f_2\rangle = = \overline{\alpha}\langle x, f_1\rangle + \overline{\beta}\langle x, f_2\rangle$, $f_1, f_2 \in \mathfrak{X}^*$, $\alpha, \beta \in \mathbb{K}$. As a consequence, we obtain a *sesquilinear* functional (linear in the first variable and semilinear in the second) which maps $\mathfrak{X} \times \mathfrak{X}^*$ in $\mathbb{K}$. This functional $\langle x, f\rangle$ is called the *scalar* (or *inner*) *product* of $x \in \mathfrak{X}$ and $f \in \mathfrak{X}^*$.

The elements $x \in \mathfrak{X}$ and $f \in \mathfrak{X}^*$ are called *orthogonal* if $\langle x, f\rangle = 0$. The sets $\mathfrak{M} \subset \mathfrak{X}$ and $\mathfrak{M}^* \subseteq \mathfrak{X}^*$ are *orthogonal* if $\langle x, f\rangle = 0$ for any $x \in \mathfrak{M}$ and $f \in \mathfrak{M}^*$. The set $\mathfrak{M}^\perp$ of all elements in $\mathfrak{X}^*$ orthogonal to $\mathfrak{M} \subseteq \mathfrak{X}$ is called the *orthogonal complement* of $\mathfrak{M}$. The set ${}^\perp\mathfrak{M}^*$ of all elements of $\mathfrak{X}$ orthogonal to a set $\mathfrak{M}^* \subseteq \mathfrak{X}^*$ is called the *-orthogonal complement of $\mathfrak{M}^*$.

**THEOREM 9.2.** *For any $\mathfrak{M} \subseteq \mathfrak{X}$ ($\mathfrak{M}^* \subseteq \mathfrak{X}^*$), its orthogonal (*-orthogonal) complement is a subspace of $\mathfrak{X}^*$ ($\mathfrak{X}$) and ${}^\perp\mathfrak{M}^\perp = \operatorname{clin}\mathfrak{M}$ (${}^\perp\mathfrak{M}^{*\perp} = \operatorname{clin}\mathfrak{M}^*$), where $\operatorname{clin}\mathfrak{M}$ is the smallest closed subspace of $\mathfrak{X}$ containing $\mathfrak{M}$.*

## §10. Weak topology and reflexivity

**1.** WEAK TOPOLOGY. A normed space $\mathfrak{X}$ carries, along with the norm topology, topologies induced by linear functionals. The *weak topology* on $\mathfrak{X}$ is defined as follows: a directedness $\{x_d \mid d \in \mathfrak{D}\}$ is *weak convergent* if the numerical directednesses $\{\langle x_d, f\rangle \mid d \in \mathfrak{D}\}$ are convergent for every $f \in \mathfrak{X}^*$. In general, the weak topology is weaker than the norm topology. However, in the case of finite dimensional spaces these topologies coincide.

**THEOREM 10.1.** *In a finite dimensional space the weak topology coincides with the norm topology.*

We return to the general case of normed space.

**THEOREM 10.2.** *If a sequence $\{x_n \mid n \in \mathbb{N}\}$ is weakly convergent to $x_0$, then the* *sequence is norm bounded. In addition*

$$\|x_0\| \leqslant \underline{\lim}\, \|x_n\|.$$

**2.** Let $\mathfrak{X} = \mathfrak{Y}^*$ be the conjugate space of a normed space $\mathfrak{Y}$. The *ultraweak* (*-weak) *topology* on $\mathfrak{X}$ is by definition the topology induced on $\mathfrak{X}$ by the topology of the direct product $\prod\limits_{y \in \mathfrak{Y}} \mathbb{K}$. Thus, a directedness $\{x_d \mid d \in \mathfrak{D}\}$ is ultraweakly convergent if any directedness $\{\langle y, x_d\rangle \mid d \in \mathfrak{D}\}$ is convergent for each $y \in \mathfrak{Y}$. In other words, the ultraweak convergence is the pointwise convergence.

**THEOREM 10.3.** *The ultraweak topology is weaker than the weak topology, and, in the case of a reflexive space, these two topologies coincide.*

**THEOREM 10.4** (Alaoglu-Bourbaki). *The closed unit ball of the space $\mathfrak{X}^*$ (the conjugate of the normed space $\mathfrak{X}$) is compact in the ultraweak topology. This ball is compact in the weak topology if and only if the space is reflexive.*

**3.** We consider now the action of a linear operator between normed spaces endowed with the weak topology.

**THEOREM 10.5.** *Let $\mathfrak{X}$ and $\mathfrak{Y}$ be normed spaces. A linear operator $A: \mathfrak{X} \to \mathfrak{Y}$ is bounded if and only if is continuous with respect to the weak topologies on $\mathfrak{X}$ and $\mathfrak{Y}$.*

Recall that an operator $A: \mathfrak{X} \to \mathfrak{Y}$ is completely continuous (see §6.4) if a) is continuous and b) the image of any bounded set in $D(A)$ is a precompact set in $\mathfrak{Y}$. For a linear operator, condition a) follows from condition b). Indeed, by b) the image of the unit ball of $\mathfrak{X}$ is bounded in $\mathfrak{Y}$ and hence the operator $A$ is bounded, and by Theorem 8.1 it is continuous. Thus, the next result holds.

**THEOREM 10.6.** *A linear operator $A: \mathfrak{X} \to \mathfrak{Y}$ is completely continuous if and only if the image of any bounded set in $D(A)$ is precompact in $\mathfrak{Y}$.*

**THEOREM 10.7.** *The range of a completely continuous operator is separable.*

An operator $A: \mathfrak{X} \to \mathfrak{Y}$ is called *very continuous* if it maps any weak convergent sequence in $\mathfrak{X}$ into a norm convergent sequence in $\mathfrak{Y}$.

**THEOREM 10.8.** *Any completely continuous operator is very continuous. If $\mathfrak{X}$ is reflexive, the converse is also true. If $\mathfrak{X}$ is the conjugate space of a certain normed* *space then the completely continuity of the operator $A$ is equivalent to the fact that $A$ maps any ultraweakly convergent sequence into a norm convergent sequence.*

Note that in a finite dimensional space $\mathfrak{X}$ any linear operator is completely continuous. A linear operator in a Banach space is called *finite dimensional* if its range is a finite dimensional space. Obviously, any bounded finite dimensional linear operator is completely continuous.

We conclude this section with some conditions for metrizability of weak and ultraweak topologies.

**THEOREM 10.9.** *Any bounded subset of a normed space is metrizable in the weak topology. If the space is the conjugate of a normed space, the same is true for the ultraweak topology.*

**COROLLARY 10.1.** *The weak and the ultraweak convergence in any bounded set of a normed space can be characterized in terms of sequences.*

## §11. Hilbert spaces

**1.** Let $\mathfrak{H}$ be a linear space. A functional $(\cdot,\cdot)$ defined on the product $\mathfrak{H}\times\mathfrak{H}$ is called a *scalar product* (or an inner product) if it satisfies the following conditions ($x,y,z\in\mathfrak{H}$, $\lambda\in\mathbb{K}$):

a) $(x,y)=\overline{(y,x)}$ (the bar denotes the complex conjugation);

b) $(x+y,z)=(x,z)+(y,z)$;

c) $(\lambda x,y)=\lambda(x,y)$;

d) $(x,x)\geqslant 0$; $(x,x)=0$ if and only if $x=0$.

A space $\mathfrak{H}$ with a scalar product is called a *pre-Hilbert space*. Using the scalar product, it is possible to define a norm putting $\|x\|=(x,x)^{\frac{1}{2}}$. If $\mathfrak{H}$ is complete with respect to this norm, then it is called a *Hilbert space*.

**EXAMPLES. 11.1.** In the space $\mathbb{C}^n$ the formula

$$(x,y)=\sum_{i=1}^{n}x_i\overline{y}_i,$$

for $x=(x_1,\dots,x_n)$ and $y=(y_1,\dots,y_n)$, defines a scalar product. Then $\|x\|=$ $=\left(\sum\limits_{i=1}^{n}|x_i|^2\right)^{\frac{1}{2}}$, i.e., the norm given by this scalar product is the norm (1) in Example

7.1. The space $\mathbb{C}^n$ with this scalar product is a Hilbert space.

**11.2.** In the linear space of all real or complex sequences $x = (x_1, \ldots, x_n, \ldots)$ for which $\sum_{i=1}^{\infty} |x_i|^2$ is finite, with the linear operations defined coordinatewise, consider the scalar product

$$(x, y) = \sum_{i=1}^{\infty} x_i \overline{y}_i,$$

where $y = (y_1, \ldots, y_n, \ldots)$. We obtain a Hilbert space, called the space $\ell_2$.

**11.3.** The space $L_2[a, b]$ (the case $p = 2$ in Example 7.3) is a Hilbert space with respect to the scalar product

$$(x, y) = \int_a^b x(t)\overline{y(t)}dt.$$

**2.** The sets $\mathfrak{M}$ and $\mathfrak{N}$ in a Hilbert space $\mathfrak{H}$ are called *orthogonal* if $(x, y) = 0$ for all $x \in \mathfrak{M}$ and $y \in \mathfrak{N}$. The set $\mathfrak{M}^{\perp}$ of all vectors in $\mathfrak{H}$ orthogonal to $\mathfrak{M}$ is called the orthogonal complement of the set $\mathfrak{M}$ in $\mathfrak{H}$. We put $\mathfrak{M}^{\perp\perp} = (\mathfrak{M}^{\perp})^{\perp}$. As above (see Theorem 9.2), the next result holds:

**THEOREM 11.1.** *Let $\mathfrak{H}$ be a Hilbert space. For any $\mathfrak{M} \subseteq \mathfrak{H}$, its orthogonal complement $\mathfrak{M}^{\perp}$ is a linear subspace of $\mathfrak{H}$ and $\mathfrak{M}^{\perp\perp} = \operatorname{clin} \mathfrak{M}$.*

The direct sum of a subspace $\mathfrak{L}$ of a Hilbert space $\mathfrak{H}$ with its orthogonal complement $\mathfrak{L}^{\perp}$ is called an *orthogonal direct sum* and is denoted by $\mathfrak{L} \oplus \mathfrak{L}^{\perp}$.

**THEOREM 11.2.** *If $\mathfrak{H}$ is a Hilbert space and $\mathfrak{L}$ is an arbitrary subspace of $\mathfrak{H}$, then $\mathfrak{H} = \mathfrak{L} \oplus \mathfrak{L}^{\perp}$.*

The decomposition $\mathfrak{H} = \mathfrak{L} \oplus \mathfrak{L}^{\perp}$ produces a pair of complementary projections $P_1, P_2$ ($P_2 = I - P_1$) on the subspaces $\mathfrak{L}$ and $\mathfrak{L}^{\perp}$, respectively. These projections are called *orthogonal projections* (since their ranges are orthogonal).

**3.** Let $\mathfrak{A}$ be an indexing set. A system of vectors $\{e_a \mid a \in \mathfrak{A}\}$ in a Hilbert space $\mathfrak{H}$ is called *orthonormal* if $(e_a, e_b) = \delta_{ab}$, where $\delta_{ab}$ are the Kronecker symbols: $\delta_{ab}$ is 1 if $a = b$ and $\delta_{ab}$ is 0 for $a \neq b$. An orthonormal system $\{e_a \mid a \in \mathfrak{A}\}$ is called *complete* if there are no nonzero elements in $\mathfrak{H}$ orthogonal to all vectors of the system, i.e., if $x \in \mathfrak{H}$ and $(x, e_a) = 0$ for all $a \in \mathfrak{A}$ then $x = 0$. For any orthonormal system $\{e_a \mid a \in \mathfrak{A}\}$ and each $x \in \mathfrak{H}$ the set of nonzero numbers of the form $(x, e_a)$ is at most countable and, with a slight change of notation, we can write

$$\sum_{i=1}^{\infty} |(x, e_i)|^2 \leqslant \|x\|^2.$$

The last relation is called the *Bessel inequality*. If the orthonormal system $\{e_a \mid a \in \mathfrak{A}\}$ is complete, then we have the relation

$$\sum_{i=1}^{\infty} |(x, e_i)|^2 = \|x\|^2,$$

called the *Parceval equality*. A complete orthonormal system is called a *basis* of $\mathfrak{H}$. If the Hilbert space $\mathfrak{H}$ is separable, then all orthonormal systems are at most countable and if, in addition, $\mathfrak{H}$ is infinite dimensional, then all bases are countable.

**4.** Theorem 8.6 asserts that all normed spaces of a fixed finite dimension are homeomorphic. More is true for Hilbert spaces.

**THEOREM 11.3.** *All Hilbert spaces of a fixed finite dimension are isometrically linearly isomorphic. All real (complex) infinite dimensional separable Hilbert spaces are isometrically linearly isomorphic with the real (complex) space $\ell_2$.*

**5.** The next theorem shows that any Hilbert space is isometrically semilinearly isomorphic with its conjugate space.

**THEOREM 11.4** (F. Riesz). *Let $\mathfrak{H}$ be a Hilbert space and $f \in \mathfrak{H}^*$. Then there exists a vector $x_f \in \mathfrak{H}$ such that $\|f\| = \|x_f\|$ and $f(x) = (x, x_f)$ for each $x \in \mathfrak{H}$.*

Riesz Theorem implies that any Hilbert space $\mathfrak{H}$ is *selfconjugate*, i.e., the conjugate space $\mathfrak{H}^*$ coincides with $\mathfrak{H}$ up to an isometrically semilinear isomorphism; consequently any Hilbert space is reflexive.

**6.** If $A$ is a densely defined linear operator in $\mathfrak{H}$ (i.e., $D(A)$ is dense in $\mathfrak{H}$), then it is possible to define the conjugate (*adjoint*) operator $A^*$ as in §9.4. So the adjoint $A^*$ verifies $(Ax, y) = (x, A^*y)$ for $x \in D(A)$ and $y \in D(A^*)$. If $A = A^*$ then $A$ is called *selfadjoint*; in this case $(Ax, y) = (x, Ay)$ for all $x, y \in D(A)$, and $D(A) = D(A^*)$. Selfadjoint operators are very important in various fields of mathematics and in applications. We will discuss them in more detail in Chapter IV.

# Chapter 1

# Differential calculus in normed spaces

## §1. The derivate and the differential of a nonlinear operator

Let $\mathfrak{X}$ and $\mathfrak{Y}$ be normed spaces over the field $\mathbb{K}$ (of complex or real numbers). We consider an operator $F$ defined on an open set $\mathfrak{D} \subseteq \mathfrak{X}$ and with values in $\mathfrak{Y}$ ($D(F) = \mathfrak{D}$, $R(F) \subseteq \mathfrak{Y}$). Accordingly, we write $F: \mathfrak{D} \to \mathfrak{Y}$ (see Chapter 0, §1).

Let $x_0$ be an element in $\mathfrak{D}$.

**DEFINITION 1.1.** Suppose that for every $h \in \mathfrak{X}$ the limit

$$\lim_{t \to 0} \frac{F(x_0 + th) - F(x_0)}{t} = \delta F(x_0, h) \tag{1.1}$$

exists, where $t \in \mathbb{K}$ is chosen such that $x_0 + th \in \mathfrak{D}$. Then the limit $\delta F(x_0, h)$ is called the *first variation* of the operator $F$ at the point $x_0 \in \mathfrak{D}$, and the operator $F$ is said to be *$\delta$-differentiable* at the point $x_0$.

**DEFINITION 1.2.** Suppose that $\delta F(x_0, h) = Ah$, where $A$ is a continuous linear operator defined everywhere on $\mathfrak{X}$ and with values in $\mathfrak{Y}$ ($A \in L(\mathfrak{X}, \mathfrak{Y})$). Then the operator $F$ is said to be *Gâteaux differentiable* ($G$-differentiable) at $x_0$. The operator $A$ above is called the *Gâteaux derivative* of $F$ at $x_0$, and is denoted by $A = F'(x_0)$. The variation $\delta F(x_0, h)$ is called then the *Gâteaux differential* and is denoted by $\mathrm{D}F(x_0, h)$.

The next example shows that the class $\mathcal{V}(x_0, \mathfrak{Y})$ of operators having the first variation at $x_0$ is strictly larger than the class $\mathcal{G}(x_0, \mathfrak{Y})$ of Gâteaux differentiable at $x_0$ operators.

**EXAMPLE 1.1.** Consider the operator $F: \mathbb{R}^2 \to \mathbb{R}$ given by

$$\begin{cases} F(x_1, x_2) = \dfrac{x_1^3}{x_1^2 + x_2^2}, & x_1^2 + x_2^2 > 0 \\ F(0,0) = 0. \end{cases}$$

Obviously, for each $h \in \mathbb{R}^2$, $h = (h_1, h_2) \neq 0$, we have

$$\frac{F(th_1, th_2) - F(0,0)}{t} = \frac{h_1^3}{h_1^2 + h_2^2} = F(h_1, h_2),$$

i.e., $\delta F(0, h)$ exists, but it is given by a nonlinear operator ($\delta F(0,h) = Fh$) and therefore $F$ is not Gâteaux differentiable.

The class $\mathcal{F}(x_0, \mathfrak{Y})$ of so-called $\mathcal{F}$-differentiable operators is even smaller.

**DEFINITION 1.3.** An operator $F: \mathfrak{D} \to \mathfrak{Y}$ is called *Fréchet differentiable* ($\mathcal{F}$-differentiable) at a point $x_0$ if there exists an operator $A \in L(\mathfrak{X}, \mathfrak{Y})$ such that for all $h$ with $x_0 + h \in \mathfrak{D}$ we have

$$F(x_0 + h) - F(x_0) = Ah + \omega(x_0, h), \tag{1.2}$$

where $\omega(x_0, h)$ satisfies the condition

$$\lim_{\|h\| \to 0} \frac{\|\omega(x_0, h)\|}{\|h\|} = 0. \tag{1.3}$$

In this case the expression $Ah$ is called the Fréchet differential of $F$ at $x_0$ and is denoted by $Ah = \mathrm{d}F(x_0, h)$.

Clearly, an operator which is Fréchet differentiable at $x_0$ is continuous at this point and also Gâteaux differentiable. Generally speaking, the converse is not true.

**EXAMPLE 1.2.** Let $F: \mathbb{R}^2 \to \mathbb{R}$ be the operator defined by the equalities

$$\begin{cases} F(x_1, x_2) = 1 & \text{if } x_2 = x_1^2,\ x_1 \neq 0 \\ F(x_1, x_2) = 0 & \text{otherwise,} \end{cases}$$

This operator is not continuous at $(0,0)$ and therefore is not Fréchet differentiable. However, for any $h \in \mathbb{R}^2$, $h = (h_1, h_2)$, we have

$$\frac{F(th_1, th_2) - F(0,0)}{t} = 0$$

for sufficiently small $t$. Consequently, the first variation $\delta F(0,h)$ exists; this is the Gâteaux differential of $F$ at 0.

So, in general, the inclusions $\mathcal{F}(x_0, \mathfrak{Y}) \subseteq \mathcal{G}(x_0, \mathfrak{Y}) \subseteq \mathcal{V}(x_0, \mathfrak{Y})$ are strict. However, the next result is true.

**THEOREM 1.1.** *Let $F: \mathfrak{D} \to \mathfrak{Y}$ be an operator which is Gâteaux differentiable at any point of a neighborhood $\mathfrak{U}$ of the point $x_0 \in \mathfrak{D}$, and such that its Gâteaux derivatives are the values of an operator-valued function*

$$F': \mathfrak{U} \to L(\mathfrak{X}, \mathfrak{Y})$$

*which is continuous on $\mathfrak{U}$. Then $F \in \mathcal{F}(x_0, \mathfrak{Y})$ and $\mathrm{D}F(x_0, h) = \mathrm{d}F(x_0, h)$.*

◁ From hypotheses, the limit (1.1) exists and $\delta(x_0, h) = F'(x_0)h = \mathrm{D}F(x_0, h)$. Choose $h$ such that $x_0 + th \in \mathfrak{U}$ for all $t \in [0,1]$ and consider the expression

$$\omega(x_0, h) = F(x_0 + h) - F(x_0) - \mathrm{D}F(x_0, h).$$

For each $y^* \in \mathfrak{Y}^*$ — the conjugate spaco of $\mathfrak{Y}$ (see Chapter 0, §4) — we have:

$$\langle \omega(x_0, h), y^* \rangle = \langle F(x_0 + h) - F(x_0), y^* \rangle - \langle \mathrm{D}F(x_0, h), y^* \rangle.$$

Consider now the real-valued function

$$\varphi(t) = \mathrm{Re}\langle F(x_0 + th), y^* \rangle.$$

We show that $\varphi$ is differentiable on $(0,1)$. Indeed,

$$\begin{aligned} \varphi'(t) &= \lim_{\Delta t \to 0} \frac{\varphi(t + \Delta t) - \varphi(t)}{\Delta t} = \\ &= \lim_{\Delta t \to 0} \mathrm{Re} \left\langle \frac{F(x_0 + th + \Delta th) - F(x_0 + th)}{\Delta t}, y^* \right\rangle = \\ &= \mathrm{Re}\langle \mathrm{D}F\left((x_0 + th), h\right), y^* \rangle. \end{aligned}$$

Applying Lagrange formula to the function $\varphi(t)$ we obtain

$$|\mathrm{Re}\langle \omega(x_0, h), y^* \rangle| = |\varphi(1) - \varphi(0) - \mathrm{Re}\langle \mathrm{D}F(x_0, h), y^* \rangle| =$$

$$= |\varphi'(\tau) - \operatorname{Re}\langle \mathrm{D}F(x_0,h), y^*\rangle| = |\operatorname{Re}\langle \mathrm{D}F(x_0+\tau h) - \mathrm{D}F(x_0,h), y^*\rangle| \leqslant$$

$$\leqslant \|\mathrm{D}F(x_0+\tau h, h) - \mathrm{D}F(x_0,h)\| \, \|y^*\| \leqslant \|F'(x_0+\tau h) - F'(x_0)\| \, \|h\| \, \|y^*\|,$$

where $\tau \in [0,1]$.

By Hahn-Banach-Suchomlynoff Theorem (see Chapter 0, §8) we may choose the functional $y^*$ such that $\langle \omega(x_0,h), y^*\rangle = \|\omega(x_0,h)\|$ and $\|y^*\| = 1$. Then

$$\frac{\|\omega(x_0,h)\|}{\|h\|} \leqslant \|F'(x_0+\tau h) - F'(x_0)\| \,.$$

Hence, using also the continuity of $F'$ on $\mathfrak{U}$, it follows that the equalities (1.2) and (1.3) hold.

Therefore, the continuous Gâteaux differentiability on a neighborhood $\mathfrak{U} \subseteq \mathfrak{D}$ implies Fréchet differentiability on the same neighborhood, and, moreover, Gâteaux and Fréchet derivatives coincide on $\mathfrak{U}$. ▶

**THEOREM 1.2.** *Let $F$ be an operator such that the equality* (1.1) *holds uniformly in $h$ ($\|h\| = 1$) and $\delta F(x_0,h) = \mathrm{D}F(x_0,h)$. Then $F \in \mathcal{F}(x_0, \mathfrak{Y})$.*

◁ Under our assumptions, for all $h$ with $\|h\| = 1$ and for each $\varepsilon > 0$, there exists $\delta > 0$ such that the inequality

$$\left\| \frac{1}{t}\left(F(x_0+th) - F(x_0)\right) - \mathrm{D}F(x_0,h) \right\| < \varepsilon$$

holds for any $0 < |t| < \delta$.

For each $h_1$ with $\|h_1\| < \delta$ there exists an $h$ with $\|h\| = 1$ such that $h_1 = th$ and $\|h_1\| = t$. We have $\|\omega(x_0,h_1)\| \, \|h_1\|^{-1} < \varepsilon$, which is equivalent to (1,3), since $\varepsilon$ was arbitrary. ▶

**REMARK 1.1.** Note that in Theorem 1.2, in contradistinction with Theorem 1.1, the condition of Gâteaux differentiability on a neighborhood of $x_0$ is not required.

**REMARK 1.2.** In the case of $\mathfrak{X}$ being a complex space, the notion of first variation can be slightly relaxed, asking the existence of the limit in (1.1) only for real values of $t$. Under the conditions of Theorems 1.1 or 1.2, this is enough for obtaining the existence of the limit (1.1) for any complex $t$. However, in the general case this problem is still open.

## §2. Lagrange formula and Lipschitz condition

Let $\mathfrak{X}$ and $\mathfrak{Y}$ be normed spaces, $\mathfrak{D}$ a set in $\mathfrak{X}$ and $F$ an operator from $\mathfrak{D}$ into $\mathfrak{Y}$.

**DEFINITION 2.1.** The operator $F$ satisfies the *Lipschitz condition* (with the constant $\ell$) on $\mathfrak{D}$, if for every $x_1, x_2 \in \mathfrak{D}$ we have

$$\|F(x_1) - F(x_2)\| \leqslant \ell \|x_1 - x_2\|. \tag{2.1}$$

**THEOREM 2.1.** *Let $\mathfrak{D}$ be a convex set in $\mathfrak{X}$ and $F: \mathfrak{D} \to \mathfrak{Y}$ an operator which is Gâteaux differentiable on $\mathfrak{D}$ with $\|F'(x)\| \leqslant \ell$ for all $x \in \mathfrak{D}$. Then $F$ satisfies the Lipschitz condition* (2.1).

◁ Since $\mathfrak{D}$ is convex, it follows that $tx_1 + (1-t)x_2 \in \mathfrak{D}$ for any $x_1, x_2 \in \mathfrak{D}$ and $0 \leqslant t \leqslant 1$. Therefore, the scalar function

$$\varphi(t) = \langle F(tx_1 + (1-t)x_2), y^* \rangle$$

is well-defined on [0,1] and continuous for any $y^* \in \mathfrak{Y}^*$.

We show that $\varphi$ is differentiable on (0,1). Indeed

$$\frac{\varphi(t+\Delta t) - \varphi(t)}{\Delta t} = \left\langle \frac{F(tx_1 + (1-t)x_2 + \Delta t(x_1 - x_2)) - F(tx_1 + (1-t)x_2)}{\Delta t}, y^* \right\rangle.$$

Therefore $\varphi'(t)$ exists and $\varphi'(t) = \langle F'(tx_1 + (1-t)x_2)(x_1 - x_2), y^* \rangle$.

By the mean-value theorem we have:

$$\|\varphi(1) - \varphi(0)\| \leqslant \sup_{0 \leqslant t \leqslant 1} |\varphi'(t)|,$$

or, equivalently,

$$\begin{aligned} &|\langle F(x_1) - F(x_2), y^* \rangle| \leqslant \\ &\leqslant \sup_{0 \leqslant t \leqslant 1} |\langle F'(tx_1 + (1-t)x_2)(x_1 - x_2), y^* \rangle| \leqslant \\ &\leqslant \sup_{0 \leqslant t \leqslant 1} \|F'(tx_1 + (1-t)x_2)\| \, \|x_1 - x_2\| \, \|y^*\|. \end{aligned}$$

Hence, using also Hahn-Banach-Suchomlynoff Theorem we obtain the so-called Lagrange formula:

$$\|F(x_1) - F(x_2)\| \leqslant \sup_{0 \leqslant t \leqslant 1} \|F'(tx_1 + (1-t)x_2)\| \, \|x_1 - x_2\|,$$

which, together with $\|F'(x)\| \leqslant \ell$ for all $x \in \mathfrak{D}$, leads to (2.1). ▶

**REMARK 2.1.** Theorem 2.1 remains true if the condition of Gâteaux differentiability is replaced with the uniform boundedness of the first variation:

$$\|\delta F(x,h)\| \leqslant \ell \|h\|, \quad h \in \mathfrak{X}.$$

We close this section with the definition of the most general notion of differentiability needed in this book.

**DEFINITION 2.2.** An operator $F\colon \mathfrak{D} \to \mathfrak{Y}$ is said to be *weakly differentiable* at $x_0 \in \mathfrak{D}$ if for any $y^* \in \mathfrak{Y}^*$ and any $h \in \mathfrak{X}$, the scalar-valued function $\varphi_h(t) = \langle F(x_0 + th), y^* \rangle$ is differentiable (in the usual sense) at $t = 0$. When, in addition, for a given $h$ there exists an element $y(x_0, h) \in \mathfrak{Y}$ such that $\varphi_h'(0) = \langle y(x_0,h), y^* \rangle$, then $y(x_0,h)$ is called the *weak variation* of the function $F$ at $x_0$ along direction $h$, and is denoted by $\delta_{\mathrm{w}} F(x_0, h)$. The class of operators having weak variation along all directions $h \in \mathfrak{X}$ is denoted by $\mathcal{V}_{\mathrm{w}}(x_0, \mathfrak{Y})$. Further, if $\delta_{\mathrm{w}} F(x_0, h) = Ah$ where $A \in L(\mathfrak{X}, \mathfrak{Y})$, then the operator $F$ is called *weakly Gâteaux differentiable*; the set of these operators $F$ is denoted by $\mathcal{G}_{\mathrm{w}}(x_0, \mathfrak{Y})$.

If, moreover,

$$\lim_{t \to 0} \frac{\varphi_h(t) - \varphi_h(0)}{t} = \langle Ah, y^* \rangle$$

is uniform relatively to all $h$ with $\|h\| = 1$, then the operator $F$ is called *weakly Fréchet differentiable* and the set of these operators is denoted by $\mathcal{F}_{\mathrm{w}}(x_0, \mathfrak{Y})$.

The following inclusions are obvious:

$$\mathcal{V}(x_0, \mathfrak{Y}) \subseteq \mathcal{V}_{\mathrm{w}}(x_0, \mathfrak{Y}) \tag{2.2}$$

$$\mathcal{G}(x_0, \mathfrak{Y}) \subseteq \mathcal{G}_{\mathrm{w}}(x_0, \mathfrak{Y}) \tag{2.3}$$

$$\mathcal{F}(x_0, \mathfrak{Y}) \subseteq \mathcal{F}_{\mathrm{w}}(x_0, \mathfrak{Y}). \tag{2.4}$$

(Let us remark that in the literature weakly differentiable operators are often called Gâteaux differentiable — see, for example, Vainberg [1], [2]. In our opinion, the terminology from Definition 2.2 seems to be more natural and appropriate. A reason for this is that it is based on the notion of weak convergence in Banach spaces. More than this, we will use below the well-known notion of weak analiticity, which is strongly related to the notion of weak differentiabily and which is also based on the notion of weak convergence.)

**REMARK 2.2.** If the space $\mathfrak{Y}$ is finite dimensional, all relations (2.2)–(2.4) become equalities, because the strong and weak convergences are equivalent. We will see below in Chapter III that this is also the case when $\mathfrak{X}$ and $\mathfrak{Y}$ are infinite dimensional if we assume, in addition, that $\mathfrak{X}$ is a complex space and the weak variation exists on a neighborhood of $x_0$.

**REMARK 2.3.** In fact, the proof of Theorem 2.1 uses a weaker hypothesis. The same arguments lead to the next more general result:

*Let $F$ be an operator defined on the convex set $\mathfrak{D}$ such that the weak variation $\delta_{\mathrm{w}} F(x,h)$ exists at any $x \in \mathfrak{D}$ and, for all $h \in \mathfrak{X}$, we have*

$$\|\delta_{\mathrm{w}} F(x,h)\| \leqslant \ell \|h\|.$$

*Then the conclusion of Theorem 2.1 is true.*

## §3. Examples of Fréchet differentiable operators

**EXAMPLE 3.1** (Urysohn operator). Let $K(t,s,x)$ be a function defined on $a \leqslant t, s \leqslant \leqslant b$, $|x| \leqslant r$, such that $K(t,s,x)$ and $K'_x(t,s,x)$ are everywhere continuous. Then the integral operator $F$ given by

$$Fx(t) = \int_a^b K(t,s,x(s))\mathrm{d}s \tag{3.1}$$

is well-defined on the ball $\|x\| \leqslant r$ of the space $C[a,b]$ of all real-valued continuous functions on $[a,b]$ with the norm $\|x\| = \sup_{a \leqslant t \leqslant b} |x(t)|$. The operator $F$ takes values in $C[a,b]$ and is Fréchet differentiable on the open ball $\|x\| < r$; moreover, for any $x_0 \in C[a,b]$ with $\|x_0\| < r$ and for every $h \in C[a,b]$ we have:

$$(F'(x_0)h)(t) = \int_a^b K'_x(t,s,x_0(s))h(s)\mathrm{d}s. \tag{3.2}$$

Indeed, first of all notice that the operator defined by (3.2) is linear and continuous on $C[a,b]$. Further

$$\|F(x_0+h)-F(x_0)-F'(x_0)h\| \leqslant$$

$$\leqslant \int_a^b \sup_{a\leqslant t\leqslant b} |K(t,s,x_0(s)+h(s))-K(t,s,x_0(s))-K'_x(t,s,x_0(s))h(s)|\,\mathrm{d}s \leqslant$$

$$\leqslant \|h\| \int_a^b \sup_{a\leqslant t\leqslant b} |K'_x(t,s,x_0(s)+\theta h(s))-K'_x(t,s,x_0(s))|\,\mathrm{d}s$$

for a certain $\theta=\theta(h)\in[0,1]$. Hence, the continuity of the function $K'_x(t,s,x)$ in the variable $x$ implies that

$$\lim_{\|h\|\to 0} \frac{1}{\|h\|} \|F(x_0+th)-F(x_0)-F'(x_0)h\| \leqslant$$

$$\leqslant \lim_{\|h\|\to 0} \int_a^b \sup_{a\leqslant t\leqslant b} |K'_x(t,s,x_0(s)+\theta h(s))-K'_x(t,s,x_0(s))|\,\mathrm{d}s = 0.$$

Note that the Urysohn operator (3.1) can be also considered on the space $\tilde{C}[a,b]$ of all complex-valued continuos functions; then $F$ is Fréchet differentiable on $\tilde{C}[a,b]$ if the kernel $K(t,s,x)$ has continuous partial derivatives with respect to the variable $x$ in the disk $|x|<r$ of the complex plane.

**EXAMPLE 3.2** (Hammerstein operator). A particular case of the operator (3.1) is the Hammerstein operator defined by the equality

$$Fx(t) = \int_a^b K(t,s)f(s,x(s))\mathrm{d}s \tag{3.3}$$

where the functions $K(t,s)$, $f(s,x)$ and $f'_x(s,x)$ are continuous with respect to all variables for $a\leqslant t,\ s\leqslant b,\ |x|\leqslant r$.

Example 3.1 implies that the Fréchet derivative of $F$ is given by

$$(F'(x_0)h)(t) = \int_a^b K(t,s)f'_x(s,x_0(s))h(s)\mathrm{d}s. \tag{3.4}$$

**EXAMPLE 3.3** (Nemytzki operator). Suppose that the functions $f(s,x)$ and $f'_x(s,x)$ are continuous with respect to all variables for $a\leqslant s\leqslant b,\ |x|\leqslant r$. Nemytzki operator is the superposition operator

$$\Phi x(s) = f(s,x(s)) \tag{3.5}$$

defined on the ball $\|x\| \leqslant r$ of the space $C[a,b]$ ($\tilde{C}[a,b]$). The operator (3.5) is Fréchet differentiable and

$$(\Phi'(x_o)h)(s) = f'_x(s, x_0(s))h(s). \tag{3.6}$$

Indeed, the operator $\Phi'(x_o)$ defined by (3.6) is linear and continuous on the space $C[a,b]$ ($\tilde{C}[a,b]$) and

$$\begin{aligned} &\frac{1}{\|h\|} \|\Phi(x_o + h) - \Phi(x_o) - \Phi'(x_o)h\| \leqslant \\ &\leqslant \frac{1}{\|h\|} \sup_{a \leqslant s \leqslant b} |f(s, x_0(s) + h(s)) - f(s, x_0(s)) - f'_x(s, x_0(s))h(s)| \leqslant \\ &\leqslant \frac{1}{\|h\|} \sup_{a \leqslant s \leqslant b} |f'_x(s, x_0(s) + \theta h(s)) - f'_x(s, x_0(s))| \, |h(s)| \leqslant \\ &\leqslant \sup_{a \leqslant s \leqslant b} |f'_x(s, x_0(s) + \theta h(s)) - f'_x(s, x_0(s))| \to 0 \end{aligned}$$

for $\|h\| \to 0$, ($\theta = \theta(h) \in [0,1]$).

**EXAMPLE 3.4.** Let $f_1(x_1, \ldots, x_n), \ldots, f_m(x_1, \ldots, x_n)$ be functions defined on a domain in the Euclidean space $\mathbb{E}^n = \{x \mid x = (x_1, \ldots, x_n)\}$ and assume that $f_1, \ldots, f_m$ and all their first order partial derivatives are continuous. Then, the operator

$$F = (f_1, \ldots, f_m) \;:\; \mathbb{E}^n \to \mathbb{E}^m$$

is Fréchet differentiable on the considered domain, and

$$F'(x)h = \begin{pmatrix} \dfrac{\partial f_1(x)}{\partial x_1} & \cdots & \dfrac{\partial f_1(x)}{\partial x_n} \\ \vdots & & \vdots \\ \dfrac{\partial f_m(x)}{\partial x_1} & \cdots & \dfrac{\partial f_m(x)}{\partial x_n} \end{pmatrix} \begin{pmatrix} h_1 \\ \vdots \\ h_n \end{pmatrix}. \tag{3.7}$$

In the right hand side of (3.7) we have the product of the Jacobi matrix of $F$ and the vector $h = (h_1, \ldots, h_n)$.

## §4. Lemmas about differentiable operators

The next lemma follows immediately from the properties of a linear operator.

**LEMMA 4.1.** *Let $A \in L(\mathfrak{X}, \mathfrak{Y})$ be a continuous linear operator. Then for any $x \in \mathfrak{X}$, the operator $A$ is Fréchet differentiable at $x$ ($A \in \mathcal{F}(x, \mathfrak{Y})$) and*

$$A'(x) \equiv A.$$

**LEMMA 4.2** (The derivative of a composition operator). *Let $\mathfrak{X}$, $\mathfrak{Y}$ and $\mathfrak{Z}$ be normed spaces. Suppose that $F: \mathfrak{X} \to \mathfrak{Y}$ is an operator defined on a neighborhood of a point $x_0 \in \mathfrak{X}$ and that $G: \mathfrak{Y} \to \mathfrak{Z}$ is defined on a neighborhood of $y_0 = f(x_0) \in \mathfrak{Y}$. Let $T = GF$ be the composition of $F$ and $G$. If $F \in \mathcal{G}(x_0, \mathfrak{Y})$ $(\mathcal{F}(x_0, \mathfrak{Y}))$ and $G \in \mathcal{G}(y_0, \mathfrak{Z})$ $(\mathcal{F}(y_0, \mathfrak{Z}))$ then $T \in \mathcal{G}(x_0, \mathfrak{Z})$ $(\mathcal{F}(x_0, \mathfrak{Z}))$ and*

$$T'(x_0) = G'(y_0)F'(x_0). \tag{4.1}$$

(The derivatives in (4.1) are understood as Fréchet or Gâteaux derivatives, depending on the class the operators $F$ and $G$ belong to. The right hand side of (4.1) is the usual product of linear operators.)

◁ For arbitrary elements $h \in \mathfrak{X}$ and $u \in \mathfrak{Y}$, define

$$\omega_1(x_0, th) = F(x_0 + th) - F(x_0) - F'(x_0)th,$$
$$\omega_2(y_0, tu) = G(y_0 + tu) - G(y_0) - G'(y_0)tu,$$

where $\omega_1$ and $\omega_2$ satisfy the conditions

$$\lim_{t\to 0} \frac{\|\omega_1(x_0, th)\|}{t} = 0, \quad \lim_{t\to 0} \frac{\|w_2(y_0, tu)\|}{t} = 0.$$

Further, take $u = F'(x_0)h + t^{-1}\omega_1(x_0, th)$ and consider the difference

$$\begin{aligned} T(x_0 + th) - T(x_0) &= GF(x_0 + th) - GF(x_0) = \\ &= G(F(x_0) + F'(x_0)th + \omega_1(x_0, th)) - G(y_0) = \\ &= G(y_0 + tu) - G(y_0) = \omega_2(y_0, tu) + G'(y_0)tu = \\ &= tG'(y_0)F'(x_0)h + G'(y_0)\omega_1(x_0, th) + \omega_2(y_0, tu). \end{aligned}$$

Consequently

$$\lim_{t\to 0} \frac{\|T(x_0 + th) - T(x_0)\|}{t} = G'(y_0)F'(x_0)h \tag{4.2}$$

a relation which is equivalent with (4.1), and which proves that $T \in \mathcal{G}(x_0, \mathfrak{Z})$. When $F \in \mathcal{F}(x_0, \mathfrak{Y})$ and $G \in \mathcal{F}(y_0, \mathfrak{Z})$, the equality (4.2) holds uniformly with respect to $h \in \mathfrak{X}$, $\|h\| = 1$, and therefore (using Theorem 1.2) $T \in \mathcal{F}(x_0, \mathfrak{Z})$. ▶

The previous two lemmas have a simple but important consequence.

**COROLLARY 4.1.** *Let $F \in \mathcal{G}(x_0, \mathfrak{Y})$ and $A \in L(\mathfrak{Y}, \mathfrak{Z})$ be two operators. Then*

$$(AF)'(x_0) = AF'(x_0). \tag{4.3}$$

The equality (4.3) shows that a linear operator "passes through" the derivation sign. Rewriting (4.3) using the differential symbol, we have

$$\mathrm{D}(AF) = A\mathrm{D}F \tag{4.4}$$

which means that the differentiation operator commutes with any linear operator.

For example, the Hammerstein operator (3.3) can be represented as the composition of the linear operator $K$, defined by the formula

$$Kx(t) = \int_a^b K(s,t)x(s)\mathrm{d}s$$

with the Nemytzki operator (3.5), namely $F = K\Phi$. Then, the relations (4.3) and (3.6) imply (3.4).

**COROLLARY 4.2.** *Let $F \in \mathcal{G}(x_0, \mathfrak{Y})$ be an operator and assume that there exists an operator $G\colon \mathfrak{Y} \to \mathfrak{X}$ defined on a neighborhood $\mathfrak{U}$ of the point $y_0 = F(x_0)$, and satisfying the relations*

$$GF = I_{\mathfrak{D}}, \quad FG = I_{\mathfrak{U}} \tag{4.6}$$

*where $\mathfrak{D}$ is a neighborhood of $x_0$ (we write in this case $G = F^{-1}$). If $F^{-1} \in \mathcal{G}(y_0, \mathfrak{X})$, then the linear operator $F'(x_0)$ has a continuous inverse and*

$$[F'(x_0)]^{-1} = (F^{-1})'(y_0). \tag{4.6}$$

◁ The equality (4.6) follows immediately from (4.1) and (4.5). ▶

**REMARK 4.1.** For a complex space $\mathfrak{X}$, Corollary 4.2 is true under weaker conditions (see Chapter IV below). The finite dimensional case of this result is a theorem of Osgood (see, for example, B. Shabat [2]).

**REMARK 4.2.** In Chapter IV it will be shown that, for complex spaces, the converse of Corollary 4.2 is also true: if the operator $F'(x_0)$ has a continuous inverse, then the operator $F$ is locally invertible and the equality (4.6) holds.

## §5. Partial derivatives

In this paragraph we suppose that the space $\mathfrak{X}$ is decomposed in a direct sum $\mathfrak{X} = \mathfrak{X}_1 \dot{+} \mathfrak{X}_2$ (see Chapter 0, §7). Assume that $F$ is an operator defined on a neighborhood of a point $a = (a_1, a_2)$, $a_i \in \mathfrak{X}_i$, $i = 1, 2$, with values in a normed space $\mathfrak{Y}$, such that $F \in \mathcal{V}(a, \mathfrak{Y})$. Choose the vectors $h^{(1)} = (h_1, 0)$ and $h^{(2)} = (0, h_2)$, where $h_i \in \mathfrak{X}_i$, $i = 1, 2$. The expressions $\delta_{h_1} F(a) = \delta F(a, h^{(1)})$ and $\delta_{h_2} F(a) = \delta F(a, h^{(2)})$ are called the *variations* of the operator $F$ at the point $a$ in the directions of the spaces $\mathfrak{X}_i$, or the *partial variations* of $F$. In other words

$$\delta_{h_1} F(a) = \lim_{t \to 0} \frac{F(a_1 + th_1, a_2) - F(a_1, a_2)}{t} \tag{5.1}$$

$$\delta_{h_2} F(a) = \lim_{t \to 0} \frac{F(a_1, a_2 + th_2) - F(a_1, a_2)}{t} \tag{5.2}$$

Note that both limits exist by Definition 1.1.

In particular, if $F \in \mathcal{G}(a, \mathfrak{Y})$ $(\mathcal{F}(a, \mathfrak{Y}))$, then the expressions (5.1) and (5.2) are linear in $h_1$ and $h_2$, respectively, and we call them the *partial differentials* in the sense of Gâteaux (Fréchet). In this case we will use the notations

$$\delta_{h_i} F(a) = F'_{x_i}(a) h_i$$

and the expressions $F'_{x_i}(a) \in L(\mathfrak{X}_i, \mathfrak{Y})$, $i = 1, 2$, are called the *partial derivatives* of the operator $F$ at the point $a$ (in the sense of Gâteaux or Fréchet, respectively).

**THEOREM 5.1.** *If $F \in \mathcal{G}(a, \mathfrak{Y})$ $(\mathcal{F}(a, \mathfrak{Y}))$ then the partial derivatives exist and*

$$F'(a)h = F'_{x_1}(a) h_1 + F'_{x_2}(a) h_2. \tag{5.3}$$

$\triangleleft$ As we have already mentioned, the existence of partial derivatives follows immediately from Definitions (1.1)–(1.3). The same definitions imply (5.3), if we notice that $F'_{x_i}(a)$ is the restriction of the linear operator $F'(a)$ on the subspace $\mathfrak{X}_i$, $i = 1, 2$. $\blacktriangleright$

In general, the converse is false: the mapping $F$ in Example 1.1 has partial derivatives with respect to both $x_1$ and $x_2$ — namely $F'_{x_1}(0) h_1 \equiv 0$ and $F'_{x_2}(0) h_2 \equiv 0$ —, but this mapping does not have a Gâteaux derivative. However, the next result is true.

**THEOREM 5.2.** *Let $F \colon \mathfrak{X} \to \mathfrak{Y}$ be an operator defined on a neighborhood $\mathfrak{D}$ of a point $a \in \mathfrak{X}$ and $\mathfrak{X} = \mathfrak{X}_1 \dot{+} \mathfrak{X}_2$. Assume that $F$ has partial Gâteaux derivatives at every point of $\mathfrak{D}$, and that these derivatives are operator-valued mappings, continuous on*

*$\mathfrak{D}$. Then $F \in \mathcal{F}(a, \mathfrak{Y})$ and the equality (5.3) holds, hence $F'(x)$ is also a continuous mapping from $\mathfrak{D}$ into $L(\mathfrak{X}, \mathfrak{Y})$.*

◁ By Theorem 1.1, the operators $F'_{x_1}(a)$ and $F'_{x_2}(a)$ are the partial Fréchet derivatives of $F$ at $a$. Consider an arbitrary vector $h = (h_1, h_2)$, such that $a + h \in \mathfrak{D}$. Then

$$\begin{gathered}F(a+h) - F(a) = \\ = F(a_1 + h_1, a_2 + h_2) - F(a_1 + h_1, a_2) + F(a_1 + h_1, a_2) - F(a_1, a_2) = \\ = F'_{x_2}(a_1 + h_1, a_2)h_2 + \omega_2(a_1 + h_1, a_2, h_2) + F'_{x_1}(a_1, a_2)h_1 + \omega_1(a_1, a_2, h_1)\end{gathered}$$

where $\omega_2 = \mathrm{o}(\|h_2\|)$, $\omega_1 = \mathrm{o}(\|h_1\|)$.

Since $F'_{x_2}\colon \mathfrak{D} \to L(\mathfrak{X}, \mathfrak{Y})$ is continuous in the uniform operator topology, we have

$$\lim_{\|h_1\| \to 0} \|F'_{x_2}(a_1 + h_1, a_2) - F'_{x_2}(a_1, a_2)\| = 0.$$

Then

$$F'_{x_2}(a_1 + h_1, a_2)h_2 = F'_{x_2}(a_1, a_2)h_2 + \omega_3(a, h_1, h_2)$$

where $\omega_3 = \mathrm{o}(\|h\|)$. Finally, we obtain

$$F(a+h) - F(a) = F'_{x_1}(a_1, a_2)h_1 + F'_{x_2}(a_1, a_2)h_2 + \omega(a, h)$$

where $\omega(a, h) = \mathrm{o}(\|h\|)$. From this, all the assertions of the theorem follow immediately. ▶

**REMARK 5.1.** Theorem 5.2 is true in the case of a complex space $\mathfrak{X}$ under a weaker hypothesis: it is sufficient to require the existence of the partial Fréchet derivatives on a neighborhood of a point $a$, without asking their continuity in the operator topology. (The finite dimensional case of this result is a theorem of Hartogs — see, for instance, B. Shabat [2].)

**REMARK 5.2.** The conclusions of Theorems 5.1 and 5.2 can be immediately extended by induction to the case when $\mathfrak{X}$ is a topological direct sum of $n$ normed spaces $\mathfrak{X}_1, \dots, \mathfrak{X}_n$. In this case, formula (5.3) becomes

$$F'(a)h = F'_{x_1}(a)h_1 + \ldots + F'_{x_n}(a)h_n. \tag{5.4}$$

This is called the formula of *complete differential.*

## §6. Multilinear operators. Duality. Homogeneous forms

Consider the space $\mathfrak{X} = \mathfrak{X}_1 \dot{+} \ldots \dot{+} \mathfrak{X}_n$, where $\mathfrak{X}_i$ are normed spaces, $i = 1, \ldots, n$.

**DEFINITION 6.1.** An operator $G: \mathfrak{X} \to \mathfrak{Y}$, everywhere defined on $\mathfrak{X}$, is called a *multilinear* ($n$-*linear*) *continuous* operator, if $G$ is linear in each of the variables $x_i \in \in \mathfrak{X}_i$, $i = 1, \ldots, n$ — by fixing the others —, and if $G$ is continuous on $\mathfrak{X}$.

**THEOREM 6.1.** *If an operator $G: \mathfrak{X} \to \mathfrak{Y}$ is linear in each of the variables $x_i \in \mathfrak{X}_i$, $i = 1, \ldots, n$ — by fixing the others —, then $G$ is a multilinear continuous operator if and only if*

$$\|G(x_1, \ldots, x_n)\| \leqslant M \|x_1\| \cdots \|x_n\| \tag{6.1}$$

*for all $x = (x_1, \ldots, x_n)$, where $0 \leqslant M < \infty$.*

◁ Let $G$ be a multilinear continuous operator. Then for all $\lambda_1, \ldots, \lambda_n > 0$ we have

$$G(\lambda_1 y_1, \ldots, \lambda_n y_n) = \lambda_1 \cdots \lambda_n G(y_1, \ldots, y_n) \tag{6.2}$$

and there exists $r > 0$ such that $\|G(y_1, \ldots, y_n)\| \leqslant 1$ if $\|y_i\| \leqslant r$, for all $i = 1, \ldots, n$. For $x_i \in \mathfrak{X}_i$, choose $\lambda_i > 0$ such that $\|\lambda_i^{-1} x_i\| = r$. From (6.2) we infer (6.1) with $M = r^{-n}$.

Conversely, suppose that (6.1) is fulfilled with a nonnegative constant $M < \infty$. We will prove the continuity of the operator $G$ at an arbitrary point $a = (a_1, \ldots, a_n) \in \in \mathfrak{X}$. For any $x = (x_1, \ldots, x_n)$, we have

$$\begin{aligned} G(x) - G(a) &= G(x_1, \ldots, x_n) - G(a_1, \ldots, a_n) = \\ &= G(x_1, \ldots, x_n) - G(a_1, x_2, \ldots, x_n) + G(a_1, x_2, \ldots, x_n) - \\ &\quad - G(a_1, a_2, x_3, \ldots, x_n) + G(a_1, a_2, x_3, \ldots, x_n) - \cdots \\ &\quad \cdots - G(a_1, \ldots, a_{n-1}, x_n) + G(a_1, \ldots, a_{n-1}, x_n) - G(a_1, \ldots, a_n) = \\ &= G(x_1 - a_1, x_2, \ldots, x_n) + G(a_1, x_2 - a_2, x_3, \ldots, x_n) + \cdots \\ &\quad \cdots + G(a_1, \ldots, a_{n-1}, x_n - a_n). \end{aligned} \tag{6.3}$$

Hence, we obtain the inequality

$$\begin{aligned} \|G(x) - G(a)\| \leqslant M \{ &\|x_1 - a_1\| \cdot \|x_2\| \cdot \cdots \cdot \|x_n\| + \|a_1\| \cdot \|x_2 - a_2\| \cdot \|x_3\| \cdot \cdots \\ &\cdots \cdot \|x_n\| + \ldots + \|a_1\| \cdot \cdots \cdot \|a_{n-1}\| \cdot \|x_n - a_n\| \}. \end{aligned}$$

The boundedness of the product $\|x_1\| \cdot \cdots \cdot \|x_n\|$, for $x$ close to $a$, implies the continuity of $G$ at $a$. ▶

**REMARK 6.1.** The proof of Theorem 6.1 implies that the inequality (6.1) holds if one requires only the continuity of $G$ at the point $(0,\ldots,0) \in \mathfrak{X}$. Therefore, the continuity of a multilinear operator at the origin implies its continuity on $\mathfrak{X}$.

**REMARK 6.2.** The equality (6.3) shows that a multilinear operator, which is separately continuous in each of its variables, is continuous on $\mathfrak{X}$.

**DEFINITION 6.2.** If $G$ is a multilinear continuous operator, then the smallest number $M$ for which the inequality (6.1) is fulfilled is called the *norm* of $G$. Thus

$$\|G\| = \sup_{\|x_i\| \leqslant 1,\ 1 \leqslant i \leqslant n} \|G(x_1,\ldots,x_n)\|. \tag{6.4}$$

The norm introduced by (6.4) defines a structure of normed space on the set $L_n(\mathfrak{X}_1,\ldots,\mathfrak{X}_n;\mathfrak{Y})$ of all multilinear continuous operators from $\mathfrak{X}_1 \dot{+} \ldots \dot{+} \mathfrak{X}_n$ into $\mathfrak{Y}$.

Let us consider in more detail the space $L_2(\mathfrak{X}_1,\mathfrak{X}_2;\mathfrak{Y})$ of bilinear continuous operators from $\mathfrak{X}_1 \dot{+} \mathfrak{X}_2$ into $\mathfrak{Y}$.

Take $G \in L(\mathfrak{X}_1,\mathfrak{X}_2;\mathfrak{Y})$ and fix a point $x_1 \in \mathfrak{X}_1$. Definition 6.1 produces a linear continuous operator in the second variable $x_2$, acting from $\mathfrak{X}_2$ into $\mathfrak{Y}$, that is, $G_{x_1} = G(x_1,\,\cdot\,)$ is an element in $L(\mathfrak{X}_2,\mathfrak{Y})$. We clearly have

$$\sup_{\|x_1\| \leqslant 1} \|G_{x_1}\| \leqslant \|G\|. \tag{6.5}$$

Taking into account the linearity of $G$ in $x_1$, the operator $G$ can be viewed as a linear operator which maps $x_1 \in \mathfrak{X}_1$ into $G_{x_1} \in L(\mathfrak{X}_2,\mathfrak{Y})$, i.e., $G \in L\left(\mathfrak{X}_1, L(\mathfrak{X}_2,\mathfrak{Y})\right)$. Then $\|G\| \leqslant \sup_{\|x_1\| \leqslant 1} \|G_{x_1}\|$. This inequality and (6.5) show that the spaces $L(\mathfrak{X}_1,\mathfrak{X}_2;\mathfrak{Y})$ and $L\left(\mathfrak{X}_1, L(\mathfrak{X}_2,\mathfrak{Y})\right)$ are linearly isometric.

Similar results are true for $n \geqslant 2$.

**THEOREM 6.2** (on duality). *Let $1 \leqslant p \leqslant n-1$ be fixed. The spaces $L_n(\mathfrak{X}_1,\ldots,\mathfrak{X}_n;\mathfrak{Y})$ and $L_p(\mathfrak{X}_1,\ldots,\mathfrak{X}_p;L_{n-p}(\mathfrak{X}_{p+1},\ldots,\mathfrak{X}_n);\mathfrak{Y}))$ are linearly isometric; their linear isometry is defined by the equalities*

$$\tilde{G}_{x_1,\ldots,x_p}(x_{p+1},\ldots,x_n) = G(x_1,\ldots,x_p,x_{p+1},\ldots,x_n)$$

*where $\tilde{G} \in L_p\left(\mathfrak{X}_1,\ldots,\mathfrak{X}_p;L_{n-p}(\mathfrak{X}_{p+1},\ldots,\mathfrak{X}_n;\mathfrak{Y})\right)$ and $G \in L(\mathfrak{X}_1,\ldots,\mathfrak{X}_n;\mathfrak{Y})$.*

Due to the symmetric role of the spaces $\mathfrak{X}_i$, $i=1,\ldots,n$, a similar result holds for any arbitrary rearrangements of the indexes.

We consider now the particular case $\mathfrak{X}_1 = \mathfrak{X}_2 = \ldots = \mathfrak{X}_n = \mathfrak{X}$. The space of all multilinear continuous operators will be denoted in this case by $L_n(\mathfrak{X};\mathfrak{Y})$.

**DEFINITION 6.3.** An operator $H: \mathfrak{X} \to \mathfrak{Y}$ is called a *homogeneous form of order n* if there exists an operator $G \in L_n(\mathfrak{X};\mathfrak{Y})$ such that

$$H(x) = G(x, x, \ldots, x). \tag{6.6}$$

We state without proof the next results (see H. Cartan [1]).

**THEOREM 6.3.** *Let $H: \mathfrak{X} \to \mathfrak{Y}$ be a homogeneous form of order $n$. Then there exists a unique operator $G \in L_n(\mathfrak{X};\mathfrak{Y})$ which is symmetric relatively to permutations of the variables and which satisfies* (6.6).

Such an operator $G$ is called the *polar form* of the operator $H$ and is denoted by $\tilde{H}$.

**THEOREM 6.4.** *Suppose that the operator $H$ is defined by* (6.6), *where $G$ acts from $\mathfrak{X}_1 \dot{+} \ldots \dot{+} \mathfrak{X}_n$ into $\mathfrak{Y}$ and is linear in each of its variables. Then $H$ is continuous if and only if its polar form $\tilde{H}$ is continuous.*

The following two relations are basic properties of a homogeneous form $H$:

$$H(\lambda x) = \lambda^n H(x), \quad x \in \mathfrak{X}, \tag{6.7}$$

for each $\lambda \in \mathbb{K}$, and

$$\|H(x)\| \leqslant \|\tilde{H}\| \, \|x\|^n, \quad x \in \mathfrak{X}. \tag{6.8}$$

It is easy to see that $\|\tilde{H}\|$ is the smallest constant for which (6.8) holds. Thus, the number $\|\tilde{H}\|$ is called the norm of the operator $H$.

The set of homogeneous forms of order $n$ will be denoted by $L^n(\mathfrak{X};\mathfrak{Y})$. Clearly, $L^n(\mathfrak{X};\mathfrak{Y})$ is a linear normed space relatively to the norm $\|\tilde{H}\|$.

## §7. Higher order derivatives

A way of defining the derivatives and differentials of higher orders is the following.

**DEFINITION 7.1.** Let $F: \mathfrak{X} \to \mathfrak{Y}$ be an operator and suppose that the first variation $\delta F(x, h_1)$ exists in a neighborhood of a point $x$, for any element $h_1 \in \mathfrak{X}$. If for each fixed $h_1$ and $h_2$ the limit

$$\lim_{t\to 0} \frac{\delta F(x+th_2,h_1)-\delta F(x,h_1)}{t} \tag{7.1}$$

exists in the space $\mathfrak{Y}$, then it is called *the second variation* of the operator $F$ and is denoted by $\delta_2 F(x,h_1,h_2)$.

The $n$-th variations are defined by induction. Namely, assume that $\delta_{n-1}(x,h_1,\ldots,h_{n-1})$ exists in a neighborhood of a point $x$ for any $h_1,h_2,\ldots,h_{n-1}\in\mathfrak{X}$. If for each fixed $h_1,h_2,\ldots,h_n$ the limit

$$\lim_{t\to 0} \frac{\delta_{n-1}F(x+th_n,h_1,\ldots,h_{n-1})-\delta_{n-1}F(x,h_1,\ldots,h_{n-1})}{t} \tag{7.2}$$

exists in the space $\mathfrak{Y}$, then it is called the *$n$-th variation* of the operator $F$ and is denoted by $\delta_n(x,h_1,\ldots,h_n)$.

Suppose now that the operator $F$ is Gâteaux differentiable in a neighborhood $\mathfrak{D}$ of a point $x$. Then, the variation $\delta F$ can be considered (on this neighborhood) as an operator acting from $\mathfrak{X}$ into $L(\mathfrak{X},\mathfrak{Y})$. Indeed, the equality

$$\delta F(x,h)=A(x)h$$

maps $x\in\mathfrak{D}$ into the operator $A(x)=F'(x)\in L(\mathfrak{X},\mathfrak{Y})$. If this operator is Gâteaux differentiable in $x_0$, that is, if the limit

$$\lim_{t\to 0} \frac{A(x+th_2)-A(x)}{t} = B(x)h_2 \tag{7.3}$$

exists, then, obviously, the value of $B(x)h_2$ at an element $h_1$ is exactly the second variation $\delta_2 F(x,h_1,h_2)$ defined by (7.1). Thus, in this case, $\delta_2 F$ can be considered as an element of the space $L\left(\mathfrak{X},L(\mathfrak{X},\mathfrak{Y})\right)$, or, according to Theorem 6.2, as an element of the space $L_2(\mathfrak{X};\mathfrak{Y})$, i.e., as a bilinear operator (relatively to $h_1$ and $h_2$) acting from $\mathfrak{X}\dot{+}\mathfrak{X}$ into $\mathfrak{Y}$.

This operator is called the *second Gâteaux derivative* and its value at an element $(h_1,h_2)\in\mathfrak{X}\dot{+}\mathfrak{X}$ is called the *second Gâteaux differential* and is denoted by $\mathrm{D}^2F(x_0,h_1,h_2)$ $(=\delta_2F(x_0,h_1,h_2))$.

The *$n$-th Gâteaux derivative* and the *$n$-th Gâteaux differential* of the operator $F$ are defined by induction, using the reccurent relations

$$\mathrm{D}^nF(x,h_1,\ldots,h_n)=\mathrm{D}\left(\mathrm{D}^{n-1}F(x,h_1,\ldots,h_{n-1});h_n\right).$$

The *n-th Fréchet derivative* and the *n-th Fréchet differential* of the operator $F$ are defined using Definition 1.3. Sometimes, it is more convenient to use Theorem 1.2. Thus, if for a Gâteaux differentiable operator $F$ the equality (7.3) is uniformly fulfilled for $h_2$ with $\|h_2\| = 1$, where $B(x) \in L(\mathfrak{X}, L(\mathfrak{X}, \mathfrak{Y}))$, then the operator $F$ is said to be twice Fréchet differentiable.

$\lhd$ Suppose that the $(n+1)$-th variations of an operator $F$ exist at a point $x_0 \in \in \mathfrak{X}$ and at each point of an open interval $]x_0, x_0 + h[$, such that

$$\|\delta_{n+1} F(x, h_1, \ldots, h_{n+1})\| \leqslant M$$

for all $x \in \,]x_0, x_0 + h[$, $\|h_1\| = \cdots = \|h_{n+1}\| = 1$, where $M < \infty$. Consider the function $\varphi(t) = \langle F(x_0 + th), y^* \rangle$ defined on $[0, 1)$ where $y^*$ is a fixed, arbitrary element of $\mathfrak{Y}^*$. The function $\varphi$ is differentiable at each point of $[0,\ 1)$, and

$$\varphi'(t) = \lim_{\Delta t \to 0} \left\langle \frac{1}{\Delta t} \left( F(x_0 + th + \Delta th) - F(x_0 + th) \right), y^* \right\rangle = \\ = \langle \delta F(x_0 + th, h), y^* \rangle.$$

Obviously, the second derivative also exists

$$\varphi''(t) = \lim_{\Delta t \to 0} \frac{\varphi'(t + \Delta t) - \varphi'(t)}{\Delta t} = \\ = \lim_{\Delta t \to 0} \frac{\langle (\delta F(x_0 + th + \Delta th, h) - \delta F(x_0 + th, h)), y^* \rangle}{\Delta t} = \\ = \langle \delta_2 F(x_0 + th, h, h), y^* \rangle.$$

Analogously, all the derivatives $\varphi^{(k)}(t)$, $k = 3, \ldots, n+1$ exist on the interval $[0, 1)$ and $\left|\varphi^{(n+1)}(t)\right| \leqslant M\|y^*\|$. Using Taylor formula for $\varphi$, we have

$$\varphi(t) = \varphi(0) + \varphi'(0) + \frac{1}{2!}\varphi''(0)t^2 + \cdots + \frac{1}{n!}\varphi^{(n)}(0)t^n + \frac{1}{(n+1)!}\varphi^{(n+1)}(\tau)t^{n+1},$$

where $\tau \in [0, t]$. The last equality is equivalent to

$$F(x_0 + th) = F(x_0) + \delta F(x_0, h)t + \frac{1}{2!}\delta_2 F(x_0, h, h)t^2 + \cdots \\ \cdots + \frac{1}{n!}\delta_n F(x_0, h, \ldots, h)t^n + \frac{1}{(n+1)!}\delta_{n+1} F(x_0 + \tau h, h, \ldots, h)t^{n+1}. \tag{7.4}$$

$\blacktriangleright$

Relation (7.4) is called the *Taylor formula* for the operator $F$. In particular, if the operator $F$ in Gâteaux or Fréchet differentiable up to the order $n+1$, then, taking $t = 1$ in (7.4) and denoting $\frac{1}{k!}\delta_k F(x_0, h, \ldots, h) = A_k(x_0, h)$, we have

$$F(x_0 + h) = F(x_0) + A_1(x_0, h) + A_2(x_0, h) + \cdots + A_n(x_0, h) + \omega(x_0, h) \tag{7.5}$$

where $A_i(x,\cdot)$ are homogeneous forms of order $i$, $1 \leqslant i \leqslant n$, and

$$\|\omega(x_0, h)\| \leqslant M\|h\|^{n+1}. \tag{7.6}$$

Note also that when the operator $F$ is Fréchet differentiable up to the order $n+1$ on a neighborhood of the point $x_0$, then the representation (7.5) is uniformly valid in $h$, $\|h\| \leqslant \varepsilon$, and the number $M$ in (7.6) does not depend on $h$. In this case, the operator $F$ is called *Taylor differentiable* and the expression $A_i(x_0, h)$ is called the *Taylor derivative* of order $i$, $1 \leqslant i \leqslant n$.

The converse assertion is also true. If the operator $F$ is differentiable up to the order $n+1$ on the interval $[x_0, x_0+h]$ and admits the representations (7.5)–(7.6), where $A_i(x_0, h)$ are homogeneous forms of order $i$, $1 \leqslant i \leqslant n$, then

$$i!\tilde{A}_i(x_0)(h_1, \dots, h_i) = \delta_i F(x_0, h_1, \dots, h_i) \tag{7.7}$$

where, of course, $\tilde{A}_i$ is the polar form of the operator $A_i$, $1 \leqslant i \leqslant n$.

A proof of this assertion is contained, for example, in H. Cartan [1].

**REMARK 7.1.** The existence of the succesive derivatives up to the order $n+1$ is an essential assumption in the previous assertion. Here is an example. Consider the operator $F$ defined by $F(x)=x^{n+1}\sin x^{-n-1}$, $x \in \mathbb{R}$, $x \neq 0$, and $F(0)=0$. The operator $F$ is Taylor differentiable at the point 0 up to order $n$, i.e., $F$ admits the representation (7.5)–(7.6) (with $A_i(x_0, h) = 0$, $1 \leqslant i \leqslant n$, and $\omega(0, h) = h^{n+1}\sin h^{-n-1}$) but $F$ does not have even the first order derivative at 0.

**DEFINITION 7.2.** An operator $F\colon \mathfrak{X} \to \mathfrak{Y}$ defined in a neighborhood of a point $x_0$ is said to be *$\delta$-analytic* at this point if for any $h \in \mathfrak{X}$ there exists a number $r(x_0, h)$ such that the representation

$$F(x_0 + th) = F(x_0) + \sum_{n=1}^{\infty} B_n(x_0, h)t^n$$

holds for all $t \in \mathbb{K}$ with $|t| < r(x_0, h)$, where the series in the right-hand side is normally convergent, i.e.,

$$\sum_{n=1}^{\infty} \|B_n(x_0, h)\| |t|^n < \infty. \tag{7.8}$$

The operator $F$ is called *$\mathcal{G}$-analytic* (or Gâteaux analytic) if the expressions $B_n(x_0, h)$ are represented as the values of some homogeneous forms from $\mathfrak{X}$ into $\mathfrak{Y}$, evaluated at $h \in \mathfrak{X}$.

Let $\rho(x_0, h)$ be the convergence radius of the power series appearing in the inequality (7.8). By Cauchy-Hadamard formula we have

$$\rho(x_0, h) = \frac{1}{\overline{\lim\limits_{n\to\infty}} \sqrt[n]{\|B_n(x_0, h)\|}}. \tag{7.9}$$

The operator $F$ is called *$\mathcal{F}$-analytic* (or, simply, *analytic*) in the point $x_0$, if it is $\mathcal{G}$-analytic and for all $h \in \mathfrak{X}$ with $\|h\| = 1$ there exists $\rho > 0$ such that

$$\rho(x_0, h) \geqslant \rho,$$

where $\rho(x_0, h)$ is defined by (7.9).

Obviously, if an operator $F$ is $\mathcal{G}$-analytic and,

$$a = \overline{\lim_{n\to\infty}} \sqrt[n]{\|B_n\|} < \infty,$$

then $F$ is analytic and for all $h$ with $\|h\| < a^{-1} = r$ we have the representation

$$F(x_0 + h) = F(x_0) + \sum_{n=1}^{\infty} B_n(x_0, h), \tag{7.10}$$

where the series in the right-hand side is uniformly convergent for all $h$ with $\|h\| \leqslant \tilde{r} < < r$. (Here $\|B_n\|$ is taken according to §6.) We denote the set of all $\delta$-analytic (respectively $\mathcal{G}$-analytic or analytic) by $\mathcal{H}_\delta(x_0, \mathfrak{Y})$ (respectively $\mathcal{H}_\mathcal{G}(x_0, \mathfrak{Y})$ or $\mathcal{H}(x_0, \mathfrak{Y})$). Using the notations in §1, we have

$$\mathcal{H}_\delta(x_0, \mathfrak{Y}) \subseteq \mathcal{V}(x_0, \mathfrak{Y})$$
$$\mathcal{H}_\mathcal{G}(x_0, \mathfrak{Y}) \subseteq \mathcal{G}(x_0, \mathfrak{Y})$$
$$\mathcal{H}(x_0, \mathfrak{Y}) \subseteq \mathcal{F}(x_0, \mathfrak{Y}).$$

For complex spaces all these inclusions are, in fact, equalities (see Chapter III below). For real spaces, these inclusions are strict.

In this chapter we constantly pointed out the special role of complex spaces in the theory of differentiability. The properties of differentiable operators on complex spaces — mentioned in some remarks — lead to far-reaching consequences. For instance, some local features of the solutions of operatorial equations in complex spaces are global. However, the methods of differentiation are not enough for proving these results; it is necessary to combine them with methods from integral calculus.

# §8. Complete continuity of operators and of their derivatives

This section will be devoted to the following question: how does the property of complete continuity of a differentiable operator influence the properties of its derivatives? First, we need the following result.

**LEMMA 8.1.** *Let $\mathfrak{X}$ and $\mathfrak{Y}$ be Banach spaces. A linear operator $A: \mathfrak{X} \to \mathfrak{Y}$ is completely continuous on $\mathfrak{X}$ if and only if it is completely continuous on a bounded ball $\mathfrak{U} \subseteq \mathfrak{X}$ centered at the origin.*

◁ Necessity follows from the definitions (see Chapter 0, §§6 and 10). For sufficiency, let $\mathfrak{D}$ be an arbitrary bounded set in $\mathfrak{X}$. Choose a sequence $\{y_n\}_{n\in\mathbb{N}} \subseteq A\mathfrak{D}$, i.e., $y_n = Ax_n$ where $\{x_n\}_{n\in\mathbb{N}} \subseteq \mathfrak{D}$. Since the ball $\mathfrak{U}$ is centered at the origin, it absorbs $\mathfrak{D}$, i.e., there exists a number $\lambda > 0$ such that $\lambda\mathfrak{U} \supseteq \mathfrak{D}$. This means that the sequence $\{x'_n\}_{n\in\mathbb{N}}$, where $x'_n = \lambda^{-1}x_n$, lies in $\mathfrak{U}$, so the sequence $\{y'_n\}_{n\in\mathbb{N}}$, where $y'_n = Ax'_n$, is precompact in $\mathfrak{Y}$. But $y_n = \lambda y'_n$, and so the sequence $\{y_n\}_{n\in\mathbb{N}}$ is also precompact. ▶

**EXAMPLE 8.1.** Let $\mathfrak{X} = \mathfrak{Y} = \tilde{C}[a,b]$. Consider the linear integral operator $A: \mathfrak{X} \to \mathfrak{Y}$ defined by

$$Ax(t) = \int_a^b K(t,s)x(s)\mathrm{d}s \tag{8.1}$$

where $K(t,s)$ is a complex-valued function, continuous for $(t,s) \in \Omega = [a,b]\times[a,b]$ (see also (3.3)). The operator $A$ is completely continuous. Indeed, $A$ is clearly continuous. According to Arzelà-Ascolli theorem and to Lemma 8.1, for proving the precompactness of $A$ it is sufficient to show the equicontinuity of any function $y(t) \in A\mathfrak{B}$, where $\mathfrak{B}$ is the unit ball in $\mathfrak{X}$. We have

$$|y(t_1) - y(t_2)| \leqslant \int_a^b |K(t_1,s) - K(t_2,s)|\,|x(s)|\mathrm{d}s.$$

Since the function $K(t,s)$ is uniformly continous on $\Omega$, then for each $\varepsilon > 0$ there exists $\delta = \delta(\varepsilon)$ such that $|K(t_1,s) - K(t_2,s)| < \varepsilon$ for any $s \in [a,b]$ and any $t_1, t_2 \in [a,b]$ with $|t_1 - t_2| < \delta$. But $y = Ax$ with $x \in \mathfrak{B}$, so taking into account that

$$|x(s)| \leqslant \|x\| = \max_{[a,b]} |x(s)| \leqslant 1,$$

we obtain

$$|y(t_1) - y(t_2)| \leqslant \varepsilon(b-a)$$

if $|t_1 - t_2| < \delta$.

**EXAMPLE 8.2.** Consider the operator $A$ defined by the formula (8.1) on the space $\mathfrak{X} = \tilde{L}_2[a,b]$. It can be shown that, in this case, if $K(t,s)$ is continuous on $\Omega = [a,b] \times \times [a,b]$ then the operator $A$ is completely continuous. (Actually, it is sufficient to assume that the kernel $K(t,s)$ is a so-called Hilbert-Schmidt kernel, that is,

$$\int_a^b \int_a^b |K(t,s)|^2 \mathrm{d}s\mathrm{d}t < \infty,$$

see, for instance, M. A. Krasnoselskii [4].) The proof follows from similar arguments as above and from Cauchy-Buniakovski inequality:

$$|y(t_1) - y(t_2)| \leqslant \int_a^b |K(t_1,s) - K(t_2,s)|\,|x(s)|\mathrm{d}s \leqslant$$

$$\leqslant \left( \int_a^b |K(t_1,s) - K(t_2,s)|^2 \,\mathrm{d}s \right)^{1/2} \left( \int_a^b |x(s)|^2 \mathrm{d}s \right)^{1/2} \leqslant \varepsilon(b-a).$$

Using Hölder inequality we can show that the operator (8.1) with a continuous kernel $K(t,s)$ is completely continuous on each $\tilde{L}_p[a,b]$ and, moreover, that it is completely continous as an operator from $\tilde{L}_p[a,b]$ into $\tilde{C}[a,b]$.

The following result is obvious:

**LEMMA 8.2.** *Let $\mathfrak{X}, \mathfrak{Y}, \mathfrak{Z}$ be topological spaces and let $F: \mathfrak{X} \to \mathfrak{Y}$ and $G: \mathfrak{Y} \to \mathfrak{Z}$ be operators defined and continous on the domains $\mathfrak{D} \subseteq \mathfrak{X}$, $\mathfrak{U} \subseteq \mathfrak{Y}$ respectively, with $F(\mathfrak{D}) \subseteq \mathfrak{U}$. If one of the operators $F$ or $G$ is compact, then the composed operator $GF: \mathfrak{X} \to \mathfrak{Y}$ is completely continuous.*

This property gives easy proofs for the complete continuity of many operators.

**EXAMPLE 8.3.** Let $\mathfrak{X} = \mathfrak{Y} = \tilde{C}[a,b]$. Consider the Hammerstein operator $F: \mathfrak{X} \to \mathfrak{Y}$,

$$Fx(t) = \int_a^b K(t,s) f(s, x(s)) \mathrm{d}s,$$

where $K(t,s)$ is continuous on the square $\Omega = [a,b] \times [a,b]$ and $f(s,x)$ is continuous on the whole domain of the variables $a \leqslant s \leqslant b$ and $x \in \mathbb{C}$, $|x| \leqslant r$. Then the operator $F$ is completely continuous on the ball $\mathfrak{D} = \{x \in \mathfrak{X} : \|x\| \leqslant r\}$. Indeed, it is sufficient to notice that the operator $F$ is the composition of the superposition operator $f(s,\cdot)$ and of the linear completely continuous operator $A$ given by

$$Au(t) = \int_a^b K(t,s)u(s)\mathrm{d}s.$$

**REMARK 8.2.** The superposition operator acting on the space $\tilde{C}[a,b]$ is not completely continuous, excepting the trivial case when $f(s,x)$ does not depend on $x$. However, the same operator may be completely continuous on other spaces. For example, a continuous superposition operator is completely continuous if the space on which it acts is a Montel space (see Chapter 0, §4). Note that Theorem 0.4.1 implies that an infinite dimensional normed space does not have this property.

Let now $\mathfrak{X}, \mathfrak{Y}$ be arbitrary Banach spaces and $\mathfrak{D}$ a domain in $\mathfrak{X}$.

**THEOREM 8.1** (M. A. Krasnoselskii). *Let $F: \mathfrak{D} \to \mathfrak{Y}$ be a completely continuous operator on $\mathfrak{D}$. If $F$ is Fréchet differentiable at $x_0 \in \mathfrak{D}$ then the operator $A = F'(x_0): \mathfrak{X} \to \mathfrak{Y}$ is also completely continuous.*

◁ Suppose that $A$ is not completely continuous. Consequently, the set of values of $A$ on the unit ball is not precompact. Then there exist a sequence $\{h_n\}_{n\in\mathbb{N}} \subseteq \mathfrak{X}$, $\|h_n\| = 1$, and a number $\delta > 0$ for which $\|A(h_n - h_m)\| > 3\delta$, $n, m \in \mathbb{N}$, $n \neq m$. Put $\omega(x_0, h) = F(x_0 + h) - F(x_0) - Ah$. Let $x_n = x_0 + \rho h_n$, where the number $\rho > 0$ is chosen such that $x_n \in \mathfrak{D}$ for all $n \in \mathbb{N}$, and

$$\frac{1}{\rho}\|\omega(x_0, \rho h_n)\| = \frac{1}{\rho}\|F(x_0 + \rho h_n) - F(x_0) - \rho A h_n\| < \delta.$$

Then, from the equality

$$F(x_n) - F(x_m) = \rho A(h_n - h_m) + \omega(x_0, \rho h_n) - \omega(x_0, \rho h_m)$$

it follows that

$$\|F(x_n) - F(x_m)\| \geqslant \rho\|A(h_n - h_m\| - \|\omega(x_0, \rho h_n)\| - \|\omega(x_0, \rho h_m)\| \geqslant \rho\delta,$$

$n, m \in \mathbb{N}$, which means — since the sequence $\{x_n\}_{n\in\mathbb{N}} \subseteq \mathfrak{D}$ is bounded — that the operator $F$ is not completely continuous. ▶

The following more general result can be proved by induction.

**THEOREM 8.2.** *Suppose that the operator $F: \mathfrak{X} \to \mathfrak{Y}$ is completely continuous on the domain $\mathfrak{D}$ and admits, about $x_0 \in \mathfrak{D}$, the Taylor representation*

$$F(x_0 + h) = F(x_0) + A_1(x_0, h) + \ldots + A_n(x_0, h) + \omega(x_0, h)$$

*where $A_i(x_0, h)$ are homogeneous forms (in $h$) of order $i$, $i = 1, 2 \ldots, n$, and*

$$\|\omega(x_0, h)\| \leqslant M\|h\|^{n+1}$$

*where $M < \infty$ does not depend on $h$. Then the forms $A_i(x_0, h)$ are completely continuous as operators from $\mathfrak{X}$ into $\mathfrak{Y}$.*

**REMARK 8.3.** Under the assumptions of Theorem 8.2, it is enough to suppose that the forms $A_i(x_0, h)$ correspond to some $i$-linear, not necessarily continuous, operators $\tilde{A}_i(x_o, h_1, \ldots, h_n)$, in order to prove their continuity in $h$.

◁ Indeed, since the operator $F$ is continuous, then for any $\varepsilon > 0$ there exists $r > 0$ such that for all $h$ with $\|h\| < r$, we have $\|F(x_0+th) - F(x_0)\| \leqslant \varepsilon$. Consequently, for all $t$, $0 \leqslant t \leqslant 1$, we obtain

$$\|\phi(x_0, th)\| = \|tA_1(x_0, h) + \cdots + t^n A_n(x_0, h)\| \leqslant \varepsilon + Mr^{n+1} = N,$$

where $\phi(x_0, th) = F(x_0 + th) - F(x_0) - \omega(x_0, th)$.

For any linear functional $y^* \in \mathfrak{Y}^*$ the functional $\varphi(t) = \langle \phi(x_0, th), y^* \rangle$ is represented as a polynomial of degree $n$ in $t$, satisfying the estimate $|\varphi(t)| < N$ for all $t$, $0 \leqslant t \leqslant 1$, and all $y^*$, $\|y^*\| \leqslant 1$. Then, there exists a number $\ell > 0$ such that the absolute value of each coefficient of the polynomial $\varphi(t)$ is less than or equal to $\ell N$, i.e.,

$$|\langle A_i(x_0, h), y^* \rangle| \leqslant N\ell.$$

Since $y^* \in \mathfrak{Y}^*$ is arbitrary, we have

$$\|A_i(x_0, h)\| \leqslant N\ell,$$

for all $h$, $\|h\| \leqslant r$, i.e., the operator $A_i(x_0, h)$ is bounded and thus, continuous. ▶

**EXAMPLE 8.4.** Consider the nonlinear Volterra type operator $F$ given by

$$Fx(t) = \int_a^t K(t,s)f(s,x(s))ds,$$

where the functions $K(t,s)$ and $f(s,u)$ are continuous for $a \leqslant t, s \leqslant b,\ u \in \mathbb{C},\ |u| \leqslant r$, and $f(s,u)$ has the representation

$$f(s,u) = \sum_{i=0}^{n} a_i(s)u^i + b(s,u),$$

where

$$\sup_{a \leqslant s \leqslant b} |b(s,u)| \leqslant m|u|^{n+1}.$$

Arguing as in the previous example, it is easy to show that the operator $F$ is completely continuous on $\tilde{C}[a,b]$ and, if the functions $a_i(s)$ are continuous on $[a,b]$ then the operator $F$ has a Taylor representation of order $n$ about zero, with completely continuous homogeneous forms

$$(A_i(0,h))(t) = \int_a^t K(t,s)a_i(s)h^i(s)ds, \quad 1 \leqslant i \leqslant n. \tag{8.2}$$

Moreover, the continuity of the functions $a_i(s)$ is not necessary for proving the complete continuity of the operators in (8.2). To see this, it is sufficient to show that the operators (8.2) act from $\tilde{C}[a,b]$ into $\tilde{C}[a,b]$. Indeed, by Remark 8.3 it follows that the functions $a_i(s)$ are bounded on the interval $[a,b]$. But then the value of the right hand side of (8.2) is a continuous function in $t$ on the interval $[a,b]$, which completes our argument.

# Chapter II

# Integration in normed spaces

## §1. Riemann - Stieltjes integrals of vector-functions

Let us remind that by a *function* we mean a mapping having the domain in the scalar field $\mathbb{K}$ of real or complex numbers (see Chapter 0, §1). In this chapter, we will consider functions with values in normed spaces — called *vector-functions* —, as well as functions with values in the field of scalars, which are called *scalar-functions*. The symbol $f(t)$ means either the function $f$ itself, or the value of $f$ at $t$; the precise meaning will follow from the context. From now on the term "vector-function" will be used in a broader sense, for any mapping $f: \mathfrak{X} \to \mathfrak{Y}$ with the domain in a finite dimensional space and with values in a normed space.

In this section we consider that $\mathbb{K} = \mathbb{R}$ and $\mathfrak{Y}$ is a Banach space, i.e., a complete normed space.

**DEFINITION 1.1.** A vector-function $y$ defined on an interval $[\alpha, \beta]$ is called a *vector-function of bounded variation*, if there exists a number $M < \infty$ such that the inequality

$$\sum_i \|y(\beta_i) - y(\alpha_i)\| \leqslant M \tag{1.1}$$

holds for any choice of a finite number of disjoint intervals $(\alpha_i, \beta_i)$ in $(\alpha, \beta)$.

The number $\mathrm{Var}[y] = \sup \sum_i \|y(\beta_i) - y(\alpha_i)\|$ is called the *complete variation* of the vector-function $y$ on the interval $[\alpha, \beta]$.

**REMARK 1.1.** If $y$ is a vector-function of bounded variation then the scalar-function $\varphi_{y^*}(t) = \langle y(t), y^* \rangle$ is also of bounded variation for any $y^* \in \mathfrak{Y}^*$. The converse is also true and was independently proved by Dunford [1] and Gelfand [1]: if $\mathrm{Var}[\varphi_{y^*}(t)] \leqslant \leqslant M(y^*) < \infty$ for any $y^* \in \mathfrak{Y}^*$ with $\|y^*\| \leqslant 1$, then $y(t)$ is a vector-function of bounded variation.

**DEFINITION 1.2.** Let $y$ be a vector-function defined on the interval $[\alpha, \beta]$ and $g$ a scalar-function with values in $\mathbb{R}$ (or $\mathbb{C}$) defined on the same interval.

The expressions

$$S_\pi(y, g) = \sum_{i=1}^{n} y(t_i)\,[g(\sigma_i) - g(\sigma_{i-1})]$$

and

$$s_\pi(y, g) = \sum_{i=1}^{n} g(t_i)\,[y(\sigma_i) - y(\sigma_{i-1})]\,,$$

where $\pi$ is a partition $\alpha = \sigma_0 \leqslant \sigma_1 \leqslant \cdots \leqslant \sigma_n = \beta$ of $[\alpha, \beta]$, $\sigma_{i-1} \leqslant t_i \leqslant \sigma_i$, $i = 1, \ldots, n$, are called *Riemann-Stieltjes integral sums*. Define also $|\pi| = \max\limits_{1 \leqslant i \leqslant n} |\sigma_i - \sigma_{i-1}|$.

**THEOREM 1.1.** *If one of the limits*

$$\lim_{|\pi| \to 0} S_n(y, g), \qquad \lim_{|\pi| \to 0} s_n(y, g)$$

*exists in the strong, weak or ultraweak topology of* $\mathfrak{Y}$ *then the other also exists. These limits are called Riemann-Stieltjes integrals and are denoted respectively by* (R-S) $\int_\alpha^\beta y(t)\mathrm{d}g(t)$ *and* (R-S) $\int_\alpha^\beta g(t)\mathrm{d}y(t)$. *Moreover, we have*

$$(\text{R-S}) \int_\alpha^\beta y(t)\mathrm{d}g(t) = y(t)g(t)\,\Big|_\alpha^\beta - (\text{R-S}) \int_\alpha^\beta g(t)\mathrm{d}y(t). \tag{1.2}$$

◁ The conclusions follow immediately from the identity

$$S_\pi(y, g) = y(\beta)g(\beta) - y(\alpha)g(\alpha) - \sum_{i=0}^{n} g(\sigma_i)\,[y(t_{i+1}) - y(t_i)]\,.$$ ▶

**THEOREM 1.2.** *Suppose that one of the functions $y$ or $g$ is of bounded variation and the other one is continuous (for $y$ the continuity is understood in the strong topology). Then both the integrals*

$$\text{(R-S)}\int_{\alpha}^{\beta} y(t)\mathrm{d}g(t) \quad \text{and} \quad \text{(R-S)}\int_{\alpha}^{\beta} g(t)\mathrm{d}y(t)$$

*exist in the strong topology of the space $\mathfrak{Y}$. In addition, if $A$ is a continuous linear operator acting from $\mathfrak{Y}$ into a Banach space $\mathfrak{Z}$, then*

$$A\left(\text{(R-S)}\int_{\alpha}^{\beta} y(t)\mathrm{d}g(t)\right) = \text{(R-S)}\int_{\alpha}^{\beta} Ay(t)\mathrm{d}g(t) \tag{1.3}$$

*and*

$$A\left(\text{(R-S)}\int_{\alpha}^{\beta} g(t)\mathrm{d}y(t)\right) = \text{(R-S)}\int_{\alpha}^{\beta} g(t)\mathrm{d}Ay(t). \tag{1.4}$$

◁ Suppose that the vector-function $y$ is continuous in the strong topology and $g$ is of bounded variation. Then for any $\varepsilon > 0$ there exists $\delta > 0$ such that $\|y(t_1) - y(t_2)\| \leqslant \varepsilon$ as soon as $|t_1 - t_2| < \delta$. Choosing $|\pi_1|, |\pi_2| < \dfrac{\delta}{2}$, we have

$$\|S_{\pi_1} - S_{\pi_2}\| \leqslant \mathrm{Var}[g] \cdot 2\varepsilon.$$

Hence the first assertion of the theorem follows. Let us prove now the equalities (1.3) and (1.4).

If the function $y$ is continuous or of bounded variation, then the linearity and the continuity of the operator $A\colon \mathfrak{Y} \to \mathfrak{Z}$ imply the same properties for the vector--function $h = Ay\colon [\alpha, \beta] \to \mathfrak{Z}$. Therefore the integrals in the right hand sides of (1.3) and (1.4) exist. Furthemore, using again the linearity of $A$, we have

$$S_\pi\,(Ay, g) \equiv AS_\pi(y, g),$$
$$s_\pi\,(Ay, g) \equiv As_\pi(y, g),$$

from which, by taking the limit, we obtain (1.3) and (1.4). ▶

**COROLLARY 1.1.** *For a vector-function $y$ and a scalar-function $g$ satisfying the conditions of Theorem 1.2, the following relations*

$$\left\langle \text{(R-S)}\int_{\alpha}^{\beta} y(t)\mathrm{d}g(t), y^* \right\rangle = \text{(R-S)}\int_{\alpha}^{\beta} \langle y(t), y^*\rangle \mathrm{d}g(t) \tag{1.5}$$

$$\left\langle (\text{R-S}) \int\limits_{\alpha}^{\beta} g(t)\mathrm{d}y(t), y^* \right\rangle = (\text{R-S}) \int\limits_{\alpha}^{\beta} g(t)\mathrm{d}\langle y(t), y^* \rangle \tag{1.6}$$

*hold for all $y^* \in \mathfrak{Y}^*$, where the integrals in the right hand sides of (1.5) and (1.6) are the usual Riemann-Stieltjes integrals of scalar-functions.*

**COROLLARY 1.2.** *If the vector-function $y$ is continuous on $[\alpha, \beta]$ and $g$ is of bounded variation then*

1)

$$\left\| (\text{R-S}) \int\limits_{\alpha}^{\beta} y(t)\mathrm{d}g(t) \right\| \leqslant \sup_{\alpha \leqslant t \leqslant \beta} \|y(t)\| \cdot \mathrm{Var}[g] = M; \tag{1.7}$$

2) *if a sequence $(y_n)_{n \in \mathbb{N}}$ of continuous vector-functions is uniformly convergent on $[\alpha, \beta]$ to the function $y$ in the norm topology of the space $\mathfrak{Y}$, then*

$$\lim_{n \to \infty} (\text{R-S}) \int\limits_{\alpha}^{\beta} y_n(t)\mathrm{d}g(t) = (\text{R-S}) \int\limits_{\alpha}^{\beta} y(t)\mathrm{d}g(t) \tag{1.8}$$

*where the limit in (1.8) is understood in the strong sense.*

◁ The inequality (1.7) follows from the estimate $\|S_\pi(y, g)\| \leqslant M$. The relation (1.8) follows from a similar estimate, since

$$\left\| (\text{R-S}) \int\limits_{\alpha}^{\beta} y_n(t)\mathrm{d}g(t) - (\text{R-S}) \int\limits_{\alpha}^{\beta} y(t)\mathrm{d}g(t) \right\| \leqslant \sup_{\alpha \leqslant t \leqslant \beta} \|y_n(t) - y(t)\| \cdot \mathrm{Var}[g]. \quad ▶$$

**REMARK 1.2.** Formula (1.2) is a generalization of the classical formula of integration by parts. Taking $g(t) \equiv 1$ in this formula, we obtain the analogue of the Newton--Leibniz formula

$$(\text{R-S}) \int\limits_{\alpha}^{\beta} \mathrm{d}y(t) = y(\beta) - y(\alpha). \tag{1.9}$$

Taking $g(t) \equiv t$ in the integral $(\text{R-S}) \int\limits_{\alpha}^{\beta} y(t)\mathrm{d}g(t)$, we obtain a generalization of the classical Riemann integral.

## §2. Pettis integral and the connection with Riemann-Stieltjes integral

The integral introduced by Pettis [1] in 1938, as a generalization of the Lebesgue integral, will play an important role in the following chapters.

**DEFINITION 2.1.** Let $y(\sigma)$ be a vector-function defined on a measurable set $\mathfrak{S}$ with values in a Banach space $\mathfrak{Y}$, such that the scalar-function $\langle y(\sigma), y^* \rangle$ is integrable in the sense of Lebesgue for each $y^* \in \mathfrak{Y}^*$. If there exists an element $z \in \mathfrak{Y}$ such that

$$\langle z, y^* \rangle = \int_{\mathfrak{S}} \langle y(\sigma), y^* \rangle \mathrm{d}\sigma$$

for all $y^* \in \mathfrak{Y}^*$, then $z$ is called the *Pettis integral* of the function $y$ and one denotes

$$z = (\mathrm{P}) \int_{\mathfrak{S}} y(\sigma) \mathrm{d}\sigma. \tag{2.1}$$

Note that in the case when $\mathfrak{Y}$ is reflexive the Pettis integral exists provided the functions $\langle y(\sigma), y^* \rangle$ are measurable and integrable in the sense of Lebesgue for all $y^* \in \mathfrak{Y}^*$ (see Hille and Phillips [1]).

The situation we are interested in is the case when $\mathfrak{Y}$ is a complex space and $\mathfrak{S}$ is a contour in $\mathbb{C}$. In this case we study the existence of Pettis integral and its connections with Riemmann-Stieltjes integral.

We take as $\mathfrak{S}$ a contour $\gamma$ in $\mathbb{C}$ given by the equation $\gamma = \{\tau \in \mathbb{C} : \tau = \tau(t), \alpha \leqslant t \leqslant \beta\}$, where $\tau(t)$ is a continuous function of bounded variation on $[\alpha, \beta]$. Let $y$ be a strongly continuous vector-function defined on $\gamma$. Then the scalar-function $\varphi_{y^*}$ given by $\varphi_{y^*}(\tau) = \langle y(\tau), y^* \rangle$, $y^* \in \mathfrak{Y}^*$, is Lebesgue integrable on $\gamma$ and

$$\int_{\gamma} \varphi_{y^*}(\tau) \mathrm{d}\tau = (\text{R-S}) \int_{\alpha}^{\beta} \varphi_{y^*}(\tau(t)) \, \mathrm{d}\tau(t) =$$

$$= \int_{\alpha}^{\beta} \langle y(\tau(t)), y^* \rangle \, \mathrm{d}\tau(t).$$

On the other hand, the vector-function $\tilde{y} = y \circ \tau$ is strongly continuous on the interval $[\alpha, \beta]$ and therefore Theorem 1.2 implies the existence of the Riemann-Stieltjes integral

$$(\text{R-S}) \int_{\alpha}^{\beta} \tilde{y}(t) \mathrm{d}\tau(t) = (\text{R-S}) \int_{\alpha}^{\beta} y(\tau(t)) \, \mathrm{d}\tau(t) = z \in \mathfrak{Y}.$$

According to (1.5), for any $y^* \in \mathfrak{Y}^*$, we have

$$\langle z, y^* \rangle = \text{(R-S)} \int_{\alpha}^{\beta} \langle y(\tau(t)), y^* \rangle \, \mathrm{d}\tau(t) = \int_{\gamma} \langle y(\tau), y^* \rangle \, \mathrm{d}\tau,$$

which proves that the function $y$ is Pettis integrable on $\gamma$:

$$z = \text{(P)} \int_{\alpha}^{\beta} y(\tau)\mathrm{d}\tau = \text{(R-S)} \int_{\alpha}^{\beta} y(\tau(t)) \, \mathrm{d}\tau(t).$$

Moreover, if $\gamma$ is a smooth contour, i.e., the derivative $\tau'(t)$ exists and is continuous on $[\alpha, \beta]$, then we have

$$\text{(P)} \int_{\gamma} y(\tau)\mathrm{d}\tau = \text{(R)} \int_{\alpha}^{\beta} y(\tau(t)) \, \tau'(t)\mathrm{d}\tau. \tag{2.2}$$

When no confusions arise, we will omit the symbol (P) in the Pettis integral, and we simply write $\int y(\tau)\mathrm{d}\tau$.

## §3. Antiderivatives of vector-functions. Integral representations

**1.** The symbol of differentiation in (1.2) is, generally speaking, only formal, since — as it follows from Theorem 1.2 — formula (1.2) is true for some vector-functions which are not differentiable; take, for example, a vector-function which has complete bounded variation.

We will show in this section that the classical Newton-Leibniz formula has a direct generalization for vector-functions which are assumed to be differentiable. To begin with, we prove, as in the classical analysis, the differentiability of an integral with respect to a variable upper limit; this result is also interesting on its own.

**LEMMA 3.1.** *Let $y$ be a vector-function defined on an interval $[\alpha, \beta]$ with values in a Banach space $\mathfrak{Y}$ which is Riemann integrable on $[\alpha, \beta]$ (see Remark 1.2) and continuous at a point $\nu \in (\alpha, \beta)$. Then the vector-function $f\colon [\alpha, \beta] \to \mathfrak{Y}$ given by*

$$f(\tau) = \text{(R)} \int_{\alpha}^{\tau} y(t)\mathrm{d}t \tag{3.1}$$

*is differentiable at* $\nu$ *and*

$$f'(\nu) = y(\nu).$$

◁ Consider the difference

$$\frac{f(\nu+\varepsilon) - f(\nu)}{\varepsilon} - y(\nu) = \frac{1}{\varepsilon}(\mathrm{R}) \int\limits_{\nu}^{\nu+\varepsilon} (y(t) - y(\nu))\,\mathrm{d}t.$$

Then, according to (1.7), we have

$$\left\| \frac{f(\nu+\varepsilon) - f(\nu)}{\varepsilon} - y(\nu) \right\| \leqslant \sup_{\nu \leqslant t \leqslant \nu+\varepsilon} \|y(t) - y(\nu)\|\,.$$

The continuity of $y(\tau)$ at $\nu$ implies that the right hand side of the previous inequality tends to zero as $\varepsilon$ approaches zero, so the conclusion of the lemma follows. ▶

**COROLLARY 3.1** (Newton-Leibniz formula). *Let* $g(t)$ *be a vector-function defined on* $[\alpha, \beta]$ *and continuously-differentiable on* $[\alpha, \beta]$. *Then the formula*

$$g(\beta) - g(\alpha) = (\mathrm{R}) \int\limits_{\alpha}^{\beta} g'(t)\mathrm{d}t \tag{3.2}$$

*holds.*

◁ Consider the function $f$ given by (3.1), where $y(t) = g'(t)$. Then $f(\beta)$ is the integral (3.2). Furthermore,

$$\langle f'(t), y^* \rangle = \langle g'(t), y^* \rangle, \quad t \in (\alpha, \beta),$$

for every $y^* \in \mathfrak{Y}^*$. Consequently, there exists a constant $c(y^*)$ such that

$$\langle f(t), y^* \rangle = \langle g(t), y^* \rangle + c(y^*). \tag{3.3}$$

Since $f(\alpha) = 0$, we have

$$c(y^*) = -\langle g(\alpha), y^* \rangle.$$

Using this equality in (3.3) and taking $t = \beta$, we have for each $y^* \in \mathfrak{Y}^*$,

$$\langle f(\beta), y^* \rangle = \langle g(\beta) - g(\alpha), y^* \rangle,$$

and so (3.2) follows. ▶

**2.** By analogy with the classical analysis, the vector-function $f$ defined by (3.1) is called the *antiderivative* (*primitive*) of $y$. If $y$ is continuous on $[\alpha, \beta]$, then its antiderivative vector-function $f$ exists and, being differentiable on $[\alpha, \beta]$, it is continuous on $[\alpha, \beta]$. Then the function $f_1 = f$ has also a continuous antiderivative $f_2$ on $[\alpha, \beta]$. By induction, we obtain the existence of succesive continuous antiderivatives $f_m$ for each order $m$ ($f_m'(t) = f_{m-1}(t)$ for $m \geqslant 2$ and $f_1'(t) = f'(t) = y(t)$). Moreover, $f_2(t)$ is an iterated integral:

$$f_2(t) = (\mathrm{R})\int_\alpha^t f(\tau)\mathrm{d}\tau = (\mathrm{R})\int_\alpha^t \mathrm{d}\tau \cdot (\mathrm{R})\int_\alpha^\tau y(s)\mathrm{d}s,$$

$(\alpha \leqslant s \leqslant \tau \leqslant t \leqslant \beta)$. Similarly, for every $m \geqslant 2$, we have

$$f_m(t) = (\mathrm{R})\int_\alpha^t \mathrm{d}t_m (\mathrm{R})\int_\alpha^{t_m} \mathrm{d}t_{m-1} \cdots (\mathrm{R})\int_\alpha^{t_1} y(t_1)\mathrm{d}t_1$$

$(\alpha \leqslant t_1 \leqslant \cdots \leqslant t_m \leqslant t \leqslant \beta)$.

Let us show now that $f_m(t)$ has an integral representation as an one-dimensional integral.

**THEOREM 3.1.** *Let $y$ be a vector-function continuous on $[\alpha, \beta]$. Then the vector-function $f_m$ given by*

$$f_m(t) = \frac{1}{(m-1)!}(\mathrm{R})\int_\alpha^t (t-s)^{m-1} y(s)\mathrm{d}s \tag{3.3}$$

*is the m-th antiderivative vector-function of $y$ satisfying the conditions*

$$f_m(\alpha) = f_m'(\alpha) = \cdots = f_m^{(m-1)}(\alpha) = 0,$$
$$f_m^{(m)}(t) = y(t).$$

◁ We shall prove the theorem by induction on $m = 1, 2, \ldots$ . For $m = 1$, the assertion follows from Lemma 3.1. Assume that the theorem is proved for the function $f_{m-1}(t)$. Consider its antiderivative vector-function

$$F(t) = (\mathrm{R})\int_\alpha^t f_{m-1}(\tau)\mathrm{d}\tau;$$

we will show that $F(t)$ coincides with the vector-function defined by (3.3). Indeed

$$F(t) = (\mathrm{R}) \int_{\alpha}^{t} \mathrm{d}\tau \ (\mathrm{R}) \int_{\alpha}^{\tau} \frac{(\tau - s)^{m-2}}{(m-2)!} y(s) \mathrm{d}s.$$

According to Corollary 1.1, for any $y^* \in \mathfrak{Y}^*$ we have

$$\langle F(t), y^* \rangle = (\mathrm{R}) \int_{\alpha}^{t} \mathrm{d}\tau \ (\mathrm{R}) \int_{\alpha}^{\tau} \frac{(\tau - s)^{m-2}}{(m-2)!} \langle y(s), y^* \rangle \, \mathrm{d}s.$$

By Fubini Theorem we can change the order of integration in the last integral, so we get

$$\langle F(t), y^* \rangle = (\mathrm{R}) \int_{\alpha}^{t} \langle y(s), y^* \rangle \, \mathrm{d}s \ (\mathrm{R}) \int_{s}^{t} \frac{(\tau - s)^{m-2}}{(m-2)!} \mathrm{d}\tau.$$

Computing the inner integral, we obtain

$$\langle F(t), y^* \rangle = \frac{1}{(m-1)!} \int_{\alpha}^{t} (t-s)^{m-1} \langle y(s), y^* \rangle \, \mathrm{d}s$$

whence, using also Corollary 1.1, it follows that

$$\langle F(t), y^* \rangle = \langle f_m(t), y^* \rangle$$

for any $y^* \in \mathfrak{Y}^*$, so the theorem is proved. ▶

Formula (3.3) may be seen as expressing a function in terms of its derivatives of order $0, 1, \ldots, m-1$, evalued at a point, and the values of its $m$-th derivative on a neighborhood of that point. For a generalization of (3.3), we formulate the following initial value problem.

Consider the differential equation of order $m$

$$F^{(m)}(t) = y(t), \tag{3.4}$$

with the initial conditions at a point $\alpha \in \mathbb{R}$

$$F(\alpha) = y_0, \ F'(\alpha) = y_1, \ \ldots, \ F^{(m-1)}(\alpha) = y_{m-1}, \tag{3.5}$$

where $y$ is a vector-function, continuous on a neighborhood of the point $\alpha$, and with values in $\mathfrak{Y}$, and $y_0, \dots, y_{m-1}$ are some given elements in $\mathfrak{Y}$. Then the next result is true.

**COROLLARY 3.2.** *The solution of the differential equation* (3.4) *with initial conditions* (3.5) *has the form*

$$F(t) = y_0 + (t-\alpha)y_1 + \cdots + \frac{1}{(m-1)!}(t-\alpha)^{m-1}y_{m-1} + (\mathrm{R})\int_\alpha^t \frac{(t-s)^{m-1}}{(m-1)!}y(s)\mathrm{d}s. \quad (3.6)$$

◁ Note that the integral in (3.6) is a antiderivative of order $m$ of $y$, which, together with all its derivatives up to order $m-1$, have the value zero at the point $\alpha$. Moreover, the sum of the other terms in the right hand side of (3.6) is a polynomial of degree $m-1$ (so its $m$-th derivative is indentically zero) which takes at $\alpha$ the given values $y_0, \dots, y_{m-1}$. ▶

Formula (3.6) is the Taylor formula for the function $F$, with the remainder in an integral form. By contradistinction with the usual Taylor formula, the integral in (3.6) gives an explicit form of the remainder, which turns out to be very useful in many cases. However, note that we impose a restrictive condition on $F$, namely, the continuity of its $m$-th derivative on a neighborhood of the point $\alpha$.

To conclude this section, consider a vector-function $f$ that is continuous on a neighborhood $\Omega$ of a point $\zeta_0 \in \mathbb{C}$ and takes values in a Banach space $\mathfrak{Y}$. Let $\zeta_1$ be an arbitrary point in $\Omega$ and let $\gamma$ be a smooth contour lying completely in $\Omega$ and joining the points $\zeta_0$ and $\zeta_1$:

$$\gamma = \{\zeta \in \Omega \ : \ \zeta = \zeta(t), \ \alpha \leqslant t \leqslant \beta, \ \zeta(\alpha) = \zeta_0, \ \zeta(\beta) = \zeta_1\}.$$

Take the partial contour $\gamma_\tau = \{\zeta \in \gamma \ : \ \zeta = \zeta(t), \ \alpha \leqslant t \leqslant \tau\}$ and consider the integral $\int_{\gamma_\tau} f(\zeta)\mathrm{d}\zeta = \Phi(\zeta(\tau))$. The last integral is a function of $\zeta = \zeta(\tau)$, where $\zeta$ runs on the contour $\gamma_\tau$ from $\zeta_0$ up to $\zeta(\tau)$. Further, according to (2.2) we have

$$\Phi(\zeta(\tau)) = \int_\alpha^\tau f(\zeta(t))\zeta'(t)\mathrm{d}t = F(\tau),$$

where $F(\tau)$ is a primitive of the function $y(\tau) = f(\zeta(\tau))\zeta'(\tau)$, with $F(\alpha) = 0$. Thus $\Phi(\zeta(\tau)) = F(\tau)$.

Suppose that the vector-function $\Phi$ is differentiable with respect to $\zeta$ in a neighborhood of $\zeta_1$. Then, using the formula for the derivative of a composite function, we find — for $\tau$ sufficiently close to $\beta$ — that

$$\Phi'(\zeta(\tau)) = \Phi'(\zeta)\zeta'(\tau) = F'(\tau) = f(\zeta(\tau))\zeta'(\tau).$$

Hence

$$\Phi'(\zeta_1) = f(\zeta_1).$$

In conclusion, if the integral

$$\int_\gamma f(\zeta)\mathrm{d}\zeta = \Phi(\zeta_1, \gamma)$$

is considered on smooth contours joining $\zeta_0$ and $\zeta_1$, and if $\Phi$ is a vector-function differentiable at $\zeta_1$, then

$$\Phi'_\zeta(\zeta_1, \gamma) = f(\zeta_1)$$

and $\Phi(\zeta_1, \gamma)$ does not depend on $\gamma$, that is the integral on any contour $\gamma$ joining the points $\zeta_0$ and $\zeta_1$ gives the value of the antiderivative of the function $f$ at the point $\zeta_1$, which is zero at $\zeta_0$.

## §4. Integrals of operators in Banach spaces

The theory of abstract integration of vector-functions has many similarities with the integration theory for scalar functions. For example, as soon as we prove the existence of the abstract Riemann-Stieltjes integral, its properties can be derived from the similar properties of the integral for scalar-functions, using Corollary 1.1 and Hahn-Banach type theorems. Similar considerations can be made for the theory of measurable vector-functions, the theory of Riemann or Lebesgue integrals, and others (see, for example, L. Schwartz [1]). However, in what follows we will need only the results stated in §§1–3.

The theory of integration of operators acting between linear topological spaces is richer and more complicated. First, it is natural to construct the integral for operators in such a way so that to generalize the corresponding notions for scalar or vector-functions. Further, it is necessary to take into account the usefulness for applications. One of the approaches to the definition of the integral of an operator relies upon the integration of its finite dimensional sections (see E. Hille and R. Phillips [1]). Although this approach is quite natural, we found it rather inefficient in many cases. For examples, from the beginning there are difficulties with changing the variable in the integral. Further, this approach does not help in constructing the analogue of the logarithmic residue from finite dimensional theory, a fact which causes many troubles for the theory of fixed points of noncompact operators.

Having in mind the traditional schemes, we introduce in this section only two types of integrals of operators acting between spaces with special properties.

**1.** Let $\mathfrak{X}$ and $\mathfrak{Z}$ be Banach spaces and let $\mathfrak{Y} = L(\mathfrak{X}, \mathfrak{Z})$ be the space of linear operators from $\mathfrak{X}$ into $\mathfrak{Z}$. Let $F: \mathfrak{X} \to \mathfrak{Y}$ be an operator defined on a domain $\mathfrak{D}$, i.e., for any $x \in \mathfrak{D}$ the value $F(x)$ is a linear continuous operator from $\mathfrak{X}$ into $\mathfrak{Z}$, depending on $x$ as a parameter.

**DEFINITION 4.1.** Assume that the vector-function $f(t) = F(x + th)h$ is integrable on the interval [0,1]. The expression

$$\int_x^{x+h} F(x)\mathrm{d}x = (\mathrm{R}) \int_0^1 F(x + th)h\mathrm{d}t \tag{4.1}$$

is called *the integral of the operator* $F$ on the interval $[x, x + h]$.

All the properties concerning the linearity of integral (4.1) can be easily checked using the results of §1. Further, if $\mathfrak{X} = \mathfrak{Z}$, then $\mathfrak{Y} = L(\mathfrak{X}, \mathfrak{X})$ is a Banach algebra with unit $I$ (the identity mapping of $\mathfrak{X}$), and

$$\int_x^{x+h} I\mathrm{d}x = \int_x^{x+h} \mathrm{d}x = h.$$

We will prove now that a Newton-Leibniz type formula holds for the integrals defined by (4.1), i.e., a formula which recovers the values of a differentiable operator from its derivative.

**THEOREM 4.1.** *Let $\mathfrak{X}$ and $\mathfrak{Z}$ be Banach spaces and let $G: \mathfrak{X} \to \mathfrak{Z}$ be an operator defined on a neighborhood of a point $x_0$. Assume that $G$ is Gâteaux differentiable at any point of an interval $[x_0, x_0 + h]$ included in the domain $\mathfrak{D}$ of $G$, such that the derivative $G'$ is continuous in the uniform operator topology on this interval. Then* *the Newton-Leibniz formula*

$$G(x_0+h)-G(x_0)=\int\limits_{x_0}^{x_0+h} G'(x)\mathrm{d}x \tag{4.2}$$

*holds.*

◁ Consider the vector-function $g$ defined on the interval $[0,1]$ by $g(t) = G(x_0+th)$. Then $g$ is continuously differentiable on $[0,1]$ and the differentiation formula for composite functions yields

$$g'(t)=G'_x(x_0+th)h.$$

From (4.1) we have

$$(\mathrm{R})\int\limits_0^1 G'_x(x_0+th)h\mathrm{d}t=\int\limits_{x_0}^{x_0+h} G'(x)\mathrm{d}x.$$

The equality (4.2) follows from the previous one and from the relations

$$G(x_0+h)-G(x_0)=g(1)-g(0)=(\mathrm{R})\int\limits_0^1 g'(t)\mathrm{d}t=\int\limits_0^1 G'(x_0+th)\mathrm{d}t. \quad ▶$$

On the other hand, a perfect analogue of the Taylor formula with the remainder term represented by an integral of the form (4.1) fails to be obtained. More precisely, for an operator $F$ which is $m$-times continuously differentiable on a neighborhood of the point $x_0$, an application of formula (3.6) gives the representation

$$\begin{aligned}F(x_0+h)&=F(x_0)+F'(x_0)h+\cdots+\frac{1}{(m-1)!}F^{(m-1)}(x_0)h^{m-1}+\\&+\frac{1}{(m-1)!}(\mathrm{R})\int\limits_0^1(1-t)^{m-1}F^{(m)}(x_0+th)h^m\mathrm{d}t\end{aligned} \tag{4.3}$$

where the symbol $F^{(k)}(x_0)h^k$ denotes the value of the homogeneous form $F^{(k)}(x_0)$ of order $k$, acting from $\mathfrak{X}^k$ into $\mathfrak{Y}$, at the element $(h,\ldots,h)$. The integral in the right hand side of (4.3) is to be understood as the integral of the vector-function $(1-t)^{m-1}F^{(m)}(x_0+th)h^m$.

**2.** Consider now the case when $\mathfrak{X}$ is a Banach algebra over the field $\mathbb{K}$ and $F:\mathfrak{X} \to \mathfrak{X}$ is an operator defined on a neighborhood $\mathfrak{D}$ of the point $x_0$. If $S$ is a rectifiable contour in $\mathfrak{D}$, i.e., $S$ is defined by an equation $x = x(t)$, $\alpha \leqslant t \leqslant \beta$, where $x(t)$ is a continuous vector-function of bounded variation, then by definition we put

$$(\mathrm{T}) \int_S F(x)\mathrm{d}x = \lim_{|\pi| \to 0} \sum_{i=1}^{n} F\left(x(\tau_i)\right)\left(x(\sigma_i) - x(\sigma_{i-1})\right),$$

where $\sigma_0 = \alpha \leqslant \sigma_1 \leqslant \ldots \leqslant \sigma_n = \beta$, $\sigma_{i-1} \leqslant \tau_i \leqslant \sigma_i$.

The existence of this integral can be proved as in §1. Its linearity, an estimate of its norm, and the commutation of the integral with linear continuous operators acting from $\mathfrak{X}$ into $\mathfrak{X}$ follow too. This integral is called the *generalized Riemann-Stieldjes-Taylor integral*, or shortly, the *T-integral*.

# Chapter III

# Holomorphic (analytic) operators and vector-functions on complex Banach spaces

In Chapter I we introduced the notions of $\delta$-differentiability, $\mathcal{G}$-differentiability and $\mathcal{F}$-differentiability of an operator $F : \mathfrak{X} \to \mathfrak{Y}$ at a given point. For each of these notions we defined the classes of $m$-times differentiable operators, $1 \leqslant m \leqslant \infty$, as well as the class of analytic operators. In the general case, each of these classes is a proper part of the previous one. However, when $\mathfrak{X}$ and $\mathfrak{Y}$ are complex spaces and $F$ is differentiable on a nieghborhood of the given point, the above mentioned classes coincide. This gives the possibility to extend the well-established theory of complex-differentiable operators, a theory with meany deep results. (We will see below that in many cases the study of the Fréchet derivative of an operator at a point has not only local consequences, but also provides indications about the global features of the behavior of the given operator.) On the other hand, many operatorial equations in a real space can be studied by considering the complexification of the space and by extending the operator such that the given equation reduces to an equation in a complex space. In order to emphasize the important role of the complex spaces in the theory of differentiable operators, we will use in what follows, for the complex case, the term "holomorphic" instead of the term "differentiable". The definitions of holomorphic operators, holomorphic vector-functions, and so on, will be given below.

There are many similarities between the theory of holomorhic operators and the theory of holomorphic vector-functions; the only differences appear when it is essential to suppose that either the domain or the range of the operator are in a finite dimensional space. Therefore we develop these two theories in the same general frame.

## §1. Differentiability in complex and real sense. Cauchy-Riemann conditions

In what follows, we suppose that $\mathfrak{X}$ and $\mathfrak{Y}$ are *complex* normed spaces. Let $F: \mathfrak{X} \to \mathfrak{Y}$ be an operator which is differentiable at a point $x_0$, according to one of the definitions I.1.1 – I.1.3 (we will say that $F$ is *complex-differentiable*). Then, the operator $F$ is also differentiable relatively to the field $\mathbb{R}$ of real numbers, according to the same definitions (in the case of Definition I.1.3, this means that the operator has a real-linear Fréchet derivative). In this situation we say that the operator is real-differentiable, according to the given definitions. When an operator $F$ is real-differentiable at a point $x_0$, the first variation of $F$ will be denoted by $\delta_{\mathbb{R}} F(x_0, h)$.

**THEOREM 1.1.** *If an operator $F: \mathfrak{X} \to \mathfrak{Y}$ is continuously real-differentiable with respect to one of the definitions* I.1.1 – I.1.3 *on a neighborhood $\mathfrak{D} \subseteq \mathfrak{X}$ of $x_0$, then the operator $F$ is complex-differentiable at $x_0$ if and only if its first variation $\delta_{\mathbb{R}} F(x_0, h)$ satisfies the condition*

$$\delta_{\mathbb{R}} F(x_0, \mathrm{i}h) = \mathrm{i}\delta_{\mathbb{R}} F(x_0, h) \tag{1.1}$$

*where* i *is the imaginary unit.*

◁ The necessity of the condition (1.1) is obvious, since the variation $\delta F(x_0, h)$, being the derivative of the vector-function $f(\lambda) = F(x_0 + \lambda h)$, is homogeneous in $\lambda$.

The conditions is also sufficient. Assume that (1.1) is verified. We represent $\lambda \in \mathbb{C}$ as $\lambda = \tau + \mathrm{i}\eta$, where $\tau, \eta \in \mathbb{R}$. For a fixed $h$, the vector-function $f(\lambda)$ defined above obviously has a continuous derivative with respect to $\tau$ at the point $\lambda = 0$. It remains to establish the existence of the continuous partial derivative of $f(\lambda)$ at $\lambda = 0$ along the direction $\mathrm{i}\eta$. Indeed, this will actually prove the existence of the complete derivative $f'(0)$ of the vector-function $f(\lambda)$, as an operator acting from the space $\mathbb{C} = \mathbb{R} + \mathrm{i}\mathbb{R}$ into the space $\mathfrak{Y}$ (see Theorem I.5.2). But

$$f'(0) = \lim_{\lambda \to 0} \frac{F(x_0 + \lambda h) - F(x_0)}{\lambda} = \delta F(x_0, h);$$

since $h$ is an arbitrary element, the existence of the first variation of the operator $F$ at the point $x_0$ will follow.

For $f'_{\mathrm{i}\eta}(0)$, we have

$$f'_{\mathrm{i}\eta}(0) = \lim_{\mathrm{i}\eta \to 0} \frac{F(x_0 + \mathrm{i}\eta h) - F(x_0)}{\mathrm{i}\eta} =$$

$$= \frac{1}{\mathrm{i}} \lim_{\eta \to 0} \frac{F(x_0 + \eta(\mathrm{i}h)) - F(x_0)}{\eta} = -\mathrm{i}\delta_{\mathbb{R}} F(x_0, \mathrm{i}h) = \delta_{\mathbb{R}} F(x_0, h).$$

To complete the proof of the theorem, notice that if the operator $F$ is Gâteaux (resp. Fréchet) differentiable, then the variation $\delta_{\mathbb{R}}F(x_0,h)$ is real-linear in $h$. By (1.1) it is also complex-linear in $h$, i.e., it is the Gâteaux (resp. Fréchet) derivative in the complex sense. ▶

Assume now that the space $\mathfrak{Y}$ is the complexification of a real Banach space $\mathfrak{V}$ (see §4 of Chapter 0), i.e., $\mathfrak{Y} = \mathfrak{V} + \mathrm{i}\mathfrak{V}$ and the topology of $\mathfrak{Y}$ coincides with the topology of the product $\mathfrak{V} \times \mathfrak{V}$. Then, any operator $F: \mathfrak{X} \to \mathfrak{Y}$ can be represented as $F = P + \mathrm{i}Q$, where $P, Q: \mathfrak{X} \to \mathfrak{V}$. Suppose further that the operators $P$ and $Q$ are real-differentiable on a domain $\mathfrak{D}$ containing the point $x_0$. Then the operator $F$ is also real-differentiable, and

$$\delta_{\mathbb{R}}F(x_0,h) = \lim_{t\to 0, t\in\mathbb{R}} \frac{P(x_0+th)-P(x_0)}{t} + \mathrm{i}\cdot \lim_{t\to 0, t\in\mathbb{R}} \frac{Q(x_0+th)-Q(x_0)}{t} = \\ = \delta_{\mathbb{R}}P(x_0,h) + \mathrm{i}\delta_{\mathbb{R}}Q(x_0,h).$$

If condition (1.1) is fulfilled, then

$$\delta_{\mathbb{R}}P(x_0,\mathrm{i}h) + \mathrm{i}\delta_{\mathbb{R}}Q(x_0,\mathrm{i}h) = \\ = \mathrm{i}\delta_{\mathbb{R}}P(x_0,h) - \delta_{\mathbb{R}}Q(x_0,h).$$

Comparing the real and the imaginary parts, this infers the equalities

$$\begin{aligned} \delta_{\mathbb{R}}P(x_0,\mathrm{i}h) &= -\delta_{\mathbb{R}}Q(x_0,h) \\ \delta_{\mathbb{R}}Q(x_0,\mathrm{i}h) &= \delta_{\mathbb{R}}P(x_0,h) \end{aligned} \tag{1.2}$$

which are generalizations of the well-known Cauchy-Riemann conditions. Indeed, if $\mathfrak{X} = \mathbb{C} = \{\lambda \mid \lambda = \tau + \mathrm{i}\eta\}$, then $P$ and $Q$ can be considered as vector-functions of two real variables $\tau, \eta \in \mathbb{R}$.

Writing $x_0 = \tau_0 + \mathrm{i}\eta_0$, $h = \Delta\tau + \mathrm{i}\Delta\eta$, we obtain

$$\delta_{\mathbb{R}}P(x_0,\mathrm{i}h) = \lim_{t\to 0, t\in\mathbb{R}} \frac{P(x_0+t\mathrm{i}h)-P(x_0)}{t} = \\ = \lim_{t\to 0, t\in\mathbb{R}} \frac{P(\tau_0 - t\Delta\eta + \mathrm{i}(\eta_0 + t\Delta\tau)) - P(\tau_0+\mathrm{i}\eta_0)}{t}.$$

For $h = (\Delta\tau, 0) \in \mathbb{R}^2$, we have

$$\delta_{\mathbb{R}}P(x_0,\mathrm{i}h) = \left.\frac{\partial P}{\partial \eta}\right|_{x_0=(\tau_0,\eta_0)}.$$

On the other hand, the first equality in (1.2) implies, for the same $h$, that

$$\delta_{\mathbb{R}}P(x_0, \mathrm{i}h) = -\delta_{\mathbb{R}}Q(x_0, h) =$$
$$= \lim_{t\to 0, t\in\mathbb{R}} \frac{Q(\tau_0 + t\Delta\tau + \mathrm{i}\eta_0) - Q(\tau_0 + \mathrm{i}\eta_0)}{t} = \frac{\partial Q}{\partial \tau}\bigg|_{x_0=(\tau_0,\eta_0)}$$

Thus we obtain the first Cauchy-Riemann condition

$$\frac{\partial P}{\partial \eta} = -\frac{\partial Q}{\partial \tau}. \tag{1.3}$$

Analogously we obtain the second Cauchy-Riemann condition

$$\frac{\partial Q}{\partial \eta} = \frac{\partial P}{\partial \tau}. \tag{1.4}$$

**REMARK 1.1.** Notice that in order to prove (1.3) and (1.4) the condition that $\mathfrak{X}$ is an one-dimensional space is not essential. It is sufficient to assume that $\mathfrak{X}$, like $\mathfrak{Y}$, is the complexification of a real space $\mathfrak{U}$. In this case, Theorem 1.1 has the following consequence.

**COROLLARY 1.1.** *Let $\mathfrak{U}$, $\mathfrak{V}$ be two real Banach spaces and let $P, Q: \mathfrak{U}\times\mathfrak{U} \to \mathfrak{V}$ be two operators defined on a neighborhood of the point $x_0 = (p_0, q_0) \in \mathfrak{U}\times\mathfrak{U}$. Assume that $P$ and $Q$ have continuous partial Fréchet derivatives on that neighborhood. Then the operator $F = P + \mathrm{i}Q: \mathfrak{X} \to \mathfrak{Y}$, where $\mathfrak{X} = \mathfrak{U} + \mathrm{i}\mathfrak{U}$ and $\mathfrak{Y} = \mathfrak{V} + \mathrm{i}\mathfrak{V}$, is Fréchet complex-differentiable if and only if the equalities*

$$\begin{aligned} P'_p(x_0) &= Q'_q(x_0) \\ P'_q(x_0) &= -Q'_p(x_0) \end{aligned} \tag{1.5}$$

*are fulfilled.*

◁ The necessity can be established as above, in the same way in which we deduced (1.3)–(1.4). For sufficiency, choose an arbitrary vector $h = \Delta p + \mathrm{i}\Delta q$ and consider $\delta_{\mathbb{R}}P(x_0, \mathrm{i}h)$. Notice that in $\mathfrak{U}\times\mathfrak{U}$ the vector $\mathrm{i}h$ has the form $\mathrm{i}h = (-\Delta q, \Delta p)$. Using (1.3), we have

$$\begin{aligned} \delta_{\mathbb{R}}P(x_0, \mathrm{i}h) &= \delta_{\mathbb{R}}P\left(p_0, q_0, (-\Delta q, \Delta p)\right) = \\ &= P'_p(x_0)(-\Delta q) + P'_q(x_0)\Delta p = \\ &= -Q'_q(x_0)\Delta q - Q'_p(x_0)\Delta p = -\delta_{\mathbb{R}}Q(x_0, h), \end{aligned}$$

which is the first equality in (1.2). The second one can be proved analogously. ▶

**COROLLARY 1.2.** *If, in addition to the conditions of the previous corollary, one assumes that $P$ and $Q$ are twice continuously differentiable and satisfy the conditions* (1.5), *then*

$$\Delta F = 0 \quad \text{and} \quad \Delta P = \Delta Q = 0$$

*where the symbol $\Delta$ denotes here the Laplace operator, i.e., for any $G: \mathfrak{U} \times \mathfrak{U} \to \mathfrak{V} \times \mathfrak{V}$,*

$$\Delta G = G''_{pp} + G''_{qq}.$$

(In fact the requirement that $P$ and $Q$ are twice continuosly differentiable is superflous in the previous corollary; as we have already mentioned, the complex-differentiability of $F$ implies its infinite real-differentiability, and this means that $P$ and $Q$ are also real-differentiable of any order.)

At the end of this section we consider the notion of the complex-conjugate differential, just by mimicing the classical constructions. To this aim, for any element $x = p + \mathrm{i}q$ in the space $\mathfrak{X} = \mathfrak{U} + \mathrm{i}\mathfrak{U}$ we introduce the conjugate element $\overline{x} = p - \mathrm{i}q$. In this case, $p = \frac{1}{2}(x + \overline{x})$, $q = \frac{1}{2\mathrm{i}}(x - \overline{x})$. Therefore, an operator $F: \mathfrak{U} \times \mathfrak{U} \to \mathfrak{Y}$ formally gives rise to an operator $\tilde{F}(x, \overline{x}) = F(p, q)$ by replacing $p$ and $q$ with their expressions in terms of $x$ and $\overline{x}$. (The problem is that $\tilde{F}$ can not be considered as a function of two independent variables, since the change of $x$ implies a simultaneous change of $\overline{x}$.) Further, if the operator $F$ is Fréchet differentiable on a neighborhood of a point $x_0 = (p_0, q_0)$ (in the real sense), then, by the formula of the complete differential, we have

$$F'(x_0)h = F'_p(x_0)\Delta p + F'_q(x_0)\Delta q$$

where $h = (\Delta p, \Delta q)$, so

$$F'(x_0)h = F'_p(x_0)\left[\frac{1}{2}(h + \overline{h})\right] + F'_q\left[\frac{1}{2\mathrm{i}}(h - \overline{h})\right] =$$
$$= \frac{1}{2}\left[F'_p(x_0) - \mathrm{i}F'_q(x_0)\right]h + \frac{1}{2}\left[F'_p(x_0) + \mathrm{i}F'_q(x_0)\right]\overline{h} =$$
$$= \tilde{F}'_x(x_0)h + \tilde{F}'_{\overline{x}}(x_0)\overline{h},$$

where $\tilde{F}'_x(x_0)h$ and $\tilde{F}'_{\overline{x}}(x_0)\overline{h}$ are the formal derivatives of the operator $\tilde{F}$ with respect to $x$ and $\overline{x}$, defined by

$$\tilde{F}'_x(x_0) = \frac{1}{2}\left[F'_p(x_0) - \mathrm{i}F'_q(x_0)\right]$$
$$\tilde{F}'_{\overline{x}}(x_0) = \frac{1}{2}\left[F'_p(x_0) + \mathrm{i}F'_q(x_0)\right].$$

Conditions (1.5) for the operator $F = P + iQ$ lead directly to the following conclusion: if the operator $F: \mathfrak{U} \times \mathfrak{U} \to \mathfrak{Y}$ is Fréchet differentiable in the real sense, then it is Fréchet differentiable in the complex sense as an operator from $\mathfrak{X} = \mathfrak{U} + i\mathfrak{U}$ into $\mathfrak{Y} = \mathfrak{V} + i\mathfrak{V}$ if and only if $\tilde{F}'_{\bar{x}}(x) = 0$.

## §2. The $\rho$-topology and holomorphy

As we have already noted, the class of operators having the first variation, the class of Gâteaux differentiable operators, and the class of Fréchet differentiable operators form a *strictly* decreasing chain with respect to the inclusion. Nevertheless, from the point of view of the complex analysis, these classes have many properties in common. It is therefore interesting to consider a characteristic of these classes which permits to recover these common properties in a unified manner, and, on the other hand, which permits to distinguish between these classes of operators. For the Gâteaux and the Fréchet differentiable operators, such a characteristic seams to be a "connection" between the domain of definition of the operator and the operator itself. More precisely, we consider, on the domain of an operator, the so-called $\rho$-topology, which is weaker than the norm topology; in this topology both classes of differentiable operators have the same properties.

Note that the interest in this approach arises not only from its generality, but also from the possibility of obtaining new results. As an example see §6 of Chapter VI, where some degenerate problems for implicite Fréchet differentiable operators are transferred to nondegenerate problems for operators differentiable relatively to the $\rho$-topology.

Our presentation and terminology mainly follow E. Hille and R. Phillips [1].

**1.** For a set $\mathfrak{D} \subseteq \mathfrak{X}$ and for elements $x \in \mathfrak{D},\ h \in \mathfrak{X}$, we introduce the following notion:

$$\rho(x,h) = \sup\{\rho \ :\ x + \tau h \in \mathfrak{D},\ \tau \in \mathbb{C},\ |\tau| \leqslant \rho\}. \tag{2.1}$$

Clearly, $\rho(x,h) \geqslant 0$, for all $x \in \mathfrak{D},\ h \in \mathfrak{X}$ and for any set $\mathfrak{D} \subseteq \mathfrak{X}$.

We will need the following property of $\rho(x,h)$.

**LEMMA 2.1.** *Let $\mathfrak{D}$ be a set in $\mathfrak{X}$. Then for any $x \in \mathfrak{D},\ h \in \mathfrak{X},\ \lambda \in \mathbb{C}\ (\lambda \neq 0)$, we have*

$$\rho(x, \lambda^{-1}h) = \rho(x,h)|\lambda|.$$

◁ The equality is obvious if $\rho(x,h) = 0$. Assume that $\rho(x,h) > 0$ for some $x \in \mathfrak{D}$ and $h \in \mathfrak{X}$. Denote $h_1 = \lambda^{-1}h$ and take an arbitrary $\rho_0$, $0 < \rho_0 \leqslant \rho(x,h_1)$. Then $x + \tau h_1 \in \mathfrak{D}$ for any $|\tau| < \rho_0$. Denoting $t = \tau\lambda^{-1}$, we have that $x + th \in \mathfrak{D}$ for any $t$ with $|t| < \rho_0|\lambda|^{-1} = \rho_1$. Consequently, $\rho_1 \leqslant \rho(x,h)$, or $\rho_0 \leqslant \rho(x,h)|\lambda|$. Hence $\rho(x,h_1) = \rho(x,\lambda^{-1}h) \leqslant \rho(x,h)|\lambda|$. Conversely, let $\rho_2 \leqslant \rho(x,h)|\lambda|$. We have $x + \tau h_1 \in \mathfrak{D}$ for any $\tau$ with $|\tau| < \rho_2$ (since $x + \tau h_1 = x + th$, where $t = \tau\lambda^{-1}$ and $|t| < \rho(x,h)$). Then $\rho_2 \leqslant \rho(x,h_1) = \rho(x,\lambda^{-1}h)$. ▶

**DEFINITION 2.1.** A set $\mathfrak{D}$ is called *ρ-open* (*one-dimensional open*) if $\rho(x,h) > 0$ for any $x \in \mathfrak{D}$ and $h \in \mathfrak{X}$.

It is clean that any open set in the norm topology of $\mathfrak{X}$ is a $\rho$-open set. The converse is false in general. To see this it is sufficient to consider the set $\mathfrak{D} \subseteq \mathbb{C}^2$ defined as $\mathfrak{D} = \mathfrak{D}_1 \setminus \mathfrak{D}_2$, where $\mathfrak{D}_1$ is an open ball in $\mathbb{C}^2$, centered at the origin, and

$$\mathfrak{D}_2 = \left\{x \in \mathbb{C}^2 \setminus \{0\},\ x = (x_1, x_2)\ :\ x_2 = x_1^2\right\}.$$

On the other hand, the following easy result is true.

**PROPOSITION 2.1.** *A set $\mathfrak{D}$ is open in the norm topology of $\mathfrak{X}$ if and only if*

$$\rho(x,h) \geqslant \delta(x) > 0 \tag{2.2}$$

*for all $x \in \mathfrak{D}$, $h \in \mathfrak{X}$, $\|h\| = 1$.*

Let $\mathfrak{D}$ be a bounded $\rho$-open set in $\mathfrak{X}$. Using (2.1) we define a convergence in $\mathfrak{D}$ as follows:

**DEFINITION 2.2.** We say that a directedness $\{x_\alpha\}_{\alpha\in\mathfrak{A}}$ in $\mathfrak{D}$ is *ρ-convergent* to an element $x \in \mathfrak{D}$ if for any positive integer $N$ there exists $\alpha_0 \in \mathfrak{A}$ such that $\rho(x, x_\alpha - x) >$ $> N$ for all $\alpha \in \mathfrak{A}$ with $\alpha > \alpha_0$. The topology defined by the $\rho$-convergence will be called the *ρ-topology*.

From Lemma 2.1 and from relation (2.2) we have:

**COROLLARY 2.1.** *If the set $\mathfrak{D}$ is open and the directedness $\{x_\alpha\}_{\alpha\in\mathfrak{A}}$ in $\mathfrak{D}$ is convergent to $x \in \mathfrak{D}$ in the norm topology, then it is also ρ-convergent to $x$.*

◁ For the proof, it is sufficient to use Lemma 2.1 with $h = x_\alpha - x$, $\lambda = \|x_\alpha - x\|$, and Proposition 2.1. ▶

**DEFINITION 2.3.** For any $x \in \mathfrak{D}$ the set

$$D_\rho(x,h) = \{z \in \mathfrak{X} \ : \ z = x + \tau h, \ \tau \in \mathbb{C}, \ |\tau| < \rho, \ h \in \mathfrak{X}, \ \|h\| = 1\}$$
$$(\overline{D}_\rho(x,h) = \{z \in \mathfrak{X} \ : \ z = x + \tau h, \ \tau \in \mathbb{C}, \ |\tau| \leqslant \rho, \ h \in \mathfrak{X}, \ \|h\| = 1\})$$

is called the *open* (*closed*) *disk of radius* $\rho$, *centered at* $x$, *and along the direction* $h$.

Clearly, a set $\mathfrak{D} \subseteq \mathfrak{X}$ is $\rho$-open if and only if for each $x \in \mathfrak{D}$ it contains an open disk of non-zero radius in any direction $h$, centered at $x$.

Let us show that a closed disk $\overline{D}_\rho(x,h)$, where $\rho < \rho(x,h)$ is compact in the $\rho$-topology of the $\rho$-open set $\mathfrak{D}$.

$\triangleleft$ Let $\{x_\alpha\}_{\alpha\in\mathfrak{A}}$ be a directedness in $\overline{D}_\rho(x,h)$. Then $x_\alpha = x + \tau_\alpha h, |\tau_\alpha| \leqslant \rho, \alpha \in \in \mathfrak{A}$. We choose a subdirectedness $\{\tau_\beta\}_{\beta\in\mathfrak{B}}$, $\mathfrak{B} \subseteq \mathfrak{A}$ convergent to a point $\tau_* \in \mathbb{C}$; we will show that the corresponding subdirectedness $\{x_\beta\}_{\beta\in\mathfrak{B}}$, $x_\beta = x + \tau_\beta h$ is $\rho$-convergent to the element $x_* = x + \tau_* h$. First of all, since $x_* \in \overline{D}_\rho(x,h)$, Proposition 2.1 implies that $\rho(x_*, h) > 0$ for any $h \in \mathfrak{X}$, $\|h\| = 1$. Then, by Lemma 2.1 we obtain

$$\rho(x_*, x_* - x_\beta) = \rho(x_*, (\tau_* - \tau_\beta)h) = \rho(x_*, h)|\tau_* - \tau_\beta|^{-1}. \tag{2.3}$$

Since for any $N > 0$ there exists $\beta_0 \in \mathfrak{B}$ such that $|\tau_* - \tau_\beta| < \rho(x_*, h)N^{-1}$ for all $\beta \in \mathfrak{B}$, $\beta > \beta_0$ (the element $h$ is fixed), the relation (2.3) implies

$$\rho(x_*, x_* - x_\beta) > N,$$

and the proof is complete. ▶

**2.** Let us turn now to the notion of holomorphy.

**DEFINITION 2.4.** Let $\mathfrak{D}$ be a $\rho$-open set in $\mathfrak{X}$. An operator $F\colon \mathfrak{D} \to \mathfrak{Y}$ is called *$\rho$-holomorphic* in $\mathfrak{D}$ (in short $F \in \mathcal{H}_\rho(\mathfrak{D}, \mathfrak{Y})$) if for any $x \in \mathfrak{D}$ and $h \in \mathfrak{X}$ the vector-function $f(\tau) = F(x + \tau h)$ is complex-differentiable in the disk $|\tau| < \rho(x,h)$ of the complex plane $\mathbb{C}$.

Thus, any $\rho$-holomorphic operator $F$ has at each point $x \in \mathfrak{D}$ the first variation $\delta F(x,h)$.

**DEFINITION 2.5.** An operator $F\colon \mathfrak{D} \to \mathfrak{Y}$ is called *$\mathcal{G}$-holomorphic* if it is $\rho$-holomorphic and for any point $x \in \mathfrak{D}$ there exists a linear operator $A(x) \in L(\mathfrak{X}, \mathfrak{Y})$ such that $\delta F(x,h) = A(x)h$. In this case we will write $F \in \mathcal{H}(\mathfrak{D}, \mathfrak{Y})$.

If the set $\mathfrak{D}$ is open, i.e., condition (2.2) is fulfilled for all $x \in \mathfrak{D}$, then a $\mathcal{G}$-holomorphic operator is called *$\mathcal{F}$-holomorphic* or, simply, *holomorphic.*

Obviously if $F \in \mathcal{H}(\mathfrak{D}, \mathfrak{Y})$ then $F$ is Gâteaux differentiable at any point $x \in \mathfrak{D}$. We will see in §3 below that if an operator $F$ is $\mathcal{F}$-holomorphic, i.e., the set $\mathfrak{D}$ is open, then $F$ is Fréchet differentiable at any point $x \in \mathfrak{D}$ (this explains the term "$\mathcal{F}$-holomorphic operator").

Since the notion of holomorphy depends on the notion of convergence, we may naturally consider the notion of weak holomorphy. However, it is not necessary to do that, because — as we have already mentioned — the strong holomorphy follows from the weak holomorphy. This result was obtained for vector-functions by N. Dunford [1]. We will prove here a sligt generalization.

An operator $F: \mathfrak{D} \to \mathfrak{Y}$ is called *locally bounded* on $\mathfrak{D}$ if for each $x \in \mathfrak{D}$ there exists a neighborhood $\mathfrak{U}$ of the point $x$ such that $F$ is bounded on $\mathfrak{U} \cap \mathfrak{D}$.

We make the supplementary assumption that there exists a Banach space $\mathfrak{Y}_*$ such that $(\mathfrak{Y}_*)^* = \mathfrak{Y}$.

**THEOREM 2.1.** *Let $\mathfrak{D}$ be a $\rho$-open set in $\mathfrak{X}$. An operator $F$, locally bounded on $\mathfrak{D}$, is $\rho$-holomorphic ($\mathcal{G}$-holomorhic) if and only if $\langle y_*, F(\cdot)\rangle \in \mathcal{H}_\rho(\mathfrak{D}, \mathbb{C})$ ($\mathcal{H}(\mathfrak{D}, \mathbb{C})$) for each $y_* \in \mathfrak{Y}^*$.*

(Maybe it is useful to explain that the value of the operator $\langle y_*, F(\cdot)\rangle$ at $x \in \mathfrak{D}$ is the number $\langle y_*, F(x)\rangle$, i.e., the value of the functional $F(x) \in \mathfrak{Y} = (\mathfrak{Y}_*)^*$ at the vector $y_* \in \mathfrak{Y}_*$.)

$\lhd$ The necessity of the condition is obvious. Let us prove the sufficiency. Consider $\langle y_*, F(\cdot)\rangle \in \mathcal{H}_\rho(\mathfrak{D}, \mathbb{C})$. This means that the function $\varphi(\tau) = \langle y_*, F(x+\tau h)\rangle$ is differentiable on the disk $|\tau| < \rho(x,h)$ for any $x \in \mathfrak{D}$, $h \in \mathfrak{X}$, $y_* \in \mathfrak{Y}_*$. We have to prove the differentiabily in the same disk of the vector-function $f(\tau) = F(x+\tau h)$ for any $x \in \mathfrak{D}$, $h \in \mathfrak{X}$. To this end, consider the sequence

$$\ell(t_n) = \frac{1}{t_n}\left[f(\tau + t_n) - f(\tau)\right] \tag{2.4}$$

for $t_n \to 0$ $(n \to \infty)$, where $|\tau|$, $|\tau + t_n| < \rho < \rho(x,h)$; we will show that this sequence is fundamental. Indeed, for any $y_* \in \mathfrak{Y}_*$, we have

$$\frac{1}{t_n - t_m}\langle y_*, \ell(t_n) - \ell(t_m)\rangle =$$

$$= \frac{1}{t_n - t_m}\left[\frac{\varphi(\tau + t_n) - \varphi(\tau)}{t_n} - \frac{\varphi(\tau + t_m) - \varphi(\tau)}{t_m}\right] =$$

$$= \frac{1}{2\pi i} \int\limits_{|\lambda|=\rho} \frac{\varphi(\lambda)}{(\lambda-\tau)(\lambda-\tau+t_n)(\lambda-\tau+t_m)} d\lambda,$$

where $\rho < \rho(x,h)$. The last equality is obtained by using the classical Cauchy integral representation and by decomposing the integrated function in partial fractions. Since $F$ is locally bounded, the function $\varphi(\lambda)(\lambda-\tau)^{-1}(\lambda-\tau+t_n)^{-1}(\lambda-\tau+t_m)^{-1}$ is bounded on the curve $|\lambda| = \rho$ by a constant $M(y_*, \rho)$. Then

$$|\langle y_*, \ell(t_n) - \ell(t_m)\rangle| \leqslant M(y_*, \rho)|t_n - t_m|,$$

for all $t_n, t_m$ with $|t_n|, |t_m| < \rho$.

Using the Banach-Steinhaus principle of uniform boundedness (Theorem 0.8.8), we obtain the inequality

$$\|\ell(t_n) - \ell(t_m)\| \leqslant M(\rho)|t_n - t_m|$$

where $M(\rho)$ is a finite constant, so the sequence (2.4) is fundamental.

Since $\mathfrak{Y}$ is complete, the limit

$$\lim_{t_n \to 0} \frac{f(\tau+t_n) - f(\tau)}{t_n} = f'(\tau) = \delta F(x+\tau h, h)$$

exists for any $\tau$ with $|\tau| < \rho(x,h)$. Thereby we proved that $F \in \mathcal{H}_\rho(\mathfrak{D}, \mathfrak{Y})$, i.e., the variation $\delta F(x,h)$ exists.

Assume now that $\langle y_*, F(\cdot)\rangle \in \mathcal{H}(\mathfrak{D}, \mathbb{C})$, that is, the variation $\delta\langle y_*, F(\cdot)\rangle(x,h)$ (for any $y_* \in \mathfrak{Y}_*$) is represented by a bounded linear operator (relatively to $h$), acting from $\mathfrak{X}$ into $\mathbb{C}$. From the equality

$$\langle y_*, \delta F(x,h)\rangle = \delta\langle y_*, F(\cdot)\rangle(x,h)$$

we infer that $F \in \mathcal{H}(\mathfrak{D}, \mathfrak{Y})$. ▶

As a consequence, we prove now an analogue of the classical Weierstrass theorem on the completeness of the spaces $\mathcal{H}_\rho(\mathfrak{D}, \mathfrak{Y})$ and $\mathcal{H}(\mathfrak{D}, \mathfrak{Y})$ for the topology of weak uniform convergence on the closed disks in $\mathfrak{D}$.

**THEOREM 2.2.** *Let $F_n \in \mathcal{H}_\rho(\mathfrak{D}, \mathfrak{Y})$ (resp. $F_n \in \mathcal{H}(\mathfrak{D}, \mathfrak{Y})$) be a sequence of operators, where $\mathfrak{D}$ is a $\rho$-open set in $\mathfrak{X}$. Assume that the sequence $F_n$ converges uniformly on the closed disks in $\mathfrak{D}$ (in the weak topology of $\mathfrak{Y}$) to an operator $F: \mathfrak{D} \to \mathfrak{Y}$. Then $F \in \mathcal{H}_\rho(\mathfrak{D}, \mathfrak{Y})$ (resp. $F \in \mathcal{H}(\mathfrak{D}, \mathfrak{Y})$).*

◁ The first assertion follows directly from the classical theorem of Weierstrass and from Theorem 2.1. Indeed, for any fixed $x \in \mathfrak{D}$ and $h \in \mathfrak{X}$, the scalar functions $\varphi_n(\tau) = \langle F_n(x+\tau h), y^*\rangle$ converge to $\varphi(\tau) = \langle F(x+\tau h), y^*\rangle$ uniformly on any disk $D_\rho(x,h)$ where $\rho < \rho(x,h)$. Hence $\langle F(\cdot), y^*\rangle \in \mathcal{H}_\rho(\mathfrak{D}, \mathbb{C})$ and consequently $F \in \mathcal{H}_\rho(\mathfrak{D}, \mathfrak{Y})$. If $F_n \in \mathcal{H}(\mathfrak{D}, \mathfrak{Y})$, then $\delta F_n(x,h)$ are linear continuous operators with respect to $h$. Thus

$$\langle \delta F_n(x,h), y^*\rangle = \varphi_n'(0) \to \varphi'(0) = \langle \delta F(x,h), y^*\rangle.$$

From this and from the principle of uniform boundedness it follows that $\delta F(x,\cdot)$ is linear and bounded, so $F \in \mathcal{H}(\mathfrak{D}, \mathfrak{Y})$. ▶

## §3. Cauchy integral theorems and their consequences

As in the classical complex analysis, the Cauchy integral theorems represent powerful tools in the study of holomorphic operators. A lot of subtle and interesting recent investigations concerning finite dimensional integral representations could be carried out for infinite dimensional spaces. Here we restrict ourselves to present the simplest results, which are enough for our further considerations.

Let $\mathfrak{D}$ be a $\rho$-open set in a Banach space $\mathfrak{X}$.

**THEOREM 3.1.** *Let $F$ be a continuous operator acting from the set $\mathfrak{D}$ endowed with the $\rho$-topology into a Banach space $\mathfrak{Y}$ with the norm topology. Then $F \in \mathcal{H}_\rho(\mathfrak{D}, \mathfrak{Y})$ if and only if the equality*

$$\text{(P)} \int\limits_{|\lambda|=\rho} F(x+\lambda h)\mathrm{d}\lambda = 0 \tag{3.1}$$

*holds for any $x \in \mathfrak{D}$, $h \in \mathfrak{X}$ and for any $\rho < \rho(x,h)$.*

◁ The vector-function $f(\tau) = F(x+\tau h)$ is continuous on the disk $|\tau| \leqslant \rho < \rho(x,h)$ for each fixed $x \in \mathfrak{D}$, $h \in \mathfrak{X}$. Therefore, the integral in the left hand side of (3.1) exists (see Chapter II, §2) and the equality

$$\left\langle \text{(P)} \int\limits_{|\lambda|=\rho} F(x+\tau h)\mathrm{d}\lambda, y^* \right\rangle = \int\limits_{|\lambda|=\rho} \langle F(x+\tau h), y^*\rangle \mathrm{d}\lambda$$

holds for any $y^* \in \mathfrak{Y}^*$.

Our theorem follows now from Theorem 2.1 and from the classical theorems of Cauchy and Morera (see, e.g., B. Shabat [1]). ▶

The analogue of the Cauchy integral representation formula can be obtained similarly.

**THEOREM 3.2.** *For an operator $F \in \mathcal{H}_\rho(\mathfrak{D}, \mathfrak{Y})$, the representation*

$$2\pi \mathrm{i} F(x+\tau h) = (\mathrm{P}) \int_{|\lambda|=\rho} F(x+\lambda h)(\lambda-\tau)^{-1} \mathrm{d}\lambda \tag{3.2}$$

*is true for any $x \in \mathfrak{D}$, $h \in \mathfrak{X}$ and $\tau \in \mathbb{C}$ with $|\tau| < \rho < \rho(x,h)$.*

*In particular*

$$2\pi \mathrm{i} F(x) = (\mathrm{P}) \int_{|\lambda|=\rho} \lambda^{-1} F(x+\lambda h) \mathrm{d}\lambda, \tag{3.3}$$

*that is, the integral in the right hand side of (3.3) does not depend on $h$ if $\rho$ is small enough, and it is equal to the value of the operator $F$ at $x$ multiplied by $2\pi\mathrm{i}$.*

**COROLLARY 3.1.** *If the operator $F$ is $\rho$-holomorphic and locally bounded on $\mathfrak{D}$, then it is $\rho$-analytic at each point of $\mathfrak{D}$, i.e., $F$ admits a decomposition as a generalized power series*

$$F(x+\tau h) = \sum_{k=0}^{\infty} \delta^k F(x,h) \tau^k \tag{3.4}$$

*which is norm convergent (in $\mathfrak{Y}$) for $|\tau| \leqslant \rho < \rho(x,h)$ and*

$$\delta^k F(x,h) = \frac{1}{2\pi \mathrm{i}} (\mathrm{P}) \int_{|\lambda|=\rho} \lambda^{-k-1} F(x+\lambda h) \mathrm{d}\lambda. \tag{3.5}$$

◁ Indeed, the kernel $(\lambda-\tau)^{-1}$ from (3.3) has the power series representation

$$\frac{1}{\lambda-\tau} = \sum_{k=0}^{\infty} \frac{\tau^k}{\lambda^{k+1}}$$

which converges absolutely for all $\tau$ with $|\tau| < |\lambda| = \rho$. Then the series

$$\sum_{k=0}^{\infty} \lambda^{-(k+1)} \tau^k F(x+\lambda h)$$

converges absolutely (in the norm of $\mathfrak{Y}$) for all $x \in \mathfrak{D}$ and $\lambda \in \mathbb{C}$ with $|\lambda| \leqslant \rho(x,h)$, because $F(x+\lambda h)$ is bounded. It remains to show that the integral and the sum "commute". For any positive integer $n$, we have

$$\begin{aligned}(\mathrm{P}) \int\limits_{|\lambda|=\rho} \sum_{k=0}^{n} \lambda^{-(k+1)} \tau^k F(x+\lambda h) \mathrm{d}\lambda = \\ = \sum_{k=0}^{n} \tau^k (\mathrm{P}) \int\limits_{|\lambda|=\rho} \lambda^{-(k+1)} F(x+\lambda h) \mathrm{d}\lambda.\end{aligned} \tag{3.6}$$

For a fixed $\tau$, denote by $P_n(\lambda)$ the expression under the integral from the left hand side of (3.6). Then $P_n(\lambda)$ converges for $n \to \infty$ to $F(x+\lambda h)(\lambda-\tau)^{-1}$, uniformly for all $\lambda$ with $|\lambda| = \rho$. Taking into account that

$$\begin{aligned}\left\| (\mathrm{P}) \int\limits_{|\lambda|=\rho} F(x+\lambda h)\lambda^{-(k+1)} \mathrm{d}\lambda \right\| \leqslant \int\limits_{|\lambda|=\rho} \|F(x+\lambda h)\| \, |\lambda^{-(k+1)}| \, |\mathrm{d}\lambda| \leqslant \\ \leqslant M \cdot \rho^{-(k+1)} 2\pi\rho = 2\pi\rho^{-k},\end{aligned} \tag{3.7}$$

where $M \geqslant \sup\limits_{z \in D_\rho(x,h)} \|F(z)\|$ and $D_\rho(x,h)$ is the open disk centered at $x$ along the direction $h$ (see Definition 2.3), it follows that the partial sums in the right hand side of (3.6) also converge for any fixed $\tau \in \mathbb{C}$ with $|\tau| < \rho$. Thus our assertion follows from Corollary II.1.2. ►

From (3.5) and (3.7) we obtain directly the next generalized Cauchy inequalities.

**COROLLARY 3.2.** *If $F \in \mathcal{H}(\mathfrak{D}, \mathfrak{Y})$ and $\|F(x)\| \leqslant M$ for all $x \in \mathfrak{D}$, then for any $x \in \mathfrak{D}$, $h \in \mathfrak{X}$ and any $\rho \leqslant \rho(x,h)$ we have*

$$\|\delta^k F(x,h)\| \leqslant M\rho^{-k}. \tag{3.8}$$

**REMARK 3.1.** Actually, from (3.5) it follows the sharper inequality

$$\|\delta^k F(x,h)\| \leqslant M(x,h)\rho^{-k}, \tag{3.8$'$}$$

where $M(x,h) \geqslant \sup\limits_{|\lambda| \leqslant \rho(x,h)} \|F(x+\lambda h)\|$, a fact which is useful in many cases.

**REMARK 3.2.** The expressions $\delta^k F(x,h)$ in (3.4) are called the *Taylor coefficients* of the operator $F$ at the point $x$ in direction $h$. Note that (3.4) can be seen as the

Taylor series of the vector-function $f(\tau) = F(x + \tau h)$ for fixed $x \in \mathfrak{D}$, $h \in \mathfrak{X}$ and $\tau$, with $|\tau| < \rho(x, h)$. Using the obvious relations

$$\frac{\mathrm{d}^k}{\mathrm{d}\tau^k}\langle f(\tau), y^*\rangle = \left\langle \frac{\mathrm{d}^k}{\mathrm{d}\tau^k} f(\tau), y^* \right\rangle, \quad k = 0, 1, 2, \dots ,$$

where $y^*$ is an arbitrary element of $\mathfrak{Y}^*$, we obtain that $\delta^k F(x, h)$ has the representation

$$\delta^k F(x, h) = \frac{1}{k!}\frac{\mathrm{d}^k}{\mathrm{d}\tau^k} F(x + \tau h)\bigg|_{\tau=0}, \quad k = 0, 1, 2, \dots .$$

If the operator $F$ is holomorphic, some other representations for $\delta^k F(x, h)$ are also valid. For example, the next result is true.

**THEOREM 3.3.** *Let $\mathfrak{D}$ be an open set in $\mathfrak{X}$. If an operator $F \in \mathcal{H}(\mathfrak{D}, \mathfrak{X})$ is locally bounded on $\mathfrak{D}$, then $F$ is indefinitely Fréchet differentiable. If $\mathrm{d}_k F(x, h_1, \dots, h_k)$ is the derivative of order $k$ of $F$, then we have*

$$\frac{1}{k!}\mathrm{d}_k F(x, h, h, \dots, h) = \delta^k F(x, h), \tag{3.9}$$

*i.e., the coefficients $\delta^k F(x, h)$ in the decomposition (3.4) are homogeneous forms in $h$ for fixed $x \in \mathfrak{D}$ (see §6 of Chapter* I*).*

(Here, for convenience, we put $\delta^0 F(x, h) = F(x)$.)

◁ Since $\mathfrak{D}$ is an open set, by Proposition 2.1, we have that $\rho(x, h) \geqslant \delta(x) > 0$ for all $h$ with $\|h\| = 1$. Assume that the inequality $\|F(z)\| \leqslant M$ is fulfilled for all points in the ball centered at $x$ and with radius $\varepsilon(x)$. Then, for any $u$ with $\|u\| \leqslant \delta =$ $= \min\{\delta(x), \varepsilon(x)\}$, and taking $h = u\tau^{-1}$, $\|h\| = 1$, relation (3.4) implies that

$$\begin{aligned} F(x + u) &= \delta F(x, h)\tau + \sum_{k=2}^{\infty} \delta^k F(x, h)\tau^k + F(x) = \\ &= A(x)u + \omega(x, u) + F(x), \end{aligned}$$

where $A(x) \in L(\mathfrak{X}, \mathfrak{Y})$. Using Cauchy inequality (3.8) and the well known formula for the geometric series, we obtain

$$\|\omega(x, u)\| = \left\| \sum_{k=2}^{\infty} \delta^k F(x, h)\tau^k \right\| \leqslant M \frac{\|u\|^2}{\delta^2\left(1 - \dfrac{\|u\|}{\delta}\right)},$$

for any $u \in \mathfrak{X}$, $\|u\| < \delta$; this estimate proves the Fréchet differentiability of the operator $F$. Hence, it follows that the vector-function $F(x+\tau_1 h_1+\ldots+\tau_n h_n)\colon \mathbb{C}^n \to \to \mathfrak{Y}$ — for any fixed $h_1,\ldots,h_n$ — has continuous partial derivatives with respect to $\tau_1,\ldots,\tau_n$ in a polydisk $\mathfrak{U}$ with a sufficiently small radius, i.e., it is holomorphic on this polydisk with respect to each of its variables. By the fundamental theorem of Hartogs (see, e.g., Shabat [2]), the function $\left\langle F\left(x+\sum_k \tau_k h_k\right), y^*\right\rangle$ is holomorphic in the polydisk $\mathfrak{U}$ with respect to all of the variables, and consequently it has partial derivatives of any order. By Theorem 2.2 it follows that the vector-function $F(x+ +\tau_1 h_1+\ldots+\tau_n h_n)$ has the same property. So, it follows, in particular, the existence of the second order derivative

$$\begin{aligned} \mathrm{d}_2 F(x,h_1,h_2) &= \mathrm{d}\,[\mathrm{d}F(x,h_1),h_2] = \mathrm{d}\left[\left.\frac{\partial F(x+\tau_1 h_1)}{\partial \tau_1}\right|_{\tau_1=0}\right] = \\ &= \frac{\partial}{\partial \tau_2}\left[\left.\frac{\partial}{\partial \tau_1}F(x+\tau_1 h_1+\tau_2 h_2)\right|_{\tau_1=0}\right]_{\tau_2=0} = \\ &= \left.\frac{\partial^2}{\partial \tau_2 \partial \tau_1}F(x+\tau_1 h_1+\tau_2 h_2)\right|_{\tau_1=0,\tau_2=0}. \end{aligned}$$

The symmetry of $\mathrm{d}_2 F$ and its linearity in $h_1$, implies the linearity in $h_2$. Analogously, we have

$$\mathrm{d}_k F(x_1,h_1,\ldots,h_k) = \left.\frac{\partial^k}{\partial \tau_1,\ldots,\partial \tau_k}F(x+\tau_1 h_1+\ldots+\tau_k h_k)\right|_{\tau_1=\cdots=\tau_k=0}, \qquad (3.10)$$

for all $k \geqslant 2$.

Taking into account the uniqueness of the Taylor decomposition, we obtain equalities (3.9). ►

Let us remark that from (3.9) and (3.10) we can infer the following representation for the form $\delta^k F(\cdot,\cdot)$:

$$\delta^k F(x,h) = \frac{1}{k!}\left.\frac{\partial^k}{\partial \tau_1,\ldots,\partial \tau_k}F(x+\tau_1 h_1+\cdots+\tau_k h_k)\right|_{\substack{\tau_1=\cdots=\tau_k=0\\ h_1=\cdots=h_k=h}}. \qquad (3.11)$$

**COROLLARY 3.3** (an analogue of Liouville Theorem). *Let $\mathfrak{X}$ and $\mathfrak{Y}$ be Banach spaces, $F \in \mathcal{H}(\mathfrak{X},\mathfrak{Y})$, and assume that for a point $x \in \mathfrak{X}$ there exists a positive integer $m > 0$ such that*

$$\|F(x+h)\| \leqslant M\|h\|^m \qquad (3.12)$$

*for all elements $h \in \mathfrak{X}$ with sufficiently large norm. Then the operator $F$ is a polynomial of degree $\leqslant m$, i.e., $F$ is a finite sum of homogeneous forms in $h$. In particular, if the value $\|F(x)\|$ is uniformly bounded for all $x \in \mathfrak{X}$, then $F(x)$ takes a constant value in $\mathfrak{Y}$.*

◁ Taking $\tau = 1$ in (3.4) (note that $\rho(x,h) = \infty$), we have

$$F(x+h) = \sum_{k=0}^{\infty} \delta^k F(x,h),$$

where, according to Theorem 3.3, $\delta^k F(x,h)$ are homogeneous forms of order $k$ in $h$. The inequalities (3.7) and (3.12) imply

$$\|\delta^k F(x,h)\| \leqslant M\|\rho\|^{m-k},$$

where $\rho = \|h\|$ is sufficiently large. For $\rho \to \infty$, we obtain that $\delta^k F(x,h) = 0$ for $k > m$, so our first assertion is proved. The second assertion follows from the first one taking $m = 0$. ▶

## §4. Uniqueness theorems and maximum principles

The next theorem easily follows from representation (3.4).

**THEOREM 4.1.** *Let $\mathfrak{D}$ be a $\rho$-open set which is a $\mathbb{C}$-star relatively to a certain point $x_0 \in \mathfrak{D}$ (i.e., $x \in \mathfrak{D}$ implies $x_0 + \lambda(x - x_0) \in \mathfrak{D}$ for all $\lambda \in \mathbb{C}$ with $|\lambda| \leqslant 1$). If a $\rho$-holomorphic operator $F$ and all its Taylor coefficients equal zero in $x_0$, then $F(x) \equiv 0$ on $\mathfrak{D}$.*

If the domain $\mathfrak{D}$ is arbitrary, but $F$ is holomorphic, then the next result is true.

**THEOREM 4.2.** *Let $\mathfrak{D}$ be a domain in $\mathfrak{X}$ and let $F \in \mathcal{H}(\mathfrak{D}, \mathfrak{Y})$ be a locally bounded operator. If $F$ and all its derivatives are zero at a certain point $x_0 \in \mathfrak{D}$, then $F(x) \equiv 0$ on $\mathfrak{D}$.*

◁ Let $\mathfrak{E}$ be the set of those points in $\mathfrak{D}$ where $F$ and all its derivatives are equal to zero. Put $\mathfrak{E}_n = \{x \in \mathfrak{D} : \delta^n F(x,h) = 0\}$, $n = 0, 1, 2, \dots$. Since $\delta^n F(x,h)$ are continuous on $\mathfrak{D}$, all the sets $\mathfrak{E}_n$ are closed. But $\mathfrak{E} = \cap \mathfrak{E}_n$, hence $\mathfrak{E}$ is also closed. We show now that $\mathfrak{E}$ is open in $\mathfrak{D}$. Let $z \in \mathfrak{E}$. By (3.9), we infer that $\delta^k F(z,h) \equiv 0$

for all $h \in \mathfrak{X}$ and $k = 0, 1, 2, \ldots$. Let $\mathfrak{W}$ be a ball centered at $z$ and included in $\mathfrak{D}$ (recall that $\mathfrak{D}$ is open). By Theorem 4.1 we have $F(x) \equiv 0$ on $\mathfrak{W}$. This means that $\mathfrak{W} \subseteq \mathfrak{E}$, so $\mathfrak{E}$ is open.

Thus we obtain that $\mathfrak{E}$ is simultaneously closed and open in the connected set $\mathfrak{D}$. Since $x_0 \in \mathfrak{E}$, it follows that $\mathfrak{E} = \mathfrak{D}$. ▶

From Theorems 3.3 and 4.1, we obtain:

**COROLLARY 4.1.** *Let $\mathfrak{D}$ be a domain in $\mathfrak{X}$ and let $F \in \mathcal{H}(\mathfrak{D}, \mathfrak{Y})$ be a bounded operator. If $F$ equals zero on a neighborhood of a point $x_0 \in \mathfrak{D}$, then $F(x) \equiv 0$ on $\mathfrak{D}$.*

One of the most fruitful features of the complex analysis is the maximum principle for the norm of a holomorphic operator. We state without proof the one-dimensional version of this principle.

**THEOREM 4.3.** *Let $\Omega$ be a bounded domain in $\mathbb{C}$ and $\varphi(\tau)$ a holomorphic function on $\Omega$ with values in $\mathbb{C}$. If the expression $|\varphi(\tau)|$ attains its greatest value at a certain point $a \in \Omega$, then $\varphi(\tau)$ is constant for all $\tau \in \Omega$. If $\varphi(\tau)$ is holomorphic on $\Omega$ and continuous on $\overline{\Omega}$, then $|\varphi(\tau)|$ attains its greatest value at a certain point $\tau \in \partial\Omega$.*

We consider now a vector-function $f(\tau)$ with values in a Banach space $\mathfrak{Y}$ which is holomorphic on a bounded set $\Omega \subseteq \mathbb{C}$. Let us suppose that the expression $\|f(\tau)\|$ attains its greatest value at a point $a \in \Omega$, i.e., $\|f(\tau)\| \leqslant \|f(a)\| = M$. Consider the function $\varphi_{y^*}(\tau) = \langle f(\tau), y^* \rangle$, where $y^* \in \mathfrak{Y}^*$; $\varphi_{y^*}(\tau)$ is also holomorphic on $\Omega$. Obviously, $|\varphi_{y^*}(\tau)| \leqslant \|f(a)\| \|y^*\|$. In particular, for any $y^*$ with $\|y^*\| = 1$, we have $|\varphi_{y^*}(\tau)| \leqslant M$. Let $M(y^*) = \sup_{\tau \in \Omega} |\varphi_{y^*}(\tau)|$; then $M(y^*) \leqslant M$. From Hahn-Banach theorem it follows that there exists an element $y_0^* \in \mathfrak{Y}^*$ with $\|y_0^*\| = 1$, such that $\varphi_{y_0^*}(a) = \langle f(a), y_0^* \rangle = \|f(a)\| = M$. Hence, by Theorem 4.3 it follows that $\varphi_{y_0^*}(\tau) = M$ for all $\tau \in \Omega$, i.e.,

$$\langle f(\tau), y_0^* \rangle = \langle f(a), y_0^* \rangle \tag{4.1}$$

Let us show now that $\|f(\tau)\| = \|f(a)\|$. Suppose it is not; then we can find a point $b \in \Omega$ so that $\|f(b)\| < M$. But then $|\varphi_{y^*}(b)| < M$, which is impossible.

If we assume, in addition, that the space $\mathfrak{Y}$ is strictly convex (see Chapter 0, §9), then from (4.1) and from the equality $\|f(\tau)\| = \|f(a)\|$, it follows that $f(\tau) = f(a)$ for all $\tau \in \Omega$.

Let us return now to the general case of a space $\mathfrak{X}$. Let $F \in \mathcal{H}_\rho(\mathfrak{D}, \mathfrak{Y})$, where $\mathfrak{D}$ is a $\rho$-open set in the Banach space $\mathfrak{X}$. Assume also that for a certain point $a \in \Omega$, the inequality $\|F(a)\| \geqslant \|F(x)\|$ for all $x \in \mathfrak{D}$ is true. Then the norm of the vector-function $f(\tau) = F(a+\tau h)\colon \mathbb{C} \to \mathfrak{Y}$ attains its greatest value at the point $\tau = 0$. From our previous results it follows that $\|f(\tau)\|$ is constant for all $\tau$ with $|\tau| < \rho(x,h)$. Since the vector $h \in \mathfrak{X}$ was taken arbitrary, the equality $\|F(x)\| = \|F(a)\|$ is valid on any $\mathbb{C}$-star set $\mathfrak{W}$ centered at the point $a$ and completely contained in $\mathfrak{D}$. Now, if the space $\mathfrak{Y}$ is strictly convex, then, from $\|F(x)\| = \|F(a)\|$, it follows that $F(x) = F(a)$ for each $x \in \mathfrak{W}$. Moreover, if $\mathfrak{D}$ is a domain, then, by Corollary 4.1, the equality $F(x) = F(a)$ follows for all $x \in \mathfrak{D}$.

Thus we obtained the following generalization of the maximum principle.

**THEOREM 4.4.** *Let $\mathfrak{D}$ be a $\rho$-open set in $\mathfrak{X}$ and let $F \in \mathcal{H}_\rho(\mathfrak{D},\mathfrak{Y})$ be an operator, where $\mathfrak{Y}$ is a Banach space. If for a certain point $a \in \mathfrak{D}$, the inequality $\|F(x)\| \leqslant \|F(a)\|$ is fulfilled for all $x \in \mathfrak{D}$, then:*

*1) the norm of the operator $F$ is constant on the $\mathbb{C}$-star component $\mathfrak{W}(a)$ of $\mathfrak{D}$, centered at $a$;*

*2) if, in addition, the space $\mathfrak{Y}$ is strictly convex, then $F(x) = F(a)$ for $x \in \mathfrak{W}(a)$.*

**COROLLARY 4.2.** *If the set $\mathfrak{D}$ is open and bounded in $\mathfrak{X}$ and $F \in \mathcal{H}(\mathfrak{D},\mathfrak{Y})$ is continuous on $\overline{\mathfrak{D}}$ then the following equalities are true:*

$$\sup_{x\in\mathfrak{D}} \|F(x)\| = \sup_{x\in\partial\mathfrak{D}} \|F(x)\| = \sup_{x\in\overline{\mathfrak{D}}} \|F(x)\|.$$

**REMARK 4.1.** The assumption in the second part of Theorem 4.4 on the space $\mathfrak{Y}$ to be strictly convex is essential. Indeed, let us take $\mathfrak{Y} = \mathbb{C}^2$ with the Tchebysheff norm $\|y\| = \max\{|y_1|,|y_2|\}$ $(y = (y_1,y_2) \in \mathbb{C}^2)$ and consider the vector-function $f(\tau) = (1,\tau)$ which is holomorphic on the unit disk $|\tau| \leqslant 1$. We obviously have $\|f(0)\| = \|f(\tau)\| = 1$ for any $\tau$ with $\|\tau\| \leqslant 1$. On the other hand, $f(\tau) \neq f(0)$ for $\tau \neq 0$.

## §5. Schwarz Lemma and its generalizations

Schwarz Lemma, which is so significant for the theory of functions of a single complex variable, plays an important role in the multi-dimensional and infinite-dimensional cases too, both for qualitative and quantitative aspects. The quantitative aspects arise in relationship with various estimates, such as, for example, the domains of existence for the solutions of a functional equation, the rate of convergence of an approximation process, and so on.

The classical version of Schwarz Lemma asserts the following:

*Let $f$ be a holomorphic function on the unit disk $\Delta = \{\lambda \in \mathbb{C} \; : \; |\lambda| < 1\}$ which transforms $\Delta$ into itself, i.e., $f(\Delta) \subseteq \Delta$. If $f(0) = 0$, then:*

a) $|f(\lambda)| \leqslant |\lambda|$, *for any* $\lambda \in \Delta$;

b) $|f'(0)| \leqslant 1$.

*Moreover,*

c) *the equalities in both* a) *and* b) *occur if and only if* $f(\lambda) = f'(0)\lambda$.

Each of the assertion a) and b) admits various generalizations. We consider some of them in the sequel.

**THEOREM 5.1.** *Let $\mathfrak{X}$ and $\mathfrak{Y}$ be Banach spaces, let $\mathfrak{D}$ be a $\rho$-open set in $\mathfrak{X}$ and let $F\colon \mathfrak{X} \to \mathfrak{Y}$ be a $\rho$-holomorphic operator on $\mathfrak{D}$. Assume that, for fixed $x \in \mathfrak{D}$ and $h \in \mathfrak{X}$, $F$ admits the representation*

$$F(x + \tau h) = \sum_{k=m}^{\infty} \tau^k \delta^k F(x, h), \tag{5.1}$$

*for a certain $m \geqslant 0$ and all $\tau \in \mathbb{C}$ with $|\tau| \leqslant \rho(x, h)$. If $M_\rho(x, h) = \sup_{z \in D_\rho(x,h)} \|F(z)\|$, where $\rho \leqslant \rho(x, h)$, then*

$$\|F(x + \tau h)\| \leqslant M_\rho(x, h)|\tau|^m \rho^{-m},$$

*for all $\tau$ with $|\tau| \leqslant \rho$.*

◁ Take an arbitrary functional $y^* \in \mathfrak{Y}^*$ with $\|y^*\| = 1$ and consider the function $\varphi(\tau) = \tau^{-m}\langle F(x+\tau h), y^* \rangle$, holomorphic on the closed disk $|\tau| \leqslant \rho < \rho(x, h)$. Then, according to the maximum principle, we infer that

$$|\varphi(\tau)| \leqslant \max_{|t|=\rho} |\varphi(t)| \leqslant M_\rho(x, h)\rho^{-m},$$

for all $\tau$ with $|\tau| \leqslant \rho$, or, equivalently

$$|\langle F(x + \tau h), y^* \rangle| \leqslant M_\rho(x, h)\rho^{-m}|\tau|^m,$$

for all $y^* \in \mathfrak{Y}^*$ with $\|y^*\| = 1$.

We are done using Corollary 0.4.1 of Hahn-Banch Theorem. ▶

In particular, an analogue of the assertion a) in the classical Schwarz Lemma follows from:

**COROLLARY 5.1.** *Let $F$ be a Fréchet differentiable operator on a ball $\mathfrak{D}$ centered at zero. Suppose that $F$ satisfies the conditions $F(\mathfrak{D}) \subseteq \mathfrak{D}$ and $F(0) = 0$. Then $\|F(x)\| \leqslant \|x\|$ for all $x \in \mathfrak{D}$.*

◁ Take $x = \tau h$, where $\|h\| = 1$ and observe that the operator $F$ admits representation (5.1) for all $\tau \in \mathbb{C}$ with $|\tau| \leqslant \|x\|$ and $m = 1$. ▶

The assertion b) can be also easily generalized for arbitrary Banach spaces. Namely, under the assumptions of Corollary 5.1, the inequality

$$\|F'(0)\| \leqslant 1 \tag{5.2}$$

is true. This follows directly from Cauchy inequality (3.8) for $k = 1$.

The case of assertion c) is considerably more complicated. All depends upon how we understand, in the general case, the right hand side of the equality in assertion c), and upon the properties of the norm of the space $\mathfrak{X}$.

If the mapping $f'(0)\lambda$ is understood as a linear operator applied to $\lambda$, then assertion c) is no longer true in the general case.

◁ Indeed, consider the space $\mathbb{C}^2$, with the Tchebysheff norm, and the operator $F: \mathbb{C}^2 \to \mathbb{C}^2$ given by the formula $F(z) = (z_1, z_2^2)$, $z = (z_1, z_2) \in \mathbb{C}^2$, which maps the unit polydisk $\mathfrak{U} = \{z \in \mathbb{C}^2,\ \|z\| < 1\}$ into itself. Clearly $F(0) = 0$ and

$$\|F'(0)\| = \left\| \begin{pmatrix} 1 & 0 \\ 0 & 0 \end{pmatrix} \right\| = 1.$$

However, the operator $F$ is obviously nonlinear. ▶

If we assume, in addition, that the operator $F'(0)$ is a linear isometry, assertion c) becomes true in the general case, too. This follows from the next generalization of a theorem due to H. Cartan for the finite-dimensional case. (Apparently, the first infinite-dimensional version of this theorem was obtained in 1969 by L. Harris [1], who supposed that $\mathfrak{D}$ is a ball. His proof heavily relies on this assumption which is a serious restriction in many cases.)

**THEOREM 5.2.** *Let $\mathfrak{D}$ be an arbitrary bounded domain in the Banach space $\mathfrak{X}$ and let $F \in \mathcal{H}(\mathfrak{D}, \mathfrak{Y})$ be an operator satisfying the conditions*

$$F(\mathfrak{D}) \subseteq \mathfrak{D} \tag{5.3}$$

*and*

$$F(z) = z, \tag{5.4}$$

*for a certain point $z \in \mathfrak{D}$. If $F'(z) = I$ (the identity operator on $\mathfrak{X}$), then $F$ is the restriction of the operator $I$ on $\mathfrak{D}$, i.e., $F(x) = x$ for all $x \in \mathfrak{D}$.*

◁ From (5.3) it follows that the operators $F^n = F \circ F^{n-1}$, $n = 1, 2, \dots$, $F^0 = I$, are defined on $\mathfrak{D}$, too; they satisfy $F^n \in \mathcal{H}(\mathfrak{D}, \mathfrak{X})$ and $F^n(\mathfrak{D}) \subseteq \mathfrak{D}$. Moreover, from (5.4) it follows that $F^n(z) = z$ and $(F^n)'(z) = I$. Therefore, the operators $F^n$ admit the representation

$$F^n(x+h) = z + Ih + \sum_{k=m_n}^{\infty} \delta^k F^n(z,h), \quad m_n \geqslant 2, \tag{5.5}$$

for all $h$ with $\|h\| < \operatorname{dist}(z, \partial\mathfrak{D}) = \inf\limits_{y \in \partial\mathfrak{D}} \|z - y\| = r$.

Let us show that $\delta^k F(z,h) = 0$ for all $h$ with $\|h\| \leqslant r$ and all $k = 2, 3, \dots$. Suppose this is not the case; then there exists a positive integer $m_1$ such that $\delta^{m_1} F(x,h)$ is the first non-identically zero form in the decomposition (5.5) of the operator $F$. Since the domain $\mathfrak{D}$ is bounded, from Cauchy inequality (3.8) we have

$$\left\|\delta^k F^n(z,h)\right\| \leqslant M(k) < \infty, \quad \|h\| \leqslant r,$$

for all $n = 1, 2, \dots$.

A direct computation using induction proves the equality

$$\delta^{m_1} F^n(z,h) = n\delta^{m_1} F(z,h)$$

whence

$$\|\delta^{m_1} F(z,h)\| \leqslant M(m_1) \cdot n^{-1}, \quad n = 1, 2, \dots .$$

But this means that $\delta^{m_1} F(z,h) \equiv 0$, a contradiction. Therefore $F(z+h) = z + h$ for all $h$ with $\|h\| \leqslant r$. By the uniqueness theorem we conclude that $F(x) = x$ for all $x \in \mathfrak{D}$. ▶

**COROLLARY 5.2.** *Let $\mathfrak{X}$ and $\mathfrak{Y}$ be complex Banach spaces and $F$ an operator which acts from the unit ball $\mathfrak{D}_{\mathfrak{X}}$ centered at zero in $\mathfrak{X}$ into the unit ball $\mathfrak{D}_{\mathfrak{Y}}$ centered at zero in $\mathfrak{Y}$. If $F \in \mathcal{H}(\mathfrak{D}_{\mathfrak{X}}, \mathfrak{Y})$, $F(0) = 0$, and $F'(0)$ is a linear isometry from $\mathfrak{X}$ into $\mathfrak{Y}$ then $F(x) = F'(0)x$ for all $x \in \mathfrak{D}_{\mathfrak{X}}$.*

◁ Denote $A = F'(0)$. Under our assumption, the operator $A$ has a continuous inverse $A^{-1}: \mathfrak{Y} \to \mathfrak{X}$, and since $\|Ax\| = \|x\|$, we infer that $\|A^{-1}y\| = \|y\|$. Hence it follows that $\Phi = A^{-1}F$ acts from $\mathfrak{D}_{\mathfrak{X}}$ into $\mathfrak{D}_{\mathfrak{X}}$ and $\Phi \in \mathcal{H}(\mathfrak{D}_{\mathfrak{X}}, \mathfrak{Y})$. Moreover, $\Phi'(0) = A^{-1}A = I_{\mathfrak{X}}$. By Theorem 5.2, we obtain $\Phi(x) = x$ for all $x \in \mathfrak{D}_{\mathfrak{X}}$. Then $F(x) = A\Phi(x) = Ax$. ▶

An operator $F: \mathfrak{D} \to \mathfrak{D}$ is called *biholomorphic* if it is holomorphic and one-to-one, and its inverse $F^{-1}$ is also holomorphic.

**COROLLARY 5.3.** *If an operator $F$ is biholomorphic on the unit ball $\mathfrak{D}$ centered at the zero element in $\mathfrak{X}$ and $F(0) = 0$ then $F(x) \equiv Ax$, $x \in \mathfrak{D}$, where $A$ is a linear isometry on the space $\mathfrak{X}$.*

◁ Consider the linear continuous operator $B = (F^{-1})'(0)$. Since $F^{-1} \circ F = I$, from (I.4.1) we obtain $B \circ F'(0) = I$. This means that the operator $A = F'(0)$ has a continuous inverse $B = A^{-1}$. Moreover, from (5.2) it follows that $\|A\| \leqslant 1$ and $\|A^{-1}\| \leqslant 1$. Thus $\|Ax\| = \|x\|$ for any $x \in \mathfrak{X}$, i.e., $A$ is a linear isometry. Our assertion follows now from Corollary 5.2. ▶

The space $\mathfrak{X}$ will be called a space *with a generalized Krein-Phillips transform*, or shortly a K-P-space, if for any point $a$ in the unit ball $\mathfrak{D}$ centered at zero in $\mathfrak{X}$ there exists a biholomorphic operator $\Phi_a$ on $\mathfrak{D}$ such that $\Phi_a(a) = a$. As examples of K-P-spaces we mention the space $\mathbb{C}^n$ with the Chebyshev or the Euclidean norm, any Hilbert space over the field $\mathbb{C}$ of complex numbers, the algebra of all linear operators on a Hilbert space, and others. In particular, for the one-dimensional space $\mathbb{C}$, the operator $\Phi_a$ has the form

$$\Phi_a(\lambda) = \frac{\lambda - a}{\lambda\overline{a} - 1}, \tag{5.6}$$

where $\overline{a}$ is the complex-conjugate of $a$.

**COROLLARY 5.4.** *Let $\mathfrak{X}$ be a K-P-space and let $F: \mathfrak{X} \to \mathfrak{X}$ be a biholomorphic operator on the ball $\mathfrak{D}$ centered at zero in the space $\mathfrak{X}$. Then the operator $F$ admits the representation*

$$F = \Phi_b^{-1} A \Phi_a, \tag{5.7}$$

*where $A$ is a linear isometry on $\mathfrak{X}$, $a$ is a point in $\mathfrak{D}$, and $b = F(a)$.*

◁ Indeed, for any point $a \in \mathfrak{D}$ the operator $\Phi_b F \Phi_a^{-1}$ is biholomorphic on $\mathfrak{D}$ and satisfies the condition $\Phi_b F \Phi_a^{-1}(0) = 0$. Now our assertion follows from Corollary 5.3. ▶

Finally, let us remark that inequality (5.2) gives an estimate for the Fréchet derivative of the operator $F$ only at a point. As we have already noted in Chapter I, a uniform estimate for the derivative on the whole domain $\mathfrak{D}$ does not exist. However, Scwarz Lemma gives the possibility of estimating the Fréchet derivatives uniformly on any closed subset of the domain $\mathfrak{D}$. The next result is implicitly contained in T. Hayden and T. Suffridge [1].

**THEOREM 5.2.** *Let $\mathfrak{D}$ be the unit ball centered at the zero element of the space $\mathfrak{X}$ and let $F \in \mathcal{H}(\mathfrak{D}, \mathfrak{X})$ be an operator satisfying the condition*

$$F(\mathfrak{D}) \subseteq \mathfrak{D}.$$

*Then the inequality*

$$\|F'(x)x\| \leqslant \frac{\|x\|}{1 - \|x\|^2}$$

*is fulfilled for any $x \in \mathfrak{D}$.*

◁ Fix $x \in \mathfrak{D}$ and choose a linear functional $x^* \in \mathfrak{X}^*$ such that $\|x^*\| = 1$ and $\langle F'(x)x, x^* \rangle = \|F'(x)x\|$. Consider the function $f(\lambda) = \langle F(\lambda x \|x\|^{-1}), x^* \rangle$, which is holomorphic on $\Delta = \{\lambda \in \mathbb{C} \ : \ |\lambda| < 1\}$. Obviously $|f(\lambda)| < 1$ for $\lambda \in \Delta$. Since $\|x\| < 1$, we have $f(\|x\|) = \langle F(x), x^* \rangle = a$, where $|a| < 1$. The functions

$$\zeta = \varphi(\lambda) = \frac{\lambda - \|x\|}{\lambda \|x\| - 1}$$

and

$$\psi(\omega) = \frac{\omega - f(\|x\|)}{\omega \overline{f(\|x\|)} - 1}$$

are holomorphic on $\Delta$ (see (5.6)), and the function $g(\zeta) = \psi(f(\varphi^{-1}(\zeta)))$ verifies the conditions of Schwarz Lemma: $|g(\zeta)| < 1$ and $g(0) = 0$. Consequently $|g(\zeta)| \leqslant |\zeta|$, or equivalently

$$\left| \frac{f(\lambda) - f(\|x\|)}{f(\lambda)\overline{f(\|x\|)} - 1} \right| \leqslant \left| \frac{\lambda - \|x\|}{\lambda \|x\| - 1} \right|.$$

Whence, we obtain the inequality

$$\left|\frac{f(\lambda) - f(\|x\|)}{\lambda - \|x\|}\right| \leqslant \left|\frac{f(\lambda)\overline{f(\|x\|)} - 1}{\lambda\|x\| - 1}\right|.$$

Taking the limit for $\lambda \to \|x\|$, we have

$$|f'(\|x\|)| \leqslant \frac{1 - |f(\|x\|)|^2}{1 - \|x\|^2} \leqslant \frac{1}{1 - \|x\|^2}.$$

But $f'(\lambda) = \langle F'(\lambda x\|x\|^{-1})x\|x\|^{-1}, x^*\rangle$, so

$$f'(\|x\|) = \langle F'(x)x, x^*\rangle\|x\|^{-1} = \|F'(x)x\| \; \|x\|^{-1},$$

hence

$$\|F'(x)x\| \leqslant \frac{\|x\|}{1 - \|x\|^2}. \qquad \blacktriangleright$$

## §6. Uniformly bounded families of $\rho$-holomorphic (holomorphic) operators. Montel property

Let $\mathfrak{D}$ be a $\rho$-open set in $\mathfrak{X}$. Until now, we did not use the fact that $\mathfrak{D}$ itself can be considered as a topological space with the $\rho$-topology defined in §2. Recall that in the initial topology of the space $\mathfrak{X}$, a $\rho$-holomorphic operator can have discontinuities (see Example I.1.2). However, for the $\rho$-topology the next assertion is true.

*Any $\rho$-holomorphic bounded operator is $\rho$-continuous.*

We prove below a stronger result.

**THEOREM 6.1.** *Let $\mathfrak{D}$ be a $\rho$-open set and let $\mathcal{F} = \{F_\alpha\}_{\alpha\in\mathfrak{A}}$, $F_\alpha \in \mathcal{H}_\rho(\mathfrak{D},\mathfrak{Y})$, be a uniformly bounded family (with respect to the norm topology of $\mathfrak{Y}$), i.e., $\|F_\alpha(x)\| \leqslant$ $\leqslant M < \infty$, for all $\alpha \in \mathfrak{A}$. Then $\mathcal{F}$ is equicontinuous as a family of operators acting from the set $\mathfrak{D}$ with the $\rho$-topology into the space $\mathfrak{Y}$ with the norm topology.*

$\triangleleft$ Given an arbitrary $\varepsilon > 0$, take $N \geqslant \max\{2\varepsilon^{-1}M, 1\}$ and let $x$ and $x'$ be such that $\rho(x, x' - x) > N$. Then, setting $x' - x = \zeta h$, where $\|h\| = 1$, $\zeta \in \mathbb{C}$, and using Lemma 2.1, we obtain

$$\rho(x, \zeta h) = \rho(x, h)|\zeta|^{-1} > N$$

or

$$|\zeta| < N^{-1}\rho(x,h) \leqslant \rho(x,h).$$

It follows that $F_\alpha(x') = F_\alpha(x+\zeta h)$ has the Taylor representation (3.4). Then, by Theorem 5.1, we have

$$\|F_\alpha(x') - F_\alpha(x)\| \leqslant \|F_\alpha(x+\zeta h) - F_\alpha(x)\| \leqslant 2M|\zeta|\,[\rho(x,h)]^{-1}. \tag{6.1}$$

As $h = \zeta^{-1}(x'-x)$, we actually have

$$\|F_\alpha(x') - F_\alpha(x)\| \leqslant 2M\,[\rho(x,x'-x)]^{-1} \leqslant 2MN^{-1} < \varepsilon. \qquad \blacktriangleright$$

**COROLLARY 6.1.** *If the set $\mathfrak{D}$ is open, then a uniformly bounded family $\mathcal{F} = \{F_\alpha\}_{\alpha\in\mathfrak{A}}$ of operators holomorphic on $\mathfrak{D}$, with values in a Banach space $\mathfrak{Y}$, satisfies a Lipschitz condition, uniformly with respect to $\alpha \in \mathfrak{A}$.*

$\lhd$ Since $\rho(x,h) \geqslant \delta(x) > 0$ (see 2.2), where $h \in \mathfrak{X}$, $\|h\| = 1$, (6.1) implies

$$\|F_\alpha(x') - F_\alpha(x)\| \leqslant 2M|\zeta|\,[\delta(x)]^{-1} \leqslant 2M\,[\delta(x)]^{-1}\,\|x - x'\|$$

for all $x'$ such that $\|x - x'\| \leqslant \delta(x)$. $\blacktriangleright$

Notice that the last relation implies the local uniformly boundedness of the Fréchet derivatives of the operators $F_\alpha$, $\alpha \in \mathfrak{A}$, too.

We return now to the general case. Let us consider again $\mathfrak{D}$ as a topological space with the $\rho$-topology. On the set of all continuous mappings from $\mathfrak{D}$ into $\mathfrak{Y}$ we introduce the topology $\tau_\rho(\mathfrak{D},\mathfrak{Y})$ of compact convergence (i.e., the uniform convergence on compact subsets in the $\rho$-topology of $\mathfrak{D}$, see §5 of Chapter 0). Using Theorem 6.1 and Theorem 0.5.4, we obtain the following analogue of the Montel property.

**THEOREM 6.2.** *Let $\mathcal{F}_M$ be a bounded subset of $\mathcal{H}_\rho(\mathfrak{D},\mathfrak{Y})$, i.e., $\mathcal{F}_M \subseteq \{F \in \mathcal{H}_\rho(\mathfrak{D},\mathfrak{Y}) : \|F(x)\| \leqslant M < \infty$ for all $x \in \mathfrak{D}\}$ and assume that for any $x \in \mathfrak{D}$, the orbit $\mathcal{F}_M(x) = \{y \in \mathfrak{Y} : y = F(x),\ F \in \mathcal{F}_M\}$ is a sequentially compact subset of $\mathfrak{Y}$. Then the family $\mathcal{F}_M$ is sequentially compact in the topology $\tau_\rho(\mathfrak{D},\mathfrak{Y})$.*

**REMARK 6.1.** In contradistinction with the classical version of Montel theorem ($\mathfrak{X} = \mathbb{C}^n$, $\mathfrak{Y} = \mathbb{C}^m$) we assume in Theorem 6.2 the compactness of the orbits; this compactness follows, for $\mathbb{C}^m$, from the boundedness of the family $\mathcal{F}_M$. The next example shows the necessity of our assumption. Let $\mathfrak{X} = \mathfrak{Y} = \ell^2$. Consider the family $\{B_k\}_{k=1}^{\infty}$ of linear bounded operators (hence holomorphic operators) defined on the set $\{x = (x_1, \ldots, x_n, \ldots) \in \ell^2 : \|x\| < R < \infty\}$ by the equalities $B_k x = (\underbrace{0, \ldots, 0}_{k}, x_1, \ldots, x_n, \ldots)$. Obviously $\|B_k x\| < R$ for all $x$ with $\|x\| < R$, $k = 1, 2, \ldots$. However the sequence $\{B_k\}_{k=1}^{\infty}$ is not compact; it is enough, for instance, to consider the values at the point $x = (1, 0, \ldots, 0, \ldots)$.

Suppose now that $\mathfrak{Y}$ is the dual space of some Banach space and that $\mathfrak{Y}$ is metrizable in the corresponding ultraweak topology (this is the case, for instance, when $\mathfrak{Y}$ is separable). Then, by Alaoglu-Bourbaki Theorem (see §10 of Chapter 0), any bounded set in $\mathfrak{Y}$ is sequentially compact in the ultraweak topology. Consequently, we can use Theorem 0.5.4 and Theorem 6.1 in order to obtain the next result.

**THEOREM 6.3.** *Suppose that $\mathfrak{Y}$ is metrizable in the ultraweak topology. Then any bounded set $\mathcal{F}_M \subseteq \mathcal{H}_\rho(\mathfrak{D}, \mathfrak{Y})$ is sequentially compact in the topology $\tau_\rho(\mathfrak{D}, \mathfrak{Y})$.*

If the space $\mathfrak{Y}$ is reflexive, the ultraweak topology in Theorem 6.3 can be replaced with the weak topology of the space $\mathfrak{Y}$. Thus, Theorems 6.1–6.3 provide a Banach-Steinhaus type theorem (see §8 of Chapter 0). (Note, once more, the similarity between the geometric properties of holomorphic bounded operators and those of linear continuous operators.)

**THEOREM 6.4.** *Assume that $\mathfrak{Y}$ is equipped with the norm (resp. ultraweak) topology. Let $\{F_n\}_{n=1}^{\infty}$, $F_n \in \mathcal{H}_\rho(\mathfrak{D}, \mathfrak{Y})$, be a sequence of bounded operators. Then the sequence $\{F_n\}_{n=1}^{\infty}$ is convergent, in the topology $\tau_\rho(\mathfrak{D}, \mathfrak{Y})$, to a bounded operator $F \in \mathcal{H}_\rho(\mathfrak{D}, \mathfrak{Y})$ if only if the following two conditions are simultaneously fulfilled:*

*1) the family $\{F_n\}_{n=1}^{\infty}$ is uniformly bounded on $\mathfrak{D}$;*

*2) the sequence $\{F_n(x)\}_{n=1}^{\infty}$ is fundamental (resp. ultraweak fundamental) for all $x \in \mathfrak{D}'$, where $\mathfrak{D}'$ is a subset of $\mathfrak{D}$, dense in the $\rho$-topology.*

We conclude this section with two generalizations of Vitali Theorem.

**THEOREM 6.5.** *Let $\mathfrak{D}$ be a $\rho$-open set which is a $\mathbb{C}$-star relatively to a point $x_0 \in \mathfrak{D}$, and let $\{F_n\}_{n=1}^{\infty}$, $F_n \in \mathcal{H}_\rho(\mathfrak{D}, \mathfrak{Y})$, be a sequence of operators, uniformly bounded on $\mathfrak{D}$. Assume that for any $h \in \mathfrak{X}$, $\|h\| = 1$, the sequence $\{F_n\}_{n=1}^{\infty}$ is fundamental on the disk $D_\rho(x_0, h) \subseteq \mathfrak{D}$ in the norm topology of the space $\mathfrak{Y}$. Then the sequence is fundamental on $\mathfrak{D}$ with respect to the norm topology of $\mathfrak{Y}$, and, consequently, it is* *strongly convergent to some operator* $F \in \mathcal{H}_\rho(\mathfrak{D}, \mathfrak{Y})$.

$\triangleleft$ Let $u$ be an arbitrary point in $\mathfrak{D}$. Using the properties of the set $\mathfrak{D}$, we find an element $h \in \mathfrak{X}$ and $\lambda \in \mathbb{C}$ with $|\lambda| < 1$, such that $x_0 + h \in \mathfrak{D}$ and $u = x_0 + \lambda h$. Moreover, $\rho(x_0, h) \geqslant 1$. Then the vector-functions $f_n(\zeta) = F_n(x_0 + \zeta h)$ are holomorphic on the unit disk $|\zeta| < 1$ and all of them are bounded on this disk, with respect to the norm of $\mathfrak{Y}$, by a constant $M < 1$. By our assumptions, there exists $d > 0$ such that the sequence $\{f_n\}_{n=1}^{\infty}$ is fundamental on the disk $|\zeta| \leqslant d$. (From now on, the proof reproduces almost verbatim the classical one. However, we may need some estimates established in the sequel, so we present a complete proof.)

By virtue of Theorem 5.4 the sequence $\{f_n\}_{n=1}^{\infty}$ is uniformly convergent, i.e., for any given positive $\delta$, there exists $N(\delta)$ such that whenever $n > N(\delta)$, we have

$$\|f_n(\zeta) - f_{n+m}(\zeta)\| < \delta, \quad \text{for} \quad |\zeta| \leqslant d, \quad \text{and} \quad m = 0, 1, \dots .$$

From these inequalities and by Cauchy inequality (3.8) we obtain

$$\|\delta_{nm}^k F(v_0, h)\| \leqslant \delta d^{-k}, \qquad k = 0, 1, 2, \dots ,$$

where $\delta_{nm}^k$ are the Taylor coefficients of the operator $F_{nm} = F_n - F_{n+m}$. Note that we also have

$$\|\delta_{nm}^k F(x_0, h)\| \leqslant 2M,$$

uniformly with respect to $k = 0, 1, 2, \dots$. Then, setting $|\lambda| = r$ $(< 1)$, we obtain

$$\|F_n(u) - F_{n+m}(u)\| = \|f_n(\lambda) - f_{n+m}(\lambda)\| \leqslant \sum_{k=0}^{\infty} \|\delta_{nm}^k F(x_0, h)\lambda^k\|.$$

Using the formula with remainder for the geometric series, we have

$$\begin{aligned} &\|F_n(u) - F_{n+m}(u)\| \leqslant \\ &\leqslant \sum_{k=0}^{p-1} \|\delta_{nm}^k F(x_0, h)\| r^k + 2M \frac{r^p}{1-r} \leqslant \\ &\leqslant \sum_{k=0}^{p-1} \delta \left(\frac{r}{d}\right)^k + 2M \frac{r^p}{1-r} = \\ &= \delta \left(1 - \frac{r}{d}\right)^{-1} \left(1 - \left(\frac{r}{p}\right)^p\right) + 2M r^p (1-r)^{-1}. \end{aligned}$$

Now, for any $\varepsilon > 0$ and $0 < q < 1$, taking

$$p > \ln\left[\varepsilon(1-r)(1-q)(2M)^{-1}\right](\ln r)^{-1} \tag{6.2}$$

and

$$\delta > \varepsilon q(r-d)d^{p-1}(r^p - d^p)^{-1}, \tag{6.3}$$

we obtain $\|F_n(u) - F_{n+m}(u)\| < \varepsilon$ as soon as $n > N(\delta)$. ▶

For the case of ultraweak convergence, the proof of the analogous assertion can be sligthly simplified; it is enough to use Montel property. However, we will be mainly interested in the case of holomorphic operators.

**THEOREM 6.6.** *Assume that the space $\mathfrak{Y}$ is metrizabile in the ultraweak topology and let $\mathfrak{D}$ be a domain in $\mathfrak{X}$. Let $\{F_n\}_{n=1}^{\infty}$ be a uniformly bounded sequence of operators holomorphic in $\mathfrak{D}$. Suppose that $\{F_n\}_{n=1}^{\infty}$ converges in the ultraweak topology to an operator $F \in \mathcal{H}(\mathfrak{D}', \mathfrak{Y})$, for any element of a certain open set $\mathfrak{D}' \subseteq \mathfrak{D}$. Then $F$ admits a holomorphic continuation on the whole $\mathfrak{D}$ and the sequence $\{F_n\}_{n=1}^{\infty}$, converges to this continuation in the ultraweak topology, for each element of $\mathfrak{D}$.*

◁ By Theorem 6.3, the sequence $\{F_n\}_{n=1}^{\infty}$ contains a subsequence $\{F_{n_k}\}_{k=1}^{\infty}$ which is convergent in the ultraweak topology, for any element of $\mathfrak{D}$, to a certain operator $\tilde{F} \in \mathcal{H}(\mathfrak{D}, \mathfrak{Y})$. The operator $\tilde{F}$ is a holomorphic continuation of the operator $F$, i.e., $\tilde{F} \mid \mathfrak{D}' = F$. The uniqueness theorem implies that any other convergent subsequence $\{F_{n_m}\}_{m=1}^{\infty}$ has the same ultraweak limit $\tilde{F}$. The proof is complete. ▶

In spite of the fact that the previous proof is very simple, it is not at all constructive; it provides no estimates analogous to those in (6.2) and (6.3), which are very useful in many applications (see §2 of Chapter VI below).

# Chapter IV

# Linear operators

This chapter, as well as Chapter 0, is mainly auxiliary. However, compared with Chapter 0, the topics considered here are more special. Some of the subsequent results were presented only in research papers, but not in monographs. Therefore, our exposition in this chapter will be quite detailed, and will include proofs.

## §1. The spectrum and the resolvent of a linear operator

**1.** CLOSED OPERATORS. Let $\mathfrak{X}$ and $\mathfrak{Y}$ be complex Banach spaces and $A: \mathfrak{X} \to \mathfrak{Y}$ a linear operator (see Chapter 0, §8 for definition). Recall that the operator $A$ is equally well described as the set of pairs $(x, Ax)$, with $x \in D(A) \subseteq \mathfrak{X}$. The operator $A$ is invertible (i.e., $A$ has an inverse operator $A^{-1}$) if and only if its kernel $N(A) = \{x \in D(A) \ : \ Ax = 0\}$ equals $\{0\}$. In this case, the operator $A^{-1}$ is defined from $R(A)$ into $D(A)$; moreover, $A^{-1}Ax = x$ for all $x \in D(A)$ and $AA^{-1}y = y$ for all $y \in R(A)$.

An operator $A$ is called *closed* if it is closed as a subset of the topological product $\mathfrak{X} \times \mathfrak{Y}$ (i.e., if $x_n \in D(A)$, $x_n \to x$ and $Ax_n \to y$ for $n \to \infty$, then $x \in D(A)$ and $Ax = y$). If $A$ is closed and invertible, then obviously $A^{-1}$ is also closed. If an operator $A$ is bounded and defined on the whole $\mathfrak{X}$, and if $B$ is a closed operator, then the operator $A + B$ — defined on $D(B)$ — is also closed.

**THEOREM 1.1.** *A bounded linear operator $A$ is closed if and only if its domain $D(A)$*

*is closed.*

◁ Since the operator $A$ is bounded then $x_n \in D(A)$ and $x_n \to x$ for $n \to \infty$ implies that $Ax_n \to Ax$. Therefore $A$ is closed if and only if $D(A)$ is closed. ▶

In particular, any operator in $L(\mathfrak{X}, \mathfrak{Y})$, i.e., any bounded operator defined on the whole $\mathfrak{X}$, is closed. The converse is also true; we give the result without proof.

**THEOREM 1.2** (Closed Graph Theorem). *Any closed linear operator $A: \mathfrak{X} \to \mathfrak{Y}$ with the domain of definition $D(A) = \mathfrak{X}$ is closed.*

**2.** SPECTRUM AND RESOLVENT. In this subsection we will consider $\mathfrak{X} = \mathfrak{Y}$. Let $A$ be a linear operator on $\mathfrak{X}$. A point $\lambda \in \mathbb{C}$ for which the operator $(A - \lambda I)^{-1}$ exists and is bounded on the whole $\mathfrak{X}$ is called a *regular point* of $A$. The set of all regular points of $A$ is denoted by $\rho(A)$.

**THEOREM 1.3.** *A linear operator $A$ with at least one regular point is closed.*

◁ If $\lambda \in \rho(A)$, then by Theorem 1.1, the operator $(A-\lambda I)^{-1}$ is closed and hence $A - \lambda I$ is also closed. Since $A = \lambda I + (A - \lambda I)$, the operator $A$ is closed, too. ▶

The complement of the set $\rho(A)$ in $\mathbb{C}$ is called the *spectrum* of the operator $A$ and is denoted by $\sigma(A)$. According to Theorem 1.3, the spectrum of a non-closed operator coincides with the set $\mathbb{C}$ of complex numbers and the set of its regular points is void. This is why, everywhere below is this section, we will consider only *closed* linear operators.

Let $A$ be a linear operator on $\mathfrak{X}$. The *resolvent* $\Gamma_A$ of $A$ is defined on the set $\rho(A)$ of its regular points by $\Gamma_A(\lambda) = (A - \lambda I)^{-1}$, $\lambda \in \rho(A)$. If no confusion may arise, the subscript $A$ in $\Gamma_A(\lambda)$ will be omitted. From the relation

$$\Gamma(\mu)(A - \lambda I)\Gamma(\lambda) - \Gamma(\mu)(A - \mu I)\Gamma(\lambda) = (\mu - \lambda)\Gamma(\mu)\Gamma(\lambda), \quad (\lambda, \mu \in \rho(A))$$

the so-called *Hilbert identity* follows:

$$\Gamma(\mu) - \Gamma(\lambda) = (\mu - \lambda)\Gamma(\mu)\Gamma(\lambda), \tag{1.1}$$

which implies the commutativity of the values of the operator-function $\Gamma(\lambda)$.

**THEOREM 1.4.** *The set $\rho(A)$ of the regular points of an operator $A$ is open and the resolvent $\Gamma(\lambda)$ is an analytic function on $\rho(A)$.*

◁ Let $\lambda \in \rho(A)$ and $\mu \in \mathbb{C}$. Then $A - \mu I = A - \lambda I - (\mu - \lambda)I$. If $|\lambda - \mu| < \|\Gamma(\lambda)\|^{-1}$, then Theorem 0.4.6 implies the existence of the operator $(A - \mu I)^{-1}$, which is defined and bounded on the whole $\mathfrak{X}$. This means that $\mu \in \rho(A)$ and $\Gamma(\mu) = (I - (\mu - \lambda)\Gamma(\lambda))^{-1}\Gamma(\lambda)$. It follows that $\Gamma(\mu) = \sum_{n=0}^{\infty}(\mu - \lambda)^n \Gamma^{n+1}(\lambda)$, where the series is convergent for all $\mu$ satisfying $|\mu - \lambda| < \|\Gamma(\mu)\|^{-1}$. Thus $\Gamma(\lambda)$ is an analytic operator-valued function. ▶

**COROLLARY 1.1.** *The spectrum $\sigma(A)$ is closed.*

◁ The conclusion follows directly from Theorem 1.4 and from de Morgan formula (see §1 of Chapter 0). ▶

Note that the conclusion of Corollary 1.1 is also true for non-closed operators, since in this case $\sigma(A) = \mathbb{C}$. Thus, the spectrum of a closed linear operator $A$ on a complex Banach space is a closed subset of $\mathbb{C}$, which can be void. If the spectrum $\sigma(A)$ is non-void, it has three parts:

1) $\sigma_1(A)$ consisting of all those $\lambda \in \sigma(A)$ for which $A - \lambda I$ has a non-zero kernel $N(A - \lambda I)$;

2) $\sigma_2(A)$ consisting of those $\lambda \in \sigma(A)$ for which $N(A - \lambda I) = \{0\}$, but there exists a sequence $\{x_n\}_{n\in\mathbb{N}}$ such that $\|x_n\| = 1$, $n \in \mathbb{N}$, and $(A-\lambda I)x_n \to 0$ as $n \to \infty$;

3) $\sigma_3(A)$ consisting of all $\lambda\in\sigma(A)$ such that $N(A-\lambda I)=\{0\}$ and $\overline{R(A-\lambda I)}= R(A - \lambda I)$.

It is obvious that $\sigma_1(A)$, $\sigma_2(A)$ and $\sigma_3(A)$ are mutually disjoint and $\sigma(A) = \bigcup_{i=1}^{3} \sigma_i(A)$.

The elements of the set $\sigma_1(A)$ are called the *eigenvalues* of $A$; if $\lambda \in \sigma_1(A)$ then there exists $x \in \mathfrak{X}$, $x \neq 0$, such that $Ax = \lambda x$. The set $\mathfrak{X}_\lambda$ of all eigenvectors $x$ corresponding to an eigenvalue $\lambda$, together with the zero element of the space $\mathfrak{X}$, form a subspace of $\mathfrak{X}$ called the *eigenspace* of $\lambda$. The dimension of the eigenspace $\mathfrak{X}_\lambda$ is called the *multiplicity* of the eigenvalue $\lambda$.

Consider now the following question: how does the spectrum of an operator looks like. When the operator is unbounded, its spectrum may be the whole complex plane $\mathbb{C}$. Let us take a bounded closed operator $A$. In this case, by Theorem 1.1, the domain $D(A)$ is closed in $\mathfrak{X}$. From now on we will consider in this section that $D(A) = \mathfrak{X}$, i.e., $A \in L(\mathfrak{X})$ $(= L(\mathfrak{X}, \mathfrak{X}))$.

**COROLLARY 1.2.** *The spectrum $\sigma(A)$ is non-void and bounded.*

◁ Assume, on the contrary, that $\rho(A) = \mathbb{C}$. Using the Hilbert identity for $\lambda = 0$ we obtain that $\Gamma(\mu) = \mu^{-1}\Gamma(0)(\mu^{-1}I - \Gamma(0))^{-1}$. It follows that $\|\Gamma(\mu)\| \to 0$ as $\mu \to \infty$, therefore the resolvent is bounded. Thus, the operator-valued function

$\Gamma\colon\mathbb{C}\to L(\mathfrak{X})$ is analytic and bounded everywhere on $\mathbb{C}$. Therefore, the function $f\circ\Gamma\colon\mathbb{C}\to\mathbb{C}$ is analytic and bounded for any $f\in(L(\mathfrak{X}))^*$. By Liouville Theorem, $f\circ\Gamma$ is constant for any $f\in(L(\mathfrak{X}))^*$, hence $\Gamma(\lambda)$ is constant. Since $\Gamma(\infty)=0$, we obtain $\Gamma(\lambda)\equiv 0$, which contradicts the definition of the resolvent.

For $|\lambda|>\|A\|$, using Theorem 0.8.4 we have that the operator $A-\lambda I = -\lambda\left(I-\frac{1}{\lambda}A\right)$ maps homeomorphically $\mathfrak{X}$ onto $\mathfrak{X}$. Therefore the spectrum $\sigma(A)$ is contained in the disk $|\lambda|\leqslant\|A\|$. ▶

**LEMMA 1.1.** *The set* $\sigma_3(A)$ *is open.*

◁ According to the definition, if $\lambda\in\sigma_3(A)$ then $R(A-\lambda I)$ is closed and $N(A-\lambda I)=\{0\}$. By Theorem 0.8.5 the operator $(A-\lambda I)^{-1}\colon R(A-\lambda I)\to\mathfrak{X}$ is bounded. Using Theorem 0.8.4 we get that the operator $I+(\lambda-\mu)(A-\lambda I)^{-1}$ is a homeomorphism from $\mathfrak{X}$ onto $\mathfrak{X}$ for any $\mu\in\mathbb{C}$ such that $|\lambda-\mu|<\|(A-\lambda I)^{-1}\|^{-1}$. It follows that

$$A-\mu I=(A-\lambda I)(I+(\lambda-\mu)(A-\lambda I)^{-1})$$

is a homeomorphism from $\mathfrak{X}$ onto $R(A-\lambda I)$, i.e., $\mu\in\sigma_3(A)$. ▶

For $\lambda\in\sigma_1(A)$, we consider the non-decreasing sequence of subspaces

$$\{0\}, N(A-\lambda I), N((A-\lambda I)^2),\dots, N((A-\lambda I)^k),\dots\,.$$

If there exists $m\geqslant 0$ such that $N((A-\lambda I)^m)=N((A-\lambda I)^{m+j})$ for any $j\in\mathbb{N}$, then the subspace $N((A-\lambda I)^m)$ is called the *root subspace* corresponding to the value $\lambda$, and the integer $m$ is called the *rank* of $\lambda$.

Let $P$ be a bounded projection on $\mathfrak{X}$, i.e., a bounded linear operator with the property $P^2=P$, and assume in addition that $P$ and $A$ commute: $PA=AP$. Then the subspaces $\mathfrak{X}_1=P\mathfrak{X}$ and $\mathfrak{X}_2=(I-P)\mathfrak{X}$ are invariant for $A$, i.e., $A\mathfrak{X}_1\subseteq\mathfrak{X}_1$ and $A\mathfrak{X}_2\subseteq\mathfrak{X}_2$. Indeed, setting $P_1=P$ and $P_2=I-P$, for any $x_i\in\mathfrak{X}_i$, we have

$$Ax_i=AP_ix_i=P_iAx_i\in\mathfrak{X}_i,\qquad i=1,2.$$

Assume that the spectrum $\sigma(A)$ of the operator $A\in L(\mathfrak{X})$ is represented as the union of two disjoint closed subsets $\sigma_1$ and $\sigma_2$. Then there exists a rectifiable, simple (i.e., without self-intersections) and closed curve $\gamma$ such that $\sigma_1$ is contained inside the curve $\gamma$ and $\sigma_2$ lies outside. In this case the next result is true.

**THEOREM 1.5.** *Suppose that the spectrum* $\sigma(A)$ *of the operator* $A$ *admits a partition (as described above) into two sets* $\sigma_1$ *and* $\sigma_2$*. Then the space* $\mathfrak{X}$ *decomposes as a* *topological direct sum $\mathfrak{X}_1 \dot{+} \mathfrak{X}_2$ of two Banach spaces $\mathfrak{X}_1$ and $\mathfrak{X}_2$ which are invariant for $A$, and such that the spectrum of the restriction $A \mid \mathfrak{X}_i$ coincides with $\sigma_i$, $i = 1, 2$.*

◁ Set

$$P_1 = -\frac{1}{2\pi i}\int_\gamma \Gamma(\xi)d\xi. \tag{1.2}$$

The operator $P_1$ is bounded and $P_1^2 = P_1$; indeed:

$$P_1^2 = \frac{1}{(2\pi i)^2}\int_\gamma\int_\gamma \Gamma(\xi)\Gamma(\zeta)d\xi d\zeta = \left(\frac{1}{2\pi i}\right)^2 \int_\gamma\int_\gamma (\xi-\zeta)^{-1}\left(\Gamma(\xi)-\Gamma(\zeta)\right)d\xi d\zeta =$$
$$= -\frac{1}{2\pi i}\int_\gamma \Gamma(\xi)d\xi.$$

(The computations above use the Hilbert identity and the Cauchy Integral Theorem). Therefore $P_1$ is a projection, and so is $P_2 = I - P_1$. Thus $\mathfrak{X} = \mathfrak{X}_1 \dot{+} \mathfrak{X}_2$ where $\mathfrak{X}_i = P_i\mathfrak{X}$, $i = 1, 2$. Further, for each $\lambda \in \rho(A)$, we have

$$P_1\Gamma(\lambda) = \Gamma(\lambda)P_1,$$

hence $AP_1 = P_1A$. Therefore, the subspaces $\mathfrak{X}_1$ and $\mathfrak{X}_2$ are invariant for $A$. Set $A_i = A \mid \mathfrak{X}_i$, $i = 1, 2$. One easily checks that $\Gamma_{A_i}(\lambda) = \Gamma(\lambda) \mid \mathfrak{X}_i$. It follows that $\rho(A_i) \supseteq \rho(A)$. In addition, $\rho(A_1) \supseteq \sigma_2$. Indeed, $\Gamma_{A_1}(\lambda)u = \Gamma(\lambda)u = \Gamma(\lambda)P_1u$ for all $u \in \mathfrak{X}_1$, $\lambda \in \rho(A)$. For any $\lambda \notin \gamma$, using (1.2) and Hilbert identity (1.1), we have

$$\Gamma(\lambda)P_1 = -\frac{1}{2\pi i}\int_\gamma \Gamma(\lambda)\Gamma(\xi)d\xi = -\frac{1}{2\pi i}\int_\gamma \left(\Gamma(\lambda)-\Gamma(\xi)\right)\frac{d\xi}{\lambda-\xi}. \tag{1.3}$$

If $\lambda$ lies outside the contour $\gamma$, then

$$\Gamma(\lambda)P_1 = \frac{1}{2\pi i}\int_\gamma \frac{1}{\lambda-\xi}\Gamma(\xi)d\xi. \tag{1.4}$$

Since the right hand side of (1.4) is holomorphic outside the contour $\gamma$, then $\Gamma(\lambda)P_1$, and consequently $\Gamma_{A_1}(\lambda)$, has an analytic continuation, holomorphic outside $\gamma$. This analytic continuation is the resolvent of $A_1$. Therefore $\rho(A_1)$ contains the exterior of the contour $\gamma$, so $\sigma(A_1) \subseteq \sigma_1$.

Analogously, from (1.3) it follows that

$$\Gamma(\lambda)P_1 = \Gamma(\lambda) + \frac{1}{2\pi i}\int_\gamma \Gamma(\xi)\frac{d\xi}{\lambda-\xi}$$

if $\lambda$ lies inside the contour $\gamma$. Thus $\Gamma(\lambda)(I - P_1)$ $(= \Gamma(\lambda)P_2)$ has an analytic continuation, holomorphic inside $\gamma$. As above, one concludes that $\sigma(A_2) \subseteq \sigma_2$.

On the other hand, a point $\lambda \in \sigma(A)$ does not belong to $\rho(A_1) \cap \rho(A_2)$. Indeed, if it does, then $\lambda \in \rho(A)$ since the operator $\Gamma_{A_1}(\lambda)P_1 + \Gamma_{A_2}(\lambda)P_2$ is an inverse of $A - \lambda I$. It follows that $\sigma(A_i) = \sigma_i$, $i = 1, 2$. ▶

## §2. Spectral radius

A useful tool in studying the spectrum of an operator $A \in L(\mathfrak{X})$ is the Neumann series

$$\sum_{i=0}^{\infty} \lambda^{-i+1} A^i, \qquad \text{where } \lambda \in \mathbb{C},\ \lambda \neq 0. \tag{2.1}$$

The convergence radius of this power series with operatorial coefficients is given by the formula

$$r(A) = \lim_{n\to\infty} \sqrt[n]{\|A^n\|}. \tag{2.2}$$

The existence of the limit in (2.2) is based on the next arguments.

◁ Set $\inf\limits_{n\geqslant 1} \|A^n\|^{\frac{1}{n}} = r$. Let us show that $\overline{\lim\limits_{n\to\infty}} \|A^n\|^{\frac{1}{n}} \leqslant r$, where the symbol $\overline{\lim}$ denotes the upper limit. For each $\varepsilon > 0$ there exists $m \in \mathbb{N}$ such that $\|A^m\|^{\frac{1}{m}} \leqslant r + \varepsilon$. For an arbitrary $n \in \mathbb{N}$ we have $n = pm + q$ where $0 \leqslant q \leqslant m - 1$, $p \in \mathbb{N}$. Then

$$\|A^n\|^{\frac{1}{n}} \leqslant \|A^m\|^{\frac{p}{n}} \|A\|^{\frac{q}{n}} \leqslant (r + \varepsilon)^{\frac{mp}{n}} \|A\|^{\frac{1}{n}}.$$

It is easy to see that $pmn^{-1} \to 1$ and $qn^{-1} \to 0$ as $n \to \infty$ ($m$ is fixed), therefore $\overline{\lim\limits_{n\to\infty}} \|A^n\|^{\frac{1}{n}} \leqslant r + \varepsilon$. Since $\varepsilon > 0$ is an arbitrary positive number, we have $\overline{\lim\limits_{n\to\infty}} \|A^n\|^{\frac{1}{n}} \leqslant r$. Thus we proved the existence of the limit

$$\lim_{n\to\infty} \sqrt[n]{\|A^n\|} = \inf_{n\geqslant 1} \sqrt[n]{\|A^n\|} = r(A).$$ ▶

**THEOREM 2.1.** *For any $A \in L(\mathfrak{X})$, we have*

$$r(A) = \sup_{\lambda\in\sigma(A)} |\lambda|. \tag{2.3}$$

◁ By the above arguments, $r(A) \geqslant \sup\limits_{\lambda\in\sigma(A)} |\lambda|$. Let us show that $r(A) \leqslant \sup\limits_{\lambda\in\sigma(A)} |\lambda|$. Set $r = \sup\limits_{\lambda\in\sigma(A)} |\lambda|$. By Theorem 1.4, the resolvent $\Gamma(\lambda)$ is analytic outside the disk of radius $r$, so it expands there in a Laurent series convergent with respect to the operatorial norm. Using the uniqueness theorem, this Laurent series coincides with the Neumann series. Consequently, $\lim_{n\to\infty} \|\lambda^{-n}A^n\| = 0$ for $|\lambda| > r$, hence, for any $\varepsilon > 0$, there exists $m$ $(= m(\varepsilon)) \in \mathbb{N}$ such that $\|A^n\| \leqslant (\varepsilon + r)^n$ for all $n \geqslant m$. It follows that

$$r(A) = \lim_{n\to\infty} \|A^n\|^{\frac{1}{n}} \leqslant r.$$ ▶

**COROLLARY 2.1.** *If $|\lambda| < r(A)$, then the Neumann series diverges.*

For $|\lambda| > r(A)$ the Neumann series is convergent in the operatorial norm and $-\sum_{i=0}^{\infty} \lambda^{1-i}A^i = \Gamma(\lambda)$. By Theorem 2.1, Corollary 1.1 and Corollary 1.2, a point from the boundary of the disk of radius $r(A)$ belongs to the spectrum of $A$. Therefore, the number $r(A)$ is called the *spectral radius* of the operator $A$, since $r(A)$ is the radius of the smallest closed disk centered at the origin of the complex plane $\mathbb{C}$, which contains the spectrum of the operator $A$. From the inequality $\|A^n\| \leqslant \|A\|^n$, $n \in \mathbb{N}$, it follows that $r(A) \leqslant \|A\|$.

It is not difficult to find an example of an operator $A$ for which $r(A) < \|A\|$.

**EXAMPLE 2.1.** Let $\mathfrak{X}$ be a two-dimensional real or complex space with a basis $\{e_1, e_2\}$, $\|e_1\| = \|e_2\| = 1$, and let $A$ be the operator given by the matrix

$$A = \begin{pmatrix} 0 & 1 \\ 0 & 0 \end{pmatrix}$$

with respect to this basis. Since $A^n = 0$ for $n \geqslant 2$, we obtain that $r(A) = 0$, while $\|A\| \geqslant 1$.

The next related result is interesting.

**THEOREM 2.2** (Ya. B. Rutitzky). *Let $A \in L(\mathfrak{X})$. Then for any $\varepsilon > 0$ there exists a norm $\|\cdot\|_\varepsilon$ on $\mathfrak{X}$, equivalent with the initial norm, such that*

$$r(A) \leqslant \|A\|_\varepsilon \leqslant r(A) + \varepsilon. \tag{2.4}$$

◁ Choose $n$ such that $\|A^n\|^{\frac{1}{n}} < r(A) + \varepsilon$, set $r = r(A)$, and let

$$\|x\|_\varepsilon = (r+\varepsilon)^{n-1}\|x\| + (r+\varepsilon)^{n-2}\|Ax\| + \cdots + \|A^{n-1}x\|.$$

From the inequalities

$$(r+\varepsilon)^{n-1}\|x\| \leqslant \|x\|_\varepsilon \leqslant \left((r+\varepsilon)^{n-1} + (r+\varepsilon)^{n-2}\|A\| + \cdots + \|A^{n-1}\|\right)\|x\|$$

it follows that the norms $\|\cdot\|$ and $\|\cdot\|_\varepsilon$ are equivalent on $\mathfrak{X}$. Further,

$$\|A\|_\varepsilon = \sup_{\|x\|_\varepsilon=1} \|Ax\|_\varepsilon = \sup_{\|x\|_\varepsilon=1} \left((r+\varepsilon)^{n-1}\|Ax\| + (r+\varepsilon)^{n-2}\|A^2x\| + \cdots + \|A^n x\|\right).$$

According to the choice of the integer $n$, it follows that $\|A^n x\| \leqslant (r+\varepsilon)^n\|x\|$. Hence, after a rearrangement of the terms, we obtain:

$$\|A\|_\varepsilon = (r+\varepsilon)\sup_{\|x\|_\varepsilon=1}\left((r+\varepsilon)^{n-1}\|x\| + (r+\varepsilon)^{n-2}\|Ax\| + \cdots + \|A^{n-1}x\|\right) =$$
$$= (r+\varepsilon)\sup_{\|x\|_\varepsilon=1}\|x\|_\varepsilon = (r+\varepsilon)$$

which proves the right hand side part of inequality (2.4).

By (2.3), $r(A)$ does not change if we replace the norm $\|\cdot\|$ on $\mathfrak{X}$ with the equivalent norm $\|\cdot\|_\varepsilon$. Setting $\|A^n\|_\varepsilon$ instead of $\|A^n\|$ in (2.3), we obtain that $r(A) \leqslant \|A\|_\varepsilon$. ▶

We conclude this paragraph by considering a family (called an *analytic operatorial sheaf* — see, for instance, A. S. Markus [1]) of linear bounded operators $A(\lambda)$ on $\mathfrak{X}$, which depends analytically on a parameter $\lambda$. Our goal is to study $r(A(\lambda)) = r(\lambda)$ as a function of the parameter $\lambda$. The next result is true.

**THEOREM 2.3** (E. Vesentini [1]). *Let $A(\lambda)$ be an analytic operatorial sheaf defined on a domain $\mathfrak{D} \subseteq \mathbb{C}$. Then the spectral radius $r(\lambda)$ is a subharmonic function.*

◁ Using Cauchy Integral Formula, we have

$$\langle A(\lambda), f\rangle = \frac{1}{2\pi \mathrm{i}}\int_0^{2\pi}\langle A(\lambda + \mathrm{e}^{\mathrm{i}\varphi}r, f\rangle\,\mathrm{d}\varphi,$$

for any $f \in L(\mathfrak{X})^*$, $\lambda \in \mathfrak{D}$ and any circle of radius $r$ centered at $\lambda$, entirely lying in $\mathfrak{D}$. Hence

$$\|A(\lambda)\| = \sup_{\|f\|=1}|\langle A(\lambda), f\rangle| \leqslant \frac{1}{2\pi}\int_0^{2\pi}\|A\langle\lambda + \mathrm{e}^{\mathrm{i}\varphi}r, f\rangle\|\,\mathrm{d}\varphi,$$

i.e., $\|A(\lambda)\|$ is a subharmonic function. Since for any $n \in \mathbb{N}$ the operatorial sheaf $A^n(\lambda)$ is also analytic, the function $\|A^n(\lambda)\|$ is subharmonic too. Therefore

$$\|A^n(\lambda)\|^{\frac{1}{n}} \leqslant \left(\frac{1}{2\pi}\int_0^{2\pi}\|A^n(\lambda + \mathrm{e}^{\mathrm{i}\varphi}r)\|\,\mathrm{d}\varphi\right)^{\frac{1}{n}} \leqslant \frac{1}{2\pi}\int_0^{2\pi}\|A^n(\lambda + \mathrm{e}^{\mathrm{i}\varphi}r)\|^{\frac{1}{n}}\,\mathrm{d}\varphi.$$

Taking the limit for $n \to \infty$ in the last inequality, we obtain the required conclusion. ▶

**COROLLARY 2.2** (Maximum Principle). *The spectral radius of an analytic operatorial sheaf attains its maximum only on the boundary of the domain* $\mathfrak{D}$.

## §3. Resolvent and spectrum of the adjoint operator

Let $A$ be a linear operator densely defined on the Banach space $\mathfrak{X}$. Then (see §9 of Chapter 0) there exists the conjugate operator $A^*$ on $\mathfrak{X}^*$.

◁ Let $\langle \cdot, \cdot \rangle$ be the canonical scalar product on $\mathfrak{X} \times \mathfrak{X}^*$ (see §9 of Chapter 0). If $A^* f = 0$ for a certain $f \in D(A^*)$ then $0 = \langle x, A^* f \rangle = \langle Ax, f \rangle$ for all $x \in D(A)$, hence $f \in R(A)^{\perp}$. Therefore, from $\overline{R(A)} = \mathfrak{X}$ it follows that $f = 0$. On the other hand, if $y \notin \overline{R(A)}$, then by Theorem 0.8.7 there exists a functional $f \in \mathfrak{X}^*$ which satisfies the conditions $f(y) = 1$ and $f \mid \overline{R(A)} = 0$. Consequently, $\langle Ax, f \rangle = 0$ for all $x \in D(A)$ so $f \in D(A^*)$ and $A^* f = 0$.

Thus we proved the next result.

**LEMMA 3.1.** *If $A$ is a linear operator on $\mathfrak{X}$ with dense domain, then the operator $(A^*)^{-1}$ exists if and only if $\overline{R(A)} = \mathfrak{X}$.* ▶

**THEOREM 3.1.** *Assume that the operator $A$ is invertible and that $\mathfrak{X} = \overline{D(A)} = \overline{R(A)}$. Then $(A^*)^{-1} = (A^{-1})^*$. In this case, the operator $A^{-1}$ is defined and bounded on the whole $\mathfrak{X}$if and only if the operator $A$ is closed and $(A^*)^{-1}$ is defined and bounded on the whole $\mathfrak{X}^*$.*

◁ The operator $(A^*)^{-1}$ exists by Lemma 3.1. The existence of the operator $(A^{-1})^*$ follows from $\overline{D(A^{-1})} = \overline{R(A)}$. If $y \in R(A)$ and $f \in D(A^*)$, then

$$\langle y, f \rangle = \langle AA^{-1}y, f \rangle = \langle A^{-1}y, A^* f \rangle,$$

so $R(A^*) \subseteq D((A^{-1})^*)$ and $(A^{-1})^* A^* f = f$ for all $f \in D(A^*)$. Therefore $R((A^{-1})^*) \subseteq D(A^*)$ and $A^*(A^{-1})^* f = f$ for all $f \in D((A^{-1})^*)$. Consequently $(A^{-1})^* \subseteq (A^*)^{-1}$, and, due to the opposite inclusion, $(A^*)^{-1} = (A^{-1})^*$.

If, in addition, the operator $A^{-1}$ is defined on the whole $\mathfrak{X}$ and is bounded, then $(A^{-1})^*$ is also bounded, hence $(A^*)^{-1}$ is bounded. By Theorem 1.1 the operator $A^{-1}$ is closed and therefore $(A^*)^{-1}$ is closed, too. Conversely, if the operator $(A^*)^{-1}$ is defined and bounded on the whole $\mathfrak{X}^*$, and if that the operator $A$ is closed, then we have

$$\left|\langle A^{-1}x, f\rangle\right| = \left|\langle x, (A^{-1})^* f\rangle\right| = \left|\langle x, (A^*)^{-1} f\rangle\right| \leqslant \|(A^*)^{-1}\| \, \|f\| \, \|x\|,$$

for all $x \in R(A)$ and $f \in \mathfrak{X}^*$, i.e., the operator $A^{-1}$ is bounded. ▶

Using these results and Theorem 1.1, we get:

**THEOREM 3.2.** *Let $A$ be a closed linear operator on $\mathfrak{X}$ with $\overline{D(A)} = \mathfrak{X}$. Then $\rho(A^*) = \{\lambda \in \mathbb{C} : \overline{\lambda} \in \rho(A)\}$ and $\Gamma_{A^*}(\lambda) = \Gamma_A(\overline{\lambda})^*$.*

◁ Let $\lambda \in \rho(A)$, i.e., the operator $(A - \lambda I)^{-1}$ is defined on the whole $\mathfrak{X}$ and is bounded. By Theorem 3.1, the operator $((A - \lambda I)^*)^{-1}$ is defined on the whole $\mathfrak{X}^*$. But $(A - \lambda I)^* = A^* - \overline{\lambda} I$, and therefore $\overline{\lambda} \in \rho(A^*)$. Conversely, let $\lambda \in \rho(A^*)$, i.e., the operator $(A^* - \lambda I)^{-1}$ is defined on the whole $\mathfrak{X}^*$ and is bounded. Let us show that $(A - \overline{\lambda} I)^{-1}$ exists. If $(A - \overline{\lambda} I)x_0 = 0$ for some $x_0 \neq 0$, then

$$\langle x_0, (A^* - \lambda I)f\rangle = \langle (A - \overline{\lambda} I)x_0, f\rangle = 0,$$

for all $f \in D(A^* - \lambda I)$. But $R(A^* - \lambda I) = \mathfrak{X}$. Thus, $x_0 \in {}^{\perp}\mathfrak{X}^* = \{0\}$, so the operator $A - \overline{\lambda} I$ has an inverse $(A - \overline{\lambda} I)^{-1}$ and both these operators are closed. By Lemma 3.1 we obtain that $\overline{D\left((A - \overline{\lambda} I)^{-1}\right)} \left(= \overline{R(A - \overline{\lambda} I)}\right) = \mathfrak{X}$. This equality and Theorem 3.1 imply that $\overline{\lambda} \in \rho(A)$. ▶

Using de Morgan formulas, we obtain:

**COROLLARY 3.1.** *Under the conditions of Theorem 3.2, the spectrum of the adjoint operator $A^*$ is the complex-conjugate of the spectrum of the operator $A$, i.e.,*

$$\sigma(A^*) = \{\lambda \in \mathbb{C} : \overline{\lambda} \in \sigma(A)\}.$$

Let us consider now an operator acting on a Hilbert space $\mathfrak{X} = \mathfrak{H}$. In this case, the conjugate operator acts on $\mathfrak{H}$, too, so the notion of a selfadjoint operator, i.e., a linear operator $A$ with $\overline{D(A)} = \mathfrak{H}$, satisfying $A^* = A$, makes sense. Then $\langle Ax, y\rangle = \langle x, Ay\rangle$ for all $x, y \in D(A)$ $(= D(A^*))$. By Corollary 3.1, the spectrum of a selfadjoint operator $A$ is real, i.e., $\sigma(A) \subseteq \mathbb{R}$, and its resolvent has the next property:

**THEOREM 3.3.** *Let $A$ be a selfadjoint operator and $\lambda = \alpha + \mathrm{i}\beta \in \mathbb{C}$, $\beta \neq 0$. Then*

$$\|\Gamma_A(\lambda)\| \leqslant \frac{1}{|\beta|}. \tag{3.1}$$

◁ We have

$$\|(A - \lambda I)x\|^2 = \|(A - \alpha I)x - \mathrm{i}\beta x\|^2 = \|(A - \alpha I)x\|^2 + |\beta|^2 \|x^2\| \geqslant |\beta|^2 \|x\|^2,$$

for all $x \in D(A)$, and thus (3.1) follows. ▶

A remarkable property of selfadjoint operators is the existence of their spectral decomposition. We state the theorem concerning the spectral decomposition without any proof. First, we need a definition. Let $B$ and $C$ be two bounded selfadjoint operators on $\mathfrak{H}$. We shall write $B \leqslant C$ if $\langle Bx, x\rangle \leqslant \langle Cx, x\rangle$ for all $x \in \mathfrak{X}$.

**THEOREM 3.4.** *Let $A$ be a selfadjoint operator on $\mathfrak{H}$ with the domain $D(A)$. Then there exists a family of orthogonal projections (see §11 of Chapter 0)$E_\lambda$, $-\infty < \lambda < < \infty$, satisfying the properties:*

*1) $E_\lambda \leqslant E_\mu$ for any $\lambda, \mu$ with $\lambda < \mu$;*

*2) $\{E_\lambda\}$ is left hand side strongly continuous, i.e., $E_{\lambda-\alpha}x \to E_\lambda x$ as $\alpha \to 0$, $\alpha > 0$, for any $x \in \mathfrak{H}$;*

*3) $E_{-\infty} = \lim_{\lambda\to-\infty} E_\lambda = 0$, $E_{+\infty} = \lim_{\lambda\to\infty} E_\lambda = I$;*

*4) $BE_\lambda = E_\lambda B$ for any bounded operator $B$ which commutes with $A$. (Here, a bounded operator $B$ on $\mathfrak{H}$ is said to commute with $A$ if $x \in D(A)$ implies $Bx \in D(A)$ and $ABx = BAx$.)*

*An element $x$ belongs to $D(A)$ if and only if*

$$\int_{-\infty}^{\infty} \lambda^2 \mathrm{d}\langle E_\lambda x, x\rangle < \infty.$$

*For such an element $x$, we have*

$$Ax = \int_{-\infty}^{\infty} \lambda \mathrm{d}E_\lambda x \quad \text{and} \quad \|Ax\|^2 = \int_{-\infty}^{\infty} \lambda^2 \mathrm{d}\langle E_\lambda x, x\rangle.$$

The integral $\int_{-\infty}^{\infty} \lambda \mathrm{d}E_\lambda x$ above is understood as the limit of the integral $\int_a^b \lambda \mathrm{d}E_\lambda x$, for $a \to -\infty$ and $b \to +\infty$.

The family $\{E_\lambda\}$ mentioned in Theorem 3.4 is called the *spectral decomposition* of the selfadjoint operator $A$.

## §4. The spectrum of a completely continuous operator

Let $\mathfrak{X}$ be a complex Banach space. Consider a completely continuous linear operator $A$ defined everywhere on $\mathfrak{X}$. According to Theorem 0.10.6, the complete continuity of $A$ is equivalent with the precompactness of the image $A\mathfrak{S}$ of the unit ball $\mathfrak{S}$ of $\mathfrak{X}$ ($=\{x \in \mathfrak{X};\ \|x\| \leqslant 1\}$). As we noticed in §1 of Chapter 0, any linear operator $A$ acting on a finite dimensional space is completely continuous. Its spectrum $\sigma(A)$ coincides with the point spectrum $\sigma_{\mathrm{p}}(A) = \sigma_1(A)$, and contains a finite number ($\leqslant$ the dimension of the space) of eigenvalues, each of these, evidently, of finite multiplicity. The spectrum of a completely continuous operator on an infinite dimensional space has a very similar structure to the one of the spectrum of an operator on a finite dimensional space. First, we establish the next preliminary result:

**THEOREM 4.1.** *The adjoint of a completely continuous operator is completely continuous.*

$\lhd$ By Theorem 0.10.7, the range of a completely continuous operator $A$ is separable. Let $\{f_n\}_{n\in\mathbb{N}}$ be a bounded sequence in $\mathfrak{X}^*$. We show that there exists a subsequence $\{f_{n_j}\}_{j\in\mathbb{N}}$ such that $\{A^*f_{n_j}\}_{j\in\mathbb{N}}$ is a Cauchy sequence in $\mathfrak{X}^*$. Let $\{v_k\}_{k\in\mathbb{N}}$ be a dense sequence in $R(A)$. For each $k$, the scalar sequence $\{\langle v_k, f_n\rangle\}_{n\in\mathbb{N}}$ is bounded. Therefore, using the diagonal method, we can choose a subsequence $\{\langle v_k, f_{n_j}\rangle\}_{j\in\mathbb{N}}$ which is Cauchy for each $k \in \mathbb{N}$. Since the sequence $\{v_k\}_{k\in\mathbb{N}}$ is dense in $R(A)$, then $\{\langle v, f_{n_j}\rangle\}_{j\in\mathbb{N}}$ is a Cauchy sequence for any $v \in R(A)$.

The functional $\lim\limits_{j\to\infty} \langle v, f_{n_j}\rangle$ is linear in $v$ on $R(A)$. By Theorem 0.8.7, there exists an extension of this functional on the whole $\mathfrak{X}$, with the same norm. Denote this extension by $f$. Then, for all $v \in R(A)$, we have $\langle v, f_{n_j}\rangle \to \langle v, f\rangle$, i.e.,

$$\langle Ax, f_{n_j}\rangle \to \langle Ax, f\rangle, \tag{4.1}$$

for all $x \in \mathfrak{X}$.

Let us show that $A^*f_{n_j} \to A^*f$ as $j \to \infty$. Assume that this is not the case. Then there exists a subsequence $\{h_\ell\}_{\ell\in\mathbb{N}}$ of the sequence $\{f_{n_j} - f\}_{j\in\mathbb{N}}$, and a number $\delta$, such that $\|A^*h_\ell\| \geqslant \delta > 0$, for all $\ell \in \mathbb{N}$. Since $\|A^*h_\ell\| = \sup\limits_{\|x\|=1} |\langle x, A^*h_\ell\rangle|$, for each $\ell \in \mathbb{N}$ there exists a vector $x_\ell \in \mathfrak{X}$, $\|x_\ell\| = 1$, such that $|\langle Ax_\ell, h_\ell\rangle| \geqslant \delta/2$. From the complete continuity of $A$ it follows that $\{Ax_\ell\}_{\ell\in\mathbb{N}}$ contains a Cauchy subsequence, so we may assume that $\{Ax_\ell\}_{\ell\in\mathbb{N}}$ is a Cauchy sequence itself, i.e., for any $\varepsilon > 0$ there exists $m = m(\varepsilon)$ such that $\|Ax_\ell - Ax_s\| < \varepsilon$ for $\ell, s > m$. Then

$$\frac{\delta}{2} \leqslant |\langle Ax_\ell, Ax_s\rangle| \leqslant M\varepsilon + |\langle Ax_s, h_\ell\rangle|,$$

where $M = \sup_{\ell \in \mathbb{N}} \|h_\ell\|$. Taking the limit for $\ell \to \infty$ and a fixed $s$, and using (4.1), we have $\delta/2 \leqslant M\varepsilon$. Since $\varepsilon$ is an arbitrary positive number, it follows that $\delta = 0$, a contradiction. ▶

**THEOREM 4.2.** a) *Let $A$ and $B$ be two completely continuous linear operators on $\mathfrak{X}$. Then any linear combination of $A$ and $B$ is a completely continuous operator, too.*

b) *If $A$ is a completely continuous linear operator on $\mathfrak{X}$ and $C \in L(\mathfrak{X})$ then $AC$ and $CA$ are completely continuous operators.*

c) *If $A = \lim_{n\to\infty} A_n$ in the norm topology of $L(\mathfrak{X})$ and $A_n$ are completely continuous linear operators on $\mathfrak{X}$, then $A$ is completely continuous.*

◁ a) The assertion is obvious.

b) Since a bounded linear operator transforms bounded sets into bounded sets, the complete continuity of $CA$ and $AC$ follows from the definition of completely continuous operators.

c) Let $A_n \in L(\mathfrak{X})$, $n \in \mathbb{N}$, and $A_n \to A$. If $\{x_j\}_{j\in\mathbb{N}}$ is a bounded sequence in $\mathfrak{X}$ then each sequence $\{A_n x_j\}_{j\in\mathbb{N}}$ is precompact. Using the diagonal method we get the precompactness of the sequence $\{Ax_j\}_{j\in\mathbb{N}}$. ▶

The main theorem of this section deals with the structure of the spectrum of a completely continuous operator. In order to prove it, we need a preliminary result.

**LEMMA 4.1.** *A bounded projection on a Banach space is a finite-dimensional operator if and only if it is completely continuous.*

◁ As we noticed in §10 of Chapter 0, any bounded linear finite-dimensional operator is completely continuous. Conversely, let $P$ be a completely continuous projection on the Banach space $\mathfrak{X}$.

Let $\dim P\mathfrak{X} = \infty$. We define recurrently a sequence $\{v_n\}_{n\in\mathbb{N}}$ such that $v_n \in P\mathfrak{X}$ and $\|v_n\| = 1 \geqslant \|v_n - v_m\|$ for $n \neq m$. Assume that the vectors $v_1, \dots, v_k$ are already defined and set $\mathfrak{M} = \operatorname{clin}\{v_1, \dots, v_k\}$. Then, there exists a vector $u \in P\mathfrak{X}$ such that $u \notin \mathfrak{M}$. Let $\mathfrak{M}_u = \operatorname{clin}\{v_1, \dots, v_k, u\}$; obviuosly we have $0 < d = \inf_{x\in\mathfrak{M}} \|u - x\|$. Choose a sequence $\{x_n\}_{n\in\mathbb{N}}$ such that $x_n \in \mathfrak{M}$ and $d = \lim_{n\to\infty} \|u - x_n\|$. Since the space $\mathfrak{M}$ is finite-dimensional, the sequence $\{x_n\}_{n\in\mathbb{N}}$ is bounded and, therefore, precompact. We may assume that $x_n \to z \in \mathfrak{M}$ as $n \to \infty$; then $d = \|u - z\|$. Put $v_{k+1} = d^{-1}(u - z)$. Clearly $\|v_{k+1}\| = 1$ and

$$\inf_{x\in\mathfrak{M}} \|v_{k+1} - x\| = \inf_{x\in\mathfrak{M}} \left\| \frac{u - z - dx}{d} \right\| = \frac{1}{d}\|u - z\| = 1.$$

Further

$$\|v_{k+1} - v_k\| \geqslant \inf_{x \in \mathfrak{M}} \|v_{k+1} - x\| \geqslant 1,$$

for $m = 1, \ldots, k$. Thus there exists a sequence $\{v_n\}_{n \in \mathbb{N}}$ satisfying all the above mentioned properties. Clearly none of its subsequences is convergent. On the other hand $Pv_n = v_n$, $n \in \mathbb{N}$, which contradicts the complete continuity of $P$. Consequently, $\dim P\mathfrak{X} < \infty$. ▶

**THEOREM 4.3.** *Let $A \in L(\mathfrak{X})$ be a completely continuous operator. Then $\sigma(A)$ is an at most countable set, with no limit points different from zero. Each $\lambda \in \sigma(A)$, $\lambda \neq 0$, is an eigenvalue of $A$ of a finite multiplicity, and $\overline{\lambda}$ is an eigenvalue of $A^*$ of the same multiplicity.*

◁ I. Let $\lambda$ be a limit point for the set of eigenvalues of the operator $A$. If we assume that $\lambda \neq 0$, then there exists a sequence $\{\lambda_n\}_{n \in \mathbb{N}}$ of eigenvalues of $A$ such that $\lambda_n \neq 0$ and $\lambda_n \to \lambda$. Let $u_n$ be an eigenvector of $A$ corresponding to the eigenvalue $\lambda_n$, and let $\mathfrak{M}_n = \operatorname{clin}\{u_1, \ldots, u_n\}$. Since the vectors $u_1, \ldots, u_n, \ldots$ are linearly independent, the subspace $\mathfrak{M}_{n-1}$ is a proper subspace of $\mathfrak{M}_n$ and $\dim \mathfrak{M}_{n-1} = \dim \mathfrak{M}_n - 1$. Therefore (see the proof of Lemma 4.1) there exists a vector $v_n \in \mathfrak{M}_n$ with $\|v_n\| = 1$ and such that $d_n = \inf\limits_{x \in \mathfrak{M}_{n-1}} \|v_n - x\| = 1$. Thus we obtain a sequence $\{v_n\}_{n \in \mathbb{N}}$ with $v_n \in \mathfrak{M}_n$, $\|v_n\| = 1$ and $d_n = 1$. Let us show that the sequence $\{\lambda_n^{-1} A v_n\}_{n \in \mathbb{N}}$ contains no Cauchy subsequences. Indeed, for $m < n$ we have

$$\lambda_n^{-1} A v_n - \lambda_m^{-1} A v_m = v_n - \left(\lambda_m^{-1} A v_m - \lambda_n^{-1} (A - \lambda_n I) v_n\right).$$

We can easily prove that the second term in the right hand side of the above equality belongs to $\mathfrak{M}_{n-1}$. Therefore the norm of the right hand side vector is greater than or equal to 1. This means that the distance between any two terms of the sequence $\{\lambda_n^{-1} A v_n\}_{n \in \mathbb{N}}$ is no less than 1. Consequently, none of the subsequences of this sequence is convergent. This contradicts the complete continuity of $A$, since the sequence $\|\lambda_n^{-1} v_n\|$ is bounded: $\|\lambda_n^{-1} v_n\| \leqslant \left(\inf\limits_{m \in \mathbb{N}} |\lambda_m|\right)^{-1} < \infty$.

Thus $\lambda = 0$ is the only possible limit point for the set of eigenvalues of $A$.

II. Let us prove now that the lineal $R(A - \lambda I)$ is closed if $\lambda \neq 0$ and $\lambda \notin \sigma_{\mathrm{p}}(A)$. Assume that $(A - \lambda I) x_n \to y$ for $n \to \infty$. We will show first that $\{x_n\}_{n \in \mathbb{N}}$ is bounded. If this is not so, then we may suppose (choosing eventually a subsequence) that $\|x_n\| \to \infty$ as $n \to \infty$. Set $u_n = \dfrac{x_n}{\|x_n\|}$. Then $\|u_n\| = 1$ and $(A - \lambda I) u_n \to 0$. Due to the complete continuity of $A$ we may assume that $A u_n \to u$, $u \neq 0$. Then $Au = \lambda u$, which contradicts our assumption. Thus $\|x_n\| \leqslant c < \infty$, and we can suppose that $A x_n \to x$. Then $x_n \to \lambda^{-1}(y - x)$, hence $y = \lambda^{-1}(A - \lambda I)(y - x) \in R(A - \lambda I)$.

III. Consider the set $\sigma'$ of all points $\lambda \in \mathbb{C}$ such that either $\lambda \in \sigma_{\mathrm{p}}(A)$ or $\overline{\lambda} \in \sigma_{\mathrm{p}}(A^*)$. By Theorem 3.2, $\sigma' \subseteq \sigma(A)$. On the other hand $\mathbb{C} \setminus \sigma' \subseteq \rho(A)$. Indeed, since $R(A - \lambda I)$ is closed for all $\lambda \in \mathbb{C} \setminus \sigma'$ it is enough to notice that $N(A^* - \lambda I) = R(A - \lambda I)^{\perp}$. So $\sigma' = \sigma(A)$. It remains to show that the eigenspace corresponding to $\lambda \in \sigma(A)$ is finite-dimensional. By Lemma 4.1 it is sufficient to establish the complete continuity of the projection $P$ onto this subspace.

As in Theorem 1.5, consider the projection $P = -(2\pi \mathrm{i})^{-1} \int_{\gamma} \Gamma(\xi) \mathrm{d}\xi$, where $\gamma$ is a circle centered at $\lambda$ and with a sufficiently small radius such that 0 is lying outside this circle.

We have $A\Gamma(\xi) = I + \xi\Gamma(\xi)$, hence $\xi^{-1} A\Gamma(\xi) = \xi^{-1} I + \Gamma(\xi)$. Since $\int_{\gamma} \xi^{-1} \mathrm{d}\xi = 0$, then $P = -(2\pi \mathrm{i})^{-1} \int_{\gamma} \xi^{-1} A\Gamma(\xi) \mathrm{d}\xi$. By Theorem 3.2, the operator $A\Gamma(\xi)$ is completely continuous for any $\xi \in \gamma$. By the same theorem, the projection $P$ is completely continuous as a uniform limit of integral sums which obviuosly are completely continuous operators.

From Theorems 3.2 and 4.1 it follows that $\sigma(A^*) \setminus \{0\}$ consists of eigenvalues of $A^*$ of finite multiplicity, and, moreover, $\sigma(A^*) \setminus \{0\} = \{\overline{\lambda} : \lambda \in \sigma(A), \lambda \neq 0\}$. The range of the projection $P^*$ — the adjoint of $P$ — is the eigenspace of $A^*$ corresponding to the eigenvalue $\overline{\lambda}$. Since the projections $P$ and $P^*$ have the same rank, the multiplicities of $\lambda$ and $\overline{\lambda}$ coincide. ►

**COROLLARY 4.1** (First Fredholm Alternative). *If $A$ is a completely continuous operator on $\mathfrak{X}$ then, either $(A - I)^{-1} \in L(\mathfrak{X})$, or there exists a non-zero solution of the equation $Ax = x$.*

## §5. Normally solvable operators

Let $\mathfrak{X}$, $\mathfrak{Y}$ be Banach spaces and $A : \mathfrak{X} \to \mathfrak{Y}$ be a linear operator with $D(A) \subseteq \mathfrak{X}$, $R(A) \subseteq \mathfrak{Y}$. Let $N(A)$ (= Ker $A$) be the kernel of the operator $A$. Let us recall (see §9 of Chapter 0) that by $\mathfrak{M}^{\perp}$ and ${}^{\perp}\mathfrak{M}^*$ we denote the orthogonal and the $*$-orthogonal complement of a set $\mathfrak{M} \subseteq \mathfrak{X}$, and $\mathfrak{M}^* \subseteq \mathfrak{X}^*$, respectively.

**THEOREM 5.1.** *Assume that the operator $A$ is densely defined, i.e., $\overline{D(A)} = \mathfrak{X}$. Then the adjoint operator $A^* : \mathfrak{Y}^* \to \mathfrak{X}^*$ exists and*

$$1) \quad N(A^*) = R(A)^{\perp}, \qquad 2) \quad \overline{R(A)} = {}^{\perp}N(A^*). \tag{5.1}$$

*If, in addition, $D(A) = \mathfrak{X}$ and the operator $A^*$ is densely defined, then $A \in L(\mathfrak{X})$ and*

$$3) \quad N(A) = {}^{\perp}R(A^*), \qquad 4) \quad \overline{R(A^*)} = N(A)^{\perp}. \tag{5.2}$$

◁ If $\overline{D(A)} = \mathfrak{X}$ then (see §9 of Chapter 0) the adjoint $A^* : \mathfrak{Y}^* \to \mathfrak{X}^*$ exists and $\langle Ax, f\rangle = \langle x, A^*f\rangle$ for all $x \in D(A)$, $f \in D(A^*)$.

1) Let $f \in N(A^*)$. Then $\langle Ax, f\rangle = \langle x, A^*f\rangle = 0$ for all $x \in D(A)$, hence $f \in R(A)^{\perp}$. So $N(A^*) \subseteq R(A)^{\perp}$.

Conversely, let $f \in R(A)^{\perp}$. Then $\langle Ax, f\rangle = 0$ for all $x \in D(A)$, hence $f \in D(A^*)$. Therefore the equality $\langle Ax, f\rangle = \langle x, A^*f\rangle$ holds and, because $D(A)$ is dense in $\mathfrak{X}$, we obtain $A^*f = 0$, i.e., $f \in N(A^*)$. So $R(A)^{\perp} \subseteq N(A^*)$ and relation 1) is proved.

By Theorem 0.9.2 the equality 2) follows from 1) taking the $*$-orthogonal complements.

Assume in addition that $D(A) = \mathfrak{X}$ and $\overline{D(A^*)} = \mathfrak{Y}$. Let us show that the operator $A$ is closed. Indeed, let $\{x_n\}_{n\in\mathbb{N}} \subset \mathfrak{X}$, such that $x_n \to x$ and $Ax_n \to y$ as $n \to \infty$. For any $f \in D(A^*)$ we have $\lim_{n\to\infty} \langle Ax_n, f\rangle = \lim_{n\to\infty} \langle x_n, A^*f\rangle = \langle x, A^*f\rangle$ and $\lim_{n\to\infty} \langle Ax_n, f\rangle = \langle y, f\rangle$. So $y = Ax$ and therefore $A$ is closed. By Theorem 1.2 the operator $A$ is bounded. Consequently, its kernel $N(A)$ is closed. The adjoint operator $A^*$ is bounded, too, and everywhere defined (see §4 of Chapter 0).

We prove now equality 3). The inclusion $N(A) \subseteq {}^{\perp}R(A^*)$ can be obtained by arguments similar to those in the proof of equality 1). Let $x \in {}^{\perp}R(A^*)$. Then $\langle x, A^*f\rangle = 0$ for all $f \in \mathfrak{Y}^*$. Therefore $Ax = 0$, i.e., $x \in N(A)$, hence ${}^{\perp}R(A^*) \subseteq N(A)$.

Equality 4) follows from 3) using Theorem 0.4.8. ▶

A densely defined linear operator $A$ is called *normally solvable* if

$$R(A) = {}^{\perp}N(A^*). \tag{5.3}$$

By equality 2) in (5.1) we have:

**THEOREM 5.2** (Hausdorff). *A densely defined operator $A$ is normally solvable if and only if its range is closed, i.e.,*

$$R(A) = \overline{R(A)}.$$

**COROLLARY 5.1.** *If an operator $B \in L(\mathfrak{X})$ is completely continuous then $A = I - B$ is normally solvable.*

◁ If $1 \in \rho(B)$ then $R(A) = \mathfrak{Y}$ $(= R(I - B))$ is closed. Assume that $1 \in \sigma(B)$. Then, by Theorem 4.3, we have that 1 is an eigenvalue of the operator $B$ of a finite multiplicity, i.e., the corresponding eigenspace $\mathfrak{X}_1$ is finite-dimensional. Let $P_1$ be the projection onto $\mathfrak{X}_1$, defined by (1.2), corresponding to the obvious spliting of the spectrum: $\sigma_1 = \{1\}$ and $\sigma_2 = \sigma(B) \setminus \sigma_1$. Then $\mathfrak{X}_1 = P_1\mathfrak{X}$ and $\mathfrak{X} = P_1\mathfrak{X} + (I - P_1)\mathfrak{X}$. Setting $\mathfrak{X}_2 = (I - P_1)\mathfrak{X}$ and $B_2 = B \mid \mathfrak{X}_2$ we obtain that $1 \in \rho(B_2)$. Therefore $(I - B_2)\mathfrak{X}_2 = \mathfrak{X}_2$ and, consequently, $R(A) = (I - B_2)\mathfrak{X} = (I - B_2)\mathfrak{X}_2 = \mathfrak{X}_2$ is a closed subspace of $\mathfrak{X}$. ▶

The next result follows from (5.3).

**THEOREM 5.3.** *If an operator $A : \mathfrak{X} \to \mathfrak{Y}$ is normally solvable then the equation*

$$Ax = y \tag{5.4}$$

*has a solution, if and only if $\langle y, f\rangle = 0$ for all solutions $f$ of the equation*

$$A^*f = 0. \tag{5.5}$$

**COROLLARY 5.2.** *If equation* (5.1) *has the only solution $f = 0$, then equation* (5.4) *is solvable for any $y \in \mathfrak{Y}$.*

## §6. Noether and Fredholm operators

The following classes of operators play an important role in the theory of linear equations with normally solvable operators.

A normally solvable operator $A : \mathfrak{X} \to \mathfrak{Y}$ is called a *Noether operator*, or an N-*operator*, if both the spaces $N(A)$ and $N(A^*)$ are finite-dimensional. For a Noether operator $A$, both $n(A) = \dim N(A)$ and $m(A) = \dim N(A^*)$ are non-negative integers, called the *number of zeros* and the *defect* of the operator $A$, respectively. The integer $\chi(A) = n(A) - m(A)$ is called the *index* of the operator $A$. A Noether operator $A$ satisfying $\chi(A) = 0$, i.e., $n(A) = m(A)$, is called a *Fredholm operator*, or an F-*operator*.

Let $A : \mathfrak{X} \to \mathfrak{Y}$ be a Noether operator. If $\{g_k\}_{k=1}^m$ is a basis of $\mathfrak{N}(A^*)$ then the condition of solvability for equation (5.4) can be written as a system of equalities:

$$\langle y, g_k\rangle = 0, \quad k = 1, 2, \dots, m.$$

If, in addition, $A$ is a Fredholm operator then, using Corollary 5.2, we obtain the next result, usually referred to as the Second Fredholm Alternative.

**THEOREM 6.1.** *If $A : \mathfrak{X} \to \mathfrak{Y}$ is an F-operator then, either the equation* (5.4) *has a unique solution for any vector in the right hand side of that equation, or equation* (5.5) *has only the zero solution.*

◁ By (5.3) the solvability of the equation (5.4) for any vector in its right hand side is equivalent with the equality $\mathfrak{Y} = {}^{\perp}N(A^*)$, which leads to $N(A^*) = \{0\}$. From $\dim N(A) = \dim N(A^*)$ we obtain $N(A) = \{0\}$.

On the other hand, let us assume that equation (5.5) has only the zero solution, i.e., $N(A) = \{0\}$. Then $N(A^*) = \{0\}$. By (5.3) we obtain $R(A) = \mathfrak{Y}$, i.e., equation (5.4) has a solution for any vector $y$ in its right hand side. Since $N(A) = 0$ such a solution is unique. ▶

Suppose now that $A \in L(\mathfrak{X}, \mathfrak{Y})$. If $A$ is a Fredholm operator and, moreover, $N(A) = \{0\}$, then by Theorem 6.1 its range $R(A)$ coincides with the whole space $\mathfrak{Y}$. By the Banach Inverse Mapping Theorem, the operator $A$ is a linear homeomorphism between the spaces $\mathfrak{X}$ and $\mathfrak{Y}$. Assume, on the contrary, that $N(A) \neq \{0\}$. In this case the operator $A$ can be perturbed by a finite rank operator $K$ such that the corresponding perturbation $B = A + K$ becomes a homeomorphism between the spaces $\mathfrak{X}$ and $\mathfrak{Y}$.

The construction of such a perturbation, which is given in the sequel, was obtained for the first time by E. Schmidt [1] in the case of integral operators.

Let $A \in L(\mathfrak{X}, \mathfrak{Y})$ be a Fredholm operator, with $n = n(A) \geqslant 1$. Let $\{a_k\}_{k=1}^n$ be a basis of $N(A)$, and let $\{g_k\}_{k=1}^n$ be a basis of $\mathfrak{N}(A^*)$. By the Hahn-Banach-Suchomlynoff Theorem we can choose two collections of elements $\{f_k\}_{k=1}^n$ and $\{b_k\}_{k=1}^n$ in the spaces $\mathfrak{X}^*$ and $\mathfrak{Y}$, respectively, such that

$$\langle a_i, f_j \rangle = \langle b_i, g_j \rangle = \delta_{ij}, \quad i, j = 1, 2, \ldots, n. \tag{6.1}$$

Define a linear operator $K$ by the formula

$$Kx = \sum_{i=1}^{n} \langle x, f_i \rangle b_i, \quad x \in \mathfrak{X}. \tag{6.2}$$

The operator $K$ acts from the space $\mathfrak{X}$ onto the finite-dimensional subspace $\mathfrak{L} = \operatorname{clin}\{b_1, \ldots, b_n\} \subseteq \mathfrak{Y}$ and satisfies the conditions

$$Ka_i = b_i, \quad i = 1, \ldots, n.$$

**LEMMA 6.1** (E. Schmidt). *Let $A \in L(\mathfrak{X}, \mathfrak{Y})$ be a Fredholm operator with $n(A) \geqslant 1$. Then the operator $B = A + K$, where $K$ is defined by* (6.2), *has a bounded inverse* *operator $B^{-1}$ defined on the whole $\mathfrak{Y}$, i.e., $B$ is a homeomorphism of the spaces $\mathfrak{X}$ and $\mathfrak{Y}$.*

◁ Assume that $Bx = 0$ for some $x \in \mathfrak{X}$. Then, by (6.2), we have

$$Ax = -\sum_{i=1}^{n} \langle x, f_i \rangle b_i. \tag{6.4}$$

From (6.4) and using the equality $N(A^*) = R(A)^{\perp}$ (see Theorem 5.1) we obtain that $\langle Ax, g_k \rangle = 0$ for any $g_k \in N(A^*)$, $k = 1, \ldots, n$. Consequently

$$0 = \sum_{i=1}^{n} \langle x, f_i \rangle \langle b_i, g_k \rangle = \langle x, f_k \rangle, \quad k = 1, \ldots, n. \tag{6.5}$$

From (6.4) and (6.3) it follows that $Ax = 0$, i.e., $x \in N(A)$, and therefore $x = \sum_{i=1}^{n} \alpha_i a_i$. By (6.5) we find $\alpha_k = \langle x, f_k \rangle = 0$, hence $x = 0$. So we conclude that $N(B) = \{0\}$.

Let us prove now that $R(B) = \mathfrak{Y}$. Let $y \in \mathfrak{Y}$ be fixed. Consider the element $\tilde{y} = y - \sum_{i=1}^{n} \langle y, g_i \rangle b_i$. By (6.1) we have $\langle \tilde{y}, g_i \rangle = 0$, hence $\tilde{y} \in {}^{\perp}N(A^*) = R(A)$. This means that there exists an element $\tilde{x} \in \mathfrak{X}$ which satisfies $\tilde{y} = A\tilde{x}$. Set

$$x = \tilde{x} + \sum_{i=1}^{n} \left( \langle y, g_i \rangle - \langle \tilde{x}, f_i \rangle \right) a_i.$$

Then $Ax = A\tilde{x}$, and using (6.3) we obtain

$$Bx = Ax + Kx = A\tilde{x} + K\tilde{x} + K\left( \sum_{i=1}^{n} \left( \langle y, g_i \rangle - \langle \tilde{x}, f_i \rangle \right) a_i \right) =$$

$$= \tilde{y} + \sum_{i=1}^{n} \langle \tilde{x}, f_i \rangle b_i + \sum_{i=1}^{n} (\langle y, g_i \rangle - \langle \tilde{x}, f_i \rangle) b_i = \tilde{y} + \sum_{i=1}^{n} \langle y, g_i \rangle b_i = y.$$

Thus the equality $R(B) = \mathfrak{Y}$ is proved. Now the conclusion of our lemma follows from the Banach Inverse Mapping Theorem. ▶

## §7. Projections. Splitable operators

**1.** PROJECTIONS. We have already dealt with projections when we defined the direct sum of spaces (see §7 of Chapter 0). Recall that by a projection on a Banach space $\mathfrak{X}$

we mean a linear operator $P$ defined on the whole space $\mathfrak{X}$, with the property $P^2 = P$. We will consider below only continuous projections which, for brevity, will be simply called projections.

If $P$ is a projection then $Q = I - P$ is a projection, too, since obviously $Q$ is linear, is defined on the whole $\mathfrak{X}$, is continuous, and $Q^2 = (I-P)^2 = I - P = Q$. The restriction of the projection $P$ on its range $R(P) = P\mathfrak{X}$ is the identity operator $P$ on this subspace, so that, if $P \neq 0$ then $\|P\| \geqslant 1$. Indeed, for any $u \in R(P)$ we find an element $x \in \mathfrak{X}$ such that $u = Px$ and, therefore, $Pu = PPx = P^2x = Px = u$. Thus $R(P)$ is closed, i.e., it is a subspace in $\mathfrak{X}$. Any element $x \in \mathfrak{X}$ has a unique representation $x = u + v$ where $u = Px$, $v = Qx$. Therefore the space $\mathfrak{X}$ decomposes into the direct sum of subspaces $R(P)$ and $R(Q)$, i.e., $\mathfrak{X} = R(P) \dot{+} R(Q)$. By Theorem 0.4.2 this direct sum is topological.

Thus any projection $P$ gives rise to a decomposition of the space $\mathfrak{X}$ into the topological direct sum $\mathfrak{X} = \mathfrak{X}_1 \dot{+} \mathfrak{X}_2$, where $\mathfrak{X}_1 = R(P)$ is the subspace onto which $P$ acts, and $\mathfrak{X}_2 = R(I-P)$ is the subspace along which the operator $P$ projects the elements of $\mathfrak{X}$. It is clear that $R(I-P) = N(P)$ and $R(P) = N(I-P)$. The projections $P$ and $Q = I - P$ are called *mutually complementary*.

According to the previous considerations, a subspace $\mathfrak{L}$ of a Banach space $\mathfrak{X}$ is called *topologically complementable* if there exists a projection $P$ on $\mathfrak{X}$ such that $R(P) = \mathfrak{L}$. Generally speaking, not all the subspaces of a Banach space are topologically complementable. There are a lot of papers and monographs devoted to this topic — the existence of topological complements (see, for instance, M. Kadec, B. Mityagin [1]).

However, in the case of a Hilbert space $\mathfrak{X} = \mathfrak{H}$, any subspace $\mathfrak{L}$ has not only a topological complement, but also an orthogonal complement (see Theorem 0.11.2). More precisely, any proper subspace $\mathfrak{L}$ can be obtained as the range of a projection $P$, with minimal norm $\|P\| = 1$, such that the subspaces $R(P)$ and $R(I-P)$ are orthogonal.

Let us return now to the general case of a Banach space $\mathfrak{X}$. A projection $P$ on $\mathfrak{X}$ will be called *proper* if $\|P\| = 1$.

Since the finite-dimensional subspaces of a Banach space have always topological complements, they play an important role in some problems concerning projections. To be more specific, let $\mathfrak{X}$ be a Banach space and $\mathfrak{E}$ a finite-dimensional subspace of $\mathfrak{X}$, with $\dim \mathfrak{E} = n$. Choose a basis $\{a_i\}_{i=1}^n$ in $\mathfrak{E}$ and define the linear functionals $f_i$, $1 \leqslant i \leqslant n$, as follows. For $u = \alpha_1 a_1 + \cdots + \alpha_n a_n \in \mathfrak{E}$ we put $f_i(u) = \alpha_i$, so that $\langle a_i, f_j \rangle = \delta_{ij}$, $i, j = 1, \ldots, n$. Obviously the functionals $f_i$ are well-defined on the subspace $\mathfrak{L}$ and are bounded. By Theorem 0.8.7 they admit extensions on the whole $\mathfrak{X}$, denoted also by $f_i$, $i = 1, \ldots, n$, with the same norms. For $x \in \mathfrak{X}$ we set

$$Px = \sum_{i=1}^{n} \langle x, f_i \rangle a_i. \tag{7.1}$$

Then the operator $P$ is a projection on $\mathfrak{X}$, with $R(P) = \mathfrak{E}$. Indeed,

$$P^2 = PPx = \sum_{i=1}^{n} \langle Px, f_i \rangle a_i = \sum_{i=1}^{n} \left\langle \sum_{k=1}^{n} \langle x, f_k \rangle a_k, f_i \right\rangle a_i =$$

$$= \sum_{i=1}^{n} \langle x, f_i \rangle a_i = Px,$$

for any $x \in \mathfrak{X}$, hence $P^2 = P$. Further, it is clear that $R(P) \subseteq \mathfrak{E}$. On the other hand, for each $u = \alpha_1 a_1 + \cdots + \alpha_n a_n$ we have $Pu = u$. Therefore $\mathfrak{E} \subseteq R(P)$ and, consequently, $\mathfrak{E} = R(P)$.

**2.** SPLITABLE OPERATORS. Assume that $P_1$ and $E_1$ are two given projections on the Banach spaces $\mathfrak{X}$ and $\mathfrak{Y}$, respectively. We denote

$$P_2 = I - P_1, \quad E_2 = I - E_1, \quad \mathfrak{L}_1 = P_1\mathfrak{X}, \quad \mathfrak{L}_2 = P_2\mathfrak{X}, \quad \mathfrak{M}_1 = E_1\mathfrak{Y}, \quad \mathfrak{M}_2 = E_2\mathfrak{Y}.$$

Then, any linear operator $A : \mathfrak{X} \to \mathfrak{Y}$ defined on the whole $\mathfrak{X}$ has the representation

$$Ax = E_1AP_1x + E_1AP_2x + E_2AP_1x + E_2AP_2x. \tag{7.2}$$

According to this representation, it is convenient to write the operator $A$ in the matrix form

$$A = \begin{pmatrix} A_{11} & A_{12} \\ A_{21} & A_{22} \end{pmatrix}, \tag{7.3}$$

where $A_{ij} : \mathfrak{L}_j \to \mathfrak{M}_i$ is the restriction of the operator $E_iAP_j$ on the subspace $\mathfrak{L}_j$, $i, j = 1, 2$.

The matrix representation (7.3) of the operator $A$ is called *non-degenerate* if the operator $A_{22}$ is a homeomorphism from the space $\mathfrak{L}_2$ onto $\mathfrak{M}_2$. Provided that this condition is fulfilled and, in addition,

$$A_{11} = A_{12}A_{22}^{-1}A_{21}, \tag{7.4}$$

then the representation (7.3) is called *regular*.

An operator $A : \mathfrak{X} \to \mathfrak{Y}$ defined on the whole space $\mathfrak{X}$ is said to be *splitable* if it admits a regular representation (7.3).

**THEOREM 7.1.** *A normally solvable operator $A \in L(\mathfrak{X}, \mathfrak{Y})$ is splitable if and only if its kernel $N(A)$ and its range $R(A)$ have topological complements.*

$\triangleleft$ Suppose that the subspaces $N(A)$ and $R(A)$ have topological complements, i.e., there exist the projections $P_1$ on $\mathfrak{X}$ and $E_1$ on $\mathfrak{Y}$ such that $R(P_1) = N(A)$ and $R(E_1) = R(A)$. Set $P_2 = I - P_1$, $E_2 = I - E_1$. Then formula (7.2) is true for all $x \in \mathfrak{X}$. Since $P_1 x \in N(A)$, then $E_1 A P_1 x = 0$, $E_2 A P_1 x = 0$ and $Ax = E_1 A P_2 x + E_2 A P_2 x$. Moreover, $y = AP_2 x \in R(A)$, so $E_2 y = 0$, hence $Ax = E_2 A P_2 x$. Thus, in our case, representation (7.3) has the particular form

$$A = \begin{pmatrix} 0 & 0 \\ 0 & A_{22} \end{pmatrix}, \tag{7.5}$$

where $A_{22}$ is the restriction of the operator $E_2 A P_2$ on $R(I - P_1)$ — the topological complement of the subspace $N(A)$. It follows that $R(A_{22}) = R(A)$, $N(A_{22}) = \{0\}$ whence by the Banach Inverse Mapping Theorem we obtain that $A_{22}$ is a homeomorphism. Obviously equality (7.4) is satisfied. So we proved that the conditions of the theorem are sufficient.

Conversely, assume the existence of the projections $P_1$ on $\mathfrak{X}$ and $E_1$ on $\mathfrak{Y}$ such that representation (7.3) is regular. Then the equation $Ax = 0$ is equivalent to the system of equations

$$\begin{cases} A_{11} P_1 x + A_{12} P_2 x = 0 \\ A_{21} P_1 x + A_{22} P_2 x = 0. \end{cases} \tag{7.6}$$

The second equation in (7.6), on its turn, is equivalent to

$$P_2 x = -A_{22}^{-1} A_{21} P_1 x. \tag{7.7}$$

From (7.7) and (7.4) it follows that the first equation in (7.6) is an identity.

Thus the kernel $N(A)$ consists exactly of those elements which satisfy (7.7). Let us define a linear operator $P$ on $\mathfrak{X}$ by

$$Px = P_1 x - A_{22}^{-1} A_{21} P_1 x, \quad x \in \mathfrak{X}. \tag{7.8}$$

We have

$$P^2 = \left(P_1 - A_{22}^{-1} A_{21} P_1\right)^2 = P_1^2 - A_{22}^{-1} A_{21} P_1^2 = P_1 - A_{22}^{-1} A_{21} P_1 = P,$$

i.e., $P$ is a projection (recall that $A_{22}^{-1}: \mathfrak{L}_2 \to \mathfrak{M}_2$ and $P_1\mathfrak{M}_2 = \{0\}$). If $x \in N(A)$, then by (7.7) we obtain $Px = (P_1 + P_2)x = x$, whence $N(A) \subseteq R(A)$. Next, let us show that $R(A) \subseteq N(A)$. For each $x \in \mathfrak{X}$ we have

$$P_1Px = P_1\left(P_1 - A_{22}^{-1}A_{21}P_1\right)x = P_1x.$$

Therefore, if we set $y = Px$, then $P_1y = P_1x$ and by (7.8) we find

$$P_2y = P_2\left(P_1 - A_{22}^{-1}A_{21}P_1\right)x = -A_{22}^{-1}A_{21}P_1x = -A_{22}^{-1}A_{21}P_1y.$$

From here and based on (7.7) it follows that $y \in N(A)$ and, consequently, $R(A) \subseteq$ $\subseteq N(A)$. Due to the inverse inclusion we obtain that $P$ is a projection onto $N(A)$, and therefore $N(A)$ has a topological complement.

Assume now that $y \in R(A)$, i.e., the equation $y = Ax$ has solutions. Equivalently, the system of equations

$$\begin{cases} A_{11}P_1x + A_{12}P_2x = E_1y \\ A_{21}P_1x + A_{22}P_2x = E_2y \end{cases} \tag{7.9}$$

has solutions. The second equation in (7.9) gives

$$P_2x = -A_{22}^{-1}A_{21}P_1x + A_{22}^{-1}E_2y.$$

Replacing $P_2x$ in the first equation of system (7.9) by the right hand side of the last equality, and according to (7.4), we obtain that $R(A)$ consists exactly of those elements $y \in \mathfrak{Y}$ for which the relation $E_1y = A_{12}A_{22}^{-1}E_2y$ is fulfilled. Put $E =$ $= E_1 - A_{12}A_{22}^{-1}E_1$. As above, we get that $E$ is a projection onto $R(A)$. Consequently $R(A)$ has a topological complement. ▶

**COROLLARY 7.1.** *Any Noether operator and, in particular, any Fredholm operator $A \in L(\mathfrak{X}, \mathfrak{Y})$ is splitable.*

At the end of this section we will establish, for a given operator $A \in L(\mathfrak{X}, \mathfrak{Y})$, the relationship between the operator $B = A + K$ considered in Lemma 6.1, where $K$ is defined by (6.2), and the representation (7.3).

Assume that $A \in L(\mathfrak{X}, \mathfrak{Y})$ is a Fredholm operator with $n = n(A) \geqslant 1$ and that $\{a_k\}_{k=1}^n$, $\{g_k\}_{k=1}^n$ are fixed bases for $N(A)$ and $N(A^*)$, respectively. Let $\{f_k\}_{k=1}^n$ and $\{b_k\}_{k=1}^n$ be the biorthogonal systems corresponding to $\{a_k\}_{k=1}^n$ and $\{g_k\}_{k=1}^n$ (see §6). Set $\mathfrak{L}_n = \operatorname{clin}\{b_1, \ldots, b_n\}$ and define the projection $P_1$ on $\mathfrak{X}$ by (7.1), and the projection $E_1$ on $\mathfrak{Y}$ by $E_1 y = \sum_{i=1}^{n} \langle y, g_i \rangle b_i$. Then the projection $E_2 = I - E_1$ maps $\mathfrak{Y}$ onto $R(A)$. Indeed, for any $y \in \mathfrak{Y}$ we have

$$E_1 y = \sum_{k=1}^{n} \langle y, g_k \rangle b_k,$$

whence $E_2 y = y - \sum_{k=1}^{n} \langle y, g_k \rangle b_k$ and

$$\langle E_2 y, g_j \rangle = \langle y, g_j \rangle - \sum_{k=1}^{n} \langle y, g_k \rangle \langle b_k, g_j \rangle = 0.$$

It follows that $E_2 y \in R(A)$. Moreover, from (5.3) and by the biorthogonality of the systems $\{b_k\}_{k=1}^{n}$ and $\{g_k\}_{k=1}^{n}$, we have $\mathfrak{L}_n \cap R(A) = \{0\}$. Therefore $\mathfrak{Y} = \mathfrak{L}_n \dot{+} R(A)$. Analogously, $\mathfrak{X} = N(A) \dot{+} \mathfrak{M}$, where $\mathfrak{M}$ consists exactly of those vectors $x \in \mathfrak{X}$ for which $\langle x, f_i \rangle = 0$, $i = 1, \ldots, n$. By Theorem 0.4.7, the $n$-dimensional spaces $\mathfrak{L}_n$ and $N(A)$ are homeomorphic.

Consider now the operator $B = A + K$, where $K$ is defined by (6.2). By (6.3), the operator $K \mid N(A)$ yields a homeomorphism between the spaces $N(A)$ and $\mathfrak{L}_n$. It is easy to see that, relatively to the four projections $P_1$, $P_2$, $E_1$, $E_2$, the operator $B$ has the matrix representation

$$B = \begin{pmatrix} K_{11} & 0 \\ 0 & A_{22} \end{pmatrix}, \tag{7.10}$$

where $K_{11} = K \mid N(A)$ and $A_{22} = A \mid \mathfrak{M}$ (see (7.5)). It follows that the operator $B$ is an extension of the operator $A_{22}$ on the whole space $\mathfrak{X}$. By Schmidt Lemma the operator $B$ is invertible. Its inverse $B^{-1}$ has the matrix representation

$$B^{-1} = \begin{pmatrix} K_{11}^{-1} & 0 \\ 0 & A_{22}^{-1} \end{pmatrix}. \tag{7.11}$$

Notice that representations (7.10) and (7.11) are not regular, since condition (7.4) is not fulfilled.

## §8. Invariant subspaces

In this section we consider operators $A \in L(\mathfrak{X})$, i.e., bounded linear operators $A : \mathfrak{X} \to \to \mathfrak{X}$ with $D(A) = \mathfrak{X}$. A subspace $\mathfrak{L} \subseteq \mathfrak{X}$ is called *invariant* relatively to $A$ if $A\mathfrak{L} \subseteq \mathfrak{L}$.

**1.** Let $A \in L(\mathfrak{X})$. Assume that the kernel $N(A)$ has a topological complement, i.e., $\mathfrak{X} = N(A)\dot{+}\mathfrak{M}$, where $\mathfrak{M}$ is a subspace of $\mathfrak{X}$. Let us consider the following question: when is the topological complement $\mathfrak{M}$ of $N(A)$ an invariant subspace relatively to $A$? If, in particular, the operator $A$ is splitable, then, by Theorem 7.1, the existence of two decompositions of the space $\mathfrak{X}$ follows:

$$\mathfrak{X} = N(A)\dot{+}\mathfrak{M}, \qquad \mathfrak{X} = \mathfrak{L}\dot{+}R(A).$$

When these two decompositions coincide, i.e., $N(A) = \mathfrak{L}$, $R(A) = \mathfrak{M}$, then the subspace $\mathfrak{M}$ is invariant relatively to the operator $A$ and $\mathfrak{X} = N(A)\dot{+}R(A)$.

Let now $A$ be an arbitrary operator in $L(\mathfrak{X})$. Consider the following two sequences of lineals:

$$\{0\}, N(A), N(A^2), \dots, N(A^k), \dots, \tag{8.1}$$

$$\mathfrak{X}, R(A), R(A^2), \dots, R(A^k), \dots . \tag{8.2}$$

Relatively to inclusion, the sequence (8.1) does not decrease and the sequence (8.2) does not increase, i.e., $N(A^k) \subseteq N(A^{k+1})$, and $R(A^k) \supseteq R(A^{k+1})$, for any $k = 0, 1, \dots$. If there exists $m \in \mathbb{N}$ such that

$$N(A^m) = N(A^{m+1}) \quad \text{and} \quad N(A^i) \neq N(A^{i+1}), \quad i = 0, \dots, m-1, \tag{8.3}$$

then we say that the operator $A$ has a finite *ascent*, equal to $m$. If there exists $r \in \mathbb{N}$ satisfying the conditions

$$R(A^r) = R(A^{r+1}), \quad \text{and} \quad R(A^j) = R(A^{j+1}), \quad j = 0, \dots, r-1, \tag{8.4}$$

then we say that the operator $A$ has a finite *descent*, equal to $r$.

It is easy to verify that condition (8.3) implies the equality $N(A^m) = N(A^k)$ for all $k \in \mathbb{N}$, $k \geqslant m$, and condition (8.4) implies the equality $R(A^l) = R(A^r)$ for all $l \in \mathbb{N}$, $l \geqslant r$. Moreover, as we will prove below, if a normally solvable operator has simultaneously a finite ascent $m$ and a finite descent $r$ then these integers coincide, i.e., $m = r$.

**THEOREM 8.1.** *Assume that a normally solvable operator $A \in L(\mathfrak{X})$ has a finite ascent $m$ and a finite descent $r$. Then*

1) *the subspace $R(A^r)$ is invariant relatively to $A$ and the restriction $A \mid R(A^r)$ is a homeomorphism;*

2) *the space $\mathfrak{X}$ decomposes into the direct sum*

$$\mathfrak{X} = N(A^r)\dot{+}R(A^r);$$

3) *the operator $A$ has the same ascent and descent, i.e., $m = r$.*

◁ 1) From $R(A^r) = R(A^{r+1})$ and $AR(A^r) = R(A^{r+1})$ we infer $AR(A^r) = R(A^r)$. Let us show now that $N(A \mid R(A^r)) = \{0\}$. Set $A \mid R(A^r) = \hat{A}$. Let $x \in R(A^r)$ with $\hat{A}x = 0$. Choose $n \in \mathbb{N}$ such that $n \geqslant \max\{m, r\}$. Then $x \in R(A^n)$, i.e., $x = A^n z$, where $z \in \mathfrak{X}$. But $A^{n+1}z = Ax = \hat{A}x = 0$, whence $z \in N(A^{n+1}) = N(A^n)$. Therefore $x\ (= A^n z) = 0$. By Theorem 5.2 it follows that $R(A^r)$ is a subspace of $\mathfrak{X}$ whence, by using Theorem 0.8.5, we obtain that $A \mid R(A^r)$ is a homeomorphism.

2) Let $x$ be an arbitrary element of $\mathfrak{X}$. Put $v = \hat{A}^{-r}A^r x$ and $u = x - v$. Then $v \in R(A^r)$ and since $A^r u = A^r x - A^r v = A^r x - A^r \hat{A}^{-r} A^r x = A^r x - A^r x = 0$ we have $u \in N(A^r)$. If we find other elements $u' \in N(A^r)$ and $v' \in R(A^r)$ such that $x = u' + v'$ then

$$A^r x = A^r u' + A^r v' = A^r v',$$

and since $v' \in R(A^r)$ it follows that $A^r v' = \hat{A}^r v'$, i.e., $v' = \hat{A}^{-r}A^r x = v$. Consequently, $u' = x - v = u$ and assertion 2) is proved.

3) Let us prove now that $m = r$. According to 2) any element $x \in N(A^{r+1})$ has the form $x = u + v$, where $u \in N(A^r)$ and $v \in R(A^r)$. Then $0 = A^{r+1}x = A^{r+1}u + A^{r+1}v$ and from $A^{r+1}u = 0$ it follows that $A^{r+1}v = 0$ hence, by virtue of 1), we obtain $v = 0$. This means that $x = u \in N(A^r)$ and therefore $N(A^{r+1}) = N(A^r)$, i.e., $m \leqslant r$.

On the other hand, if $y \in R(A^m)$, i.e., $y = A^m x$ for some $x \in \mathfrak{X}$ then, representing $x$ as $x = u + v$ where $u \in N(A^r)$, $v \in R(A^r)$, we obtain $y = A^m u + A^m v$. Since $N(A^m) = \cdots = N(A^r)$ then $u \in N(A^m)$ hence $A^m u = 0$, i.e.,

$$y = A^m v = A^{m+1}\hat{A}^{-1}v \in R(A^{m+1}).$$

So $m \geqslant r$ and the proof is complete. ▶

**REMARK 8.1.** If an operator $A$ satisfies the conditions in Theorem 8.1 and $A = I - B$, then the space $N(A^m)$ is the root subspace for the operator $B$ corresponding to the eigenvalue $\lambda = 1$ (see §1) and the integer $m$ itself is the rank of this eigenvalue.

**COROLLARY 8.1.** *Let $B \in L(\mathfrak{X})$ and assume that the operator $I - B$ is normally solvable. If $\lambda = 1$ is an eigenvalue of rank 1 for the operator $B$ and an eigenvalue of finite rank $m^*$ for the operator $B^*$ then the operator $I - B$ is splitable and*

$$\mathfrak{X} = N(I - B) \dot{+} R(I - B). \tag{8.6}$$

◁ By Remark 8.1 the operator $A = I - B$ has the finite ascent $m = 1$ and the operator $A^* = I - B^*$ has the finite ascent $m^*$. Therefore $N\left((A^*)^{m^*}\right) =$

$= N\left((A^*)^{m^*+1}\right)$. This equality and (5.1) imply that $R\left(A^{m^*}\right) = R\left(A^{m^*+1}\right)$. Consequently, the operator $A$ has the finite descent $r \leqslant m^*$. By Theorem 8.1, the equality $m = r = 1$ holds. The decomposition (8.6) follows now from this last equality and (8.5). ▶

To conclude this subsection we give an example of a normally solvable operator with a finite ascent, but without a finite descent.

**EXAMPLE 8.1.** Let $\mathfrak{H}$ be a Hilbert space with an orthonormal basis $\{e_i\}_{i=0}^{\infty}$. Define the linear operator $A$ on $\mathfrak{H}$ by the equalities: $Ae_0 = 0$, $Ae_i = e_{i+1}$, $i \geqslant 1$. Then $A \in L(\mathfrak{H})$, $N(A^k) = \text{clin}\{e_0\}$ and $R(A^k) = \text{clin}\{e_{k+1}, e_{k+2}, \dots\}$. Thus the operator $A$ has the finite ascent $m = 1$ but it has not a finite descent. At the same time the operator $A^*$ has the finite descent $r^* = 1$ but it has no finite ascent.

**2.** In this subsection we use a different approach to study the problem of decomposing the space $\mathfrak{X}$ into a direct sum of the form (8.6). A key role in this approach is played by the next condition:

(UB) *The family of the iterations of the operator $B$ is uniformly bounded,* i.e.,

$$\|B^k\| \leqslant M < \infty, \quad k = 1, 2, \dots .$$

As a preliminary step let us prove the following result.

**LEMMA 8.1.** *Assume that the operator $B \in L(\mathfrak{X})$ satisfies condition* (UB). *Then the intersection of the kernel and the range of the operator $A = I - B$ is equal to* $\{0\}$, i.e.,

$$R(A) \cap N(A) = \{0\}. \tag{8.7}$$

◁ Consider the operators

$$B_n = n^{-1} \sum_{k=1}^{n} B^k, \quad n = 1, 2, \dots . \tag{8.8}$$

From the equality $B^n A = n^{-1}\left(\sum_{k=1}^{n} B^k - \sum_{k=1}^{n} B^{k+1}\right) = n^{-1}(B - B^{n+1})$ and by condition (UB) it follows that for any $y \in R(A)$ the relation

$$\lim_{n\to\infty} B_m y = 0 \tag{8.9}$$

holds. If, on the contrary, $y \in N(A)$, then $B_n y = y$ and equality (8.9) fails for $0 \neq y \in N(A)$. ▶

**COROLLARY 8.2.** *If the operator $B \in L(\mathfrak{X})$ satisfies condition* (UB) *and the operator $A = I - B$ is Fredholm, then equality* (8.6) *is true.*

◁ Since $A$ is a Fredholm operator, the kernel $N(A)$ is a finite-dimensional subspace of $\mathfrak{X}$. Let $\dim N(A) = n$ and let $\{a_k\}_{k=1}^n$ be a basis of $N(A)$. Using Hahn-Banach-Suchomlynoff Theorem and Lemma 8.1 we may find a system of vectors $\{f_k\}_{k=1}^n \subseteq \mathfrak{X}^*$ with the properties:

$$\langle a_i, f_j \rangle = \delta_{ij}, \quad i, j = 1, \dots, n \tag{8.10}$$

$$\langle y, f_k \rangle = 0, \quad k = 1, \dots, n, \tag{8.11}$$

for any $y \in R(A)$. Condition (8.10) implies that the system of vectors $\{f_k\}_{k=1}^n$ is linearly independent and from (8.11) together with (5.1) it follows that $\{f_k\}_{k=1}^n$ is a basis of $N(A^*)$ (recall that $\dim N(A^*) = n$). The operator $P$ defined by (7.1) is a projection from $\mathfrak{X}$ onto $N(A)$, and $I - P$ is a projection onto $R(A) = {}^{\perp}N(A^*)$. ▶

Let us return now to the more general case when the operator $A$ is only normally solvable. For such an operator the next result is true.

**THEOREM 8.2.** *Assume that the operator $B \in L(\mathfrak{X})$ satisfies condition* (UB) *and the operator $A = I - B$ is normally solvable. Then equality* (8.6) *holds if and only if the limit in the strong operator topology (see* §8 *of Chapter* 0*) of the sequence of operators defined by* (8.8) *exists. The operator*

$$P = \lim_{n\to\infty} B_n = \lim_{n\to\infty} n^{-1} \sum_{m=1}^{n} B^m \tag{8.12}$$

*is a projection onto $N(A)$ and $\|P\| \leqslant M$, where $M$ is the same upper bound as in condition* (UB).

◁ Necessity. From (8.6) it follows that any element $x \in \mathfrak{X}$ can be represented as

$$x = y + z,$$

where $y \in R(I - B) = R(A)$ and $z \in N(I - B) = N(A)$. By this equality and from (8.8), (8.9) and condition (UB) it follows that $\|B_n\| \leqslant M$, $n \in \mathbb{N}$, and

$$\lim_{n\to\infty} B_n x = \lim_{n\to\infty} n^{-1} \sum_{m=1}^{n} (I - A)^m z = z. \tag{8.13}$$

As a consequence of the Uniformly Boundedness Principle, the operator $P$ defined by (8.1) is a bounded linear operator with $\|P\| \leqslant M$. According to (8.13) $P$ is a projection onto $N(A)$.

Sufficiency. Assume that the operator $P$ defined by equality (8.12) exists. Again as a consequence of the Uniformly Boundedness Principle, the operator $P$ is linear, bounded and $\|P\| \leqslant M$. For each $z \in N(A)$ we obtain (see (8.13)) $P(z) = z$, hence $N(A) \subseteq R(P)$. On the other hand, if $u \in R(P)$ then there exists $v \in \mathfrak{X}$ for which $u = Pv = \lim_{n\to\infty} n^{-1} \sum_{m=1}^{n} B^m v$. Then, by condition (UB), we have

$$Au = (I - B)u = \lim_{n\to\infty} n^{-1} \left( \sum_{m=1}^{n} B^m v - \sum_{m=1}^{n} B^{m+1} v \right) =$$
$$= \lim_{n\to\infty} n^{-1}(Bv - B^{n+1}v) = 0,$$

hence $u \in N(A)$. Thus we obtain that $R(P) = N(A)$ and that $P \mid R(P)$ is the identity operator. Consequently, $P$ is a projection onto $N(A)$. Let us show that $N(P) = R(A)$, a relation which clearly will complete the proof of (8.6).

Let $y \in N(P)$. Then, by (8.12), for any $\varepsilon > 0$ there exists $n \in \mathbb{N}$ such that $\|B_n y\| < \varepsilon$. We have $y - B_n y \in R(A)$; indeed

$$y - B_n y = \left( I - n^{-1} \sum_{m-1}^{n} B^m \right) y = n^{-1} \sum_{m=1}^{n} (I - B^m) y =$$
$$= n^{-1}(I - B) \sum_{m=1}^{n} (I + B + B^2 + \cdots + B^{m-1}) y = (I - B) y_n,$$

where $y_n = n^{-1} \sum_{m=1}^{n} (I + B + B^2 + \cdots + B^{m-1}) y$.

Since $y - B_n y \in R(A)$ and $R(A)$ is closed, taking into account that $\varepsilon$ was an arbitrary positive number, we obtain $y \in R(A)$, i.e., $N(P) \subseteq R(A)$. The converse inclusion $R(A) \subseteq N(P)$ follows directly from (8.9). ►

**REMARK 8.2.** The arguments in our proof of Theorem 8.2 above are, actually, suitable modifications of the arguments used by K. Yoshida in the proofs of the Statistical Ergodic Theorem and its consequences (see, for instance, K. Yoshida [1]). This theorem asserts the following:

*Let $B \in L(\mathfrak{X})$ be an operator satisfying condition* (UB). *Consider the operators $B_n$, $n \in \mathbb{N}$, defined by* (8.8) *and assume that the sequence $\{B_n x\}_{n\in\mathbb{N}}$ is weakly compact for $x \in \mathfrak{X}$. Then this sequence is convergent in the norm topology of the space $\mathfrak{X}$.*

This theorem and Theorem 8.2 lead to the next simple but important result.

**COROLLARY 8.3.** *Let $\mathfrak{X}$ be a reflexive space. Assume that an operator $B \in L(\mathfrak{X})$ satisfies condition* (UB) *and the operator $I - B$ is normally solvable. Then the equality* (8.6) *is true.*

Some authors (see, for example, F. Riesz [1]) refer to Corollary 8.3 as the Statistical Ergodic Theorem.

# Chapter V

# Nonlinear equations with differentiable operators

## §1. Fixed points. Banach principle

**1.** Let $\mathfrak{D}$ be a topological space. Assume that $F$ is an operator defined from $\mathfrak{D}$ into $\mathfrak{D}$, i.e.,

$$F(\mathfrak{D}) \subseteq \mathfrak{D}. \tag{1.1}$$

**DEFINITION 1.1.** A point $z \in \mathfrak{D}$ is called a *fixed point* for the operator $F$ if $F(z) = z$.

Condition (1.1), which is refered to as the invariance of the set $\mathfrak{D}$ for the operator $F$, enables us to define the iterations $F^n$ of $F$, by the recurrent relations $F^1 = F$, $F^{n+1} = F^n \circ F$, where "$\circ$" denotes the composition of operators.

**DEFINITION 1.2.** 1) A fixed point $z \in \mathfrak{D}$ of the operator $F$ is called *locally attractive* if there exists a neighborhood $\mathfrak{U} \subseteq \mathfrak{D}$ of the point $z$, such that the iterations $F^n(x)$ converge to $z$ for all $x \in \mathfrak{U}$. If $\mathfrak{U} = \mathfrak{D}$ we shall say that $z$ is an *attractive* (or *globally attractive*) fixed point.

2) The point $z$ is called an s-*fixed point* (*successful approximative fixed point*) for the operator $F$ in $\mathfrak{D}$ if it is attractive uniformly on each closed and bounded subset of $\mathfrak{D}$, i.e., the iterations $F^n(x)$ converge uniformly on each closed and bounded subset of $\mathfrak{D}$.

3) A fixed point $z$ is called *repulsive* if the sequence $F^n(x)$ converges to $z$ if and only if $x = z$.

4) A fixed point $z$ will be called *neutral* if it is neither locally attractive nor repulsive.

**DEFINITION 1.3.** A fixed point $z$ of $F$ is called *isolate* if there exists a neighborhood of this point which does not contain other fixed points for the operator $F$.

**REMARK 1.1.** A locally attractive point $z$ is an isolate fixed point. Moreover, a globally attractive point is unique. Indeed, assume that on a neighborhood $\mathfrak{U}$ of the point $z$ the iterations $F^n(x)$ converge to $z$ for all $x \in \mathfrak{U}$. If we suppose that the operator $F$ has another fixed point $y$ in $\mathfrak{U}$, then $F^n(y) \to z$. But $y$ is a fixed point for the operator $F^n$ for any $n$, since $F^n(y) = F \circ F^{n-1}(y) = F(y) = y$. Hence $z = y$. We notice also that for a continuous operator $F$ the convergence of the iterations implies the existence of fixed points.

A lot of work has been done to obtain criterions for the existence of fixed points and to study the properties of these points (uniqueness, isolation, attractiveness, and so on) for various operators. These criterions are frequently keystones in the theory of nonlinear operator equations. The theorems which establish conditions for the existence of fixed points are usually called *fixed point principles*. As a rule, the formulation of a fixed point principle starts with the condition of invariance (1.1). However, in many cases this condition is insufficient for the existence of a fixed point, and, even more, it is insufficient for the existence of an s-fixed point. Notice that the presence of an s-fixed point guarantees not only its uniqueness, but also the possibility of approximating it (by approximative computations).

Let us remark that the development of the computer techniques enables one to approximate a fixed point by iterations even for rather complicated operators. At the same time the computer methods do not insure against accidental computational errors, which practically do not matter as soon as we know in advance that the iterative proccess converges uniformly.

One of the most effective fixed point principle is the Banach principle for contractive operators.

Let $(\mathfrak{X}, \rho)$ be a complete metric space with the metric $\rho(\cdot, \cdot)$ (see §4 of Chapter 0). An operator $F$ defined from a subset $\mathfrak{D} \subseteq \mathfrak{X}$ into $\mathfrak{X}$ is called a *contractive operator* or a *contraction*, if it satisfies a Lipschitz type condition with a constant $q < 1$, i.e.,

$$\rho(Fx, Fy) \leqslant q\rho(x, y), \quad x, y \in \mathfrak{D}. \tag{1.2}$$

Obviously a contractive operator is continuous.

**THEOREM 1.1** (Banach principle for contractive operators). *Let $\mathfrak{D}$ be a closed subset of $(\mathfrak{X}, \rho)$ and let $F$ be an operator defined on $\mathfrak{D}$ which is a contraction and satisfies the invariance condition* (1.1). *Then the operator $F$ has an s-fixed point in*

*$\mathfrak{D}$. The rate of convergence for the iterations of the operator $F$ to this fixed point is described by the relations*

$$\rho(F^n x_0, F^m x_0) \leqslant \frac{q^n}{1-q}\rho(x_0, Fx_0), \tag{1.3}$$

*where $m \geqslant n$ and $x_0$ is an arbitrary element in $\mathfrak{D}$.*

◁ As a matter of fact it is sufficient to prove inequality (1.3). This inequality implies that the sequence $\{x_n\}_{n\in\mathbb{N}}$, where $x_n = F^n x_0$, is fundamental, and this property is uniformly fulfilled relatively to $x_0$ in any closed and bounded subset of $\mathfrak{D}$. Therefore, by the completeness of the space $\mathfrak{X}$ and the closedness of the set $\mathfrak{D}$, the sequence $\{x_n\}_{n\in\mathbb{N}}$ converges to a point $z \in \mathfrak{D}$, uniformly relatively to $x_0 \in \mathfrak{D}$. Indeed, by (1.2) it follows that

$$\rho(x_1, x_2) = \rho(Fx_0, Fx_1) \leqslant q\rho(x_0, x_1).$$

Analogously,

$$\rho(x_2, x_3) = \rho(Fx_1, Fx_2) \leqslant q\rho(x_1, x_2) \leqslant q^2\rho(x_0, x_1).$$

By induction we get

$$\rho(x_k, x_{k+1}) \leqslant q^k\rho(x_0, x_1),$$

for all $k = 1, 2, \ldots$. From these relations, using the triangle inequality and the formula for the goemetric series, we have

$$\begin{aligned}\rho(x_n, x_m) &\leqslant \rho(x_n, x_{n+1}) + \rho(x_{n+1}, x_{n+2}) + \cdots + \rho(x_{m-1}, x_m) \leqslant \\ &\leqslant q^n\rho(x_0, x_1)(1 + q + \cdots + q^{m-n-1}) \leqslant \frac{q^n}{1-q}\rho(x_0, x_1).\end{aligned}$$

Setting now $z = \lim\limits_{n\to\infty} F^n x_0$, where $x_0$ is an arbitrary element of $\mathfrak{D}$, we obtain that $z$ is an s-fixed point for the operator $F$ in $\mathfrak{D}$. ▶

If we weaken slightly the conditions required for the operator $F$ in Theorem 1.1, we obtain the next version of the Banach principle.

**COROLLARY 1.1.** *Let $\mathfrak{D}$ be a closed subset of a complete metric space $\mathfrak{X}$ and let $F$ be an operator satisfying the invariance condition (1.1). Assume that there exists an integer $p \geqslant 1$ such that the operator $F^p$ is a contraction on $\mathfrak{D}$. Then the operator $F$ has a globally attractive, and, consequently, a unique fixed point in $\mathfrak{D}$.*

◁ By Theorem 1.1 the operator $F^p$ has an s-fixed point $z$. Choose an arbitrary element $x_0 \in \mathfrak{D}$ and consider the sequence $\{x_n \ : \ n \in \hat{\mathbb{N}}\}$, where $x_n = F^n x_0$ and

$\hat{\mathbb{N}}$ denotes the set $\{0, 1, 2, \ldots, n, \ldots\}$. Let us show that this sequence converges to $z$. Indeed, the sequence breaks up into a finite number of subsequences

$$\left\{F^{kp}x_0 \ : \ k \in \hat{\mathbb{N}}\right\}, \quad \left\{F^{kp}x_1 \ : \ k \in \hat{\mathbb{N}}\right\}, \ldots, \left\{F^{kp}x_{p-1} \ : \ k \in \hat{\mathbb{N}}\right\}.$$

Each of these subsequences is convergent to $z$. Now we show that $z$ is a fixed point for $F$. (Notice that this does not follow from the convergence of the sequence $\{x_n \ : \ n \in$ $\in \mathbb{N}\}$, since under our assumptions the operator $F$ is not necessarily continuous.) We have $F^pFz = FF^pz = Fz$, whence it follows that $Fz$ is a fixed point for the operator $F^p$. According to the uniqueness of the fixed point for $F^p$ we obtain that $Fz = z$. ▶

**REMARK 1.2.** In the general case it is not true that an s-fixed point of the operator $F^p$, for certain $p > 1$, is an s-fixed point for $F$, too. However, as we shall prove below, this is true for holomorphic operators. In order to approximate a fixed point by iterations under the assumptions of Corollary 1.1 it is advisable, if it is possible, to start with an explicite computation of the operator $F^p$ and then to use its iterations.

**REMARK 1.3.** The condition $q < 1$ in (1.2) is not necessary for the existence of s--fixed points. For example, consider the operator $Fx = ax^2$, where $1/2 < a < 1$, which maps the interval [0, 1] into itself and has the s-fixed point 0, though $q = 2a > 1$. Nevertheless, in the general case condition $q < 1$ can not be weakened by setting $q \leqslant 1$. For example, the operator $Fx = x + a$, $a \neq 0$, which maps $\mathbb{R}$ into $\mathbb{R}$ and satisfies condition (1.2) with $q = 1$, does not have fixed points. In this example the diameter of the set on which the operator $F$ acts is infinite. However, there are examples of operators acting on domains with bounded diameter, in infinite dimensional spaces, which satisfy condition (1.1), or condition (1.2) with $q = 1$, and do not have fixed points (for more details see §2 below).

Let us return now to the concept of s-fixed points and state, without proof, a remarkable theorem due to P. R. Meyers [1], which explains the substance of this notion and represents the most complete converse of the Banach principle. (In connection with this result see also A. Levin and E. Lifsic [1].)

**THEOREM 1.2** (P. R. Meyers). *Let $\mathfrak{X}$ be a complete metric space and $\mathfrak{D}$ a closed subset of $\mathfrak{X}$. Assume that an operator $F$ satisfies the invariance condition* (1.1) *and has an attractive fixed point $z \in \mathfrak{D}$, which is an s-fixed point in a neighborhood of $z$. Then $z$ is an s-fixed point in $\mathfrak{D}$ and, moreover, for any $q$ with $0 < q < 1$ there exists a metric on the space $\mathfrak{X}$ equivalent with the original metric, such that $F$ satisfies the contraction condition* (1.2) *on $\mathfrak{D}$ with respect to this new metric.*

Although Meyers' result is merely an existence theorem, it shows that in some concrete problems, if we know a piori about the existence of an s-fixed point, then, for estimating the rate of convergence of the iterative process, it is convenient to try to find an equivalent metric, with respect to which $F$ is a contraction.

**2.** One of the most important fixed point principles is the well known principle due to J. P. Schauder, which is a generalization of the finite-dimensional fixed point principle of Brouwer.

**THEOREM 1.3** (J. P. Schauder). *Let $F$ be an operator which maps a closed convex subset $\mathfrak{D}$ of a Banach space $\mathfrak{X}$ into itself and is completely continuous on $\mathfrak{D}$. Then $F$ has at least one fixed point in $\mathfrak{D}$.*

All the proofs of this theorem we know about rely on subtle topological and geometric considerations. As a rule, these proofs use Brouwer Theorem, finite-dimensional approximations, and also some results from the degree theory for mappings and the vector fields theory (see, for instance, V. A. Trenogin [1], M. A. Krasnoselskii [1]). This is why we stated Schauder Theorem without proof.

Obviously, neither Theorem 1.1, nor Theorem 1.3, are related to each other. Let us remark also that Schauder principle has no constructive features; it fails to indicate the number of fixed points, as well as methods for their approximation.

In the following section we will consider a larger class of operators, which includes the class of contractive operators. At the same time, under the assumptions of Schauder principle, we will indicate some approximation methods for the fixed points.

## §2. Non-expansive operators

Let $(\mathfrak{X}, \rho)$ be a metric space and $\mathfrak{D}$ a closed subset of $\mathfrak{X}$.

**DEFINITION 2.1.** An operator $F : \mathfrak{D} \to \mathfrak{X}$ is called *non-expansive*, respectively *strictly non-expansive*, if the inequality

$$\rho(Fx, Fy) \leqslant \rho(x, y) \tag{2.1}$$

is fulfilled for all $x, y \in \mathfrak{D}$, respectively, if

$$\rho(Fx, Fy) < \rho(x, y), \tag{2.2}$$

for any $x, y \in \mathfrak{D}$, $x \neq y$.

A large amount of work is devoted to the theory of non-expansive operators, either on metric, or on Banach spaces (see for example Z. Opial [1]). Here we treat only those aspects which are related to the fixed points of differentiable operators.

As we have already noticed in Remark 1.2, the operators satisfying condition (2.1) may have no fixed points. In addition, let us present a few other examples.

**EXAMPLE 2.1** (T. Hayden, T. Suffridge [1]). Let $\mathbf{c}_0$ be the space of all sequences $x = (x_1, \dots, x_n, \dots)$ of real or complex numbers which converge to zero, with the norm $\|x\| = \max_n |x_n|$. Define an operator $F$ on the unit ball of this space by the equality $Fx = (\frac{1}{2}, x_1, x_2, \dots)$, for any $x = (x_1, \dots, x_n, \dots)$. Clearly $\|Fx\| \leqslant 1$ if $\|x\| \leqslant 1$, i.e., $F$ satisfies condition (1.1), and $\|Fx - Fy\| = \|(0, x_1 - y_1, \dots, x_n - y_n, \dots)\| = \|x - y\|$, i.e., $F$ satisfies condition (2.1), too. At the same time, if we assume that there exists an element $z = (z_1, \dots, z_n, \dots)$ such that $z = Fz$ then, necessarily, $z_1 = z_2 = \dots = z_n = \dots = \frac{1}{2}$, i.e., $z \notin \mathbf{c}_0$, a contradiction.

We can show that condition (2.2), as well, is not enough for the existence of fixed points.

**EXAMPLE 2.2.** Let $\mathfrak{X} = \mathbb{R}$ and $\mathfrak{D} = [1, \infty)$. The operator $F$ defined by $Fx = x + \dfrac{1}{x}$ clearly does not have fixed points, while $F(\mathfrak{D}) \subseteq \mathfrak{D}$ and

$$|Fx - Fy| = \left| x - y + \frac{1}{x} - \frac{1}{y} \right| = |x - y| \left| \frac{xy - 1}{xy} \right| < |x - y|,$$

for all $x, y \geqslant 1$, $x \neq y$.

Related to the example above we easily notice that for any $q$, with $0 < q < 1$, there exist $x, y \in [1, \infty)$ such that $|Fx - Fy| > q|x - y|$. An analogous example can be produced on a ball with a finite radius in an infinite dimensional space, too.

**EXAMPLE 2.3** Let $\mathfrak{X} = \mathbf{c}_0$ and let $\mathfrak{D}$ be the unit ball in $\mathfrak{X}$. Consider the operator $F$ defined by $Fx = (y_1, \dots, y_n, \dots)$, where $x = (x_1, \dots, x_n, \dots)$ and $y_n = \dfrac{n-1}{n} x_n + \dfrac{1}{n} \sin n$. We can easily prove that the operator $F$ satisfies both conditions (1.1) and (2.2) on $\mathfrak{D}$. At the same time, $F$ has no fixed poins in $\mathfrak{D}$. Assume, on the contrary, that $z = Fz$, where $z = (z_1, \dots, z_n) \in \mathfrak{D}$. This means that $z_n = \dfrac{n-1}{n} z_n + \dfrac{1}{n} \sin n$. Therefore, $z_n = \sin n$, which contradicts the definition of the space $\mathbf{c}_0$.

All these examples show that the existence of a fixed point for a non-expansive operator requires some additional conditions.

**THEOREM 2.1.** *Let $(\mathfrak{X}, \rho)$ be a complete metric space, $\mathfrak{D}$ a closed subset of $\mathfrak{X}$, and $F$ a strictly non-expansive operator on $\mathfrak{D}$ satisfying the invariance condition*

(1.1), *i.e.*, $F(\mathfrak{D}) \subseteq \mathfrak{D}$. *If there exists at least one point* $x_0 \in \mathfrak{D}$ *such that the orbit* $\mathfrak{F}_{x_0} = \{x \in \mathfrak{D} \,:\, x = F^n x_0, n \in \mathbb{N}\}$ *is precompact, then the operator* $F$ *has a unique fixed point* $z \in \mathfrak{D}$. *Moreover, the sequence* $\{x_n \,:\, x_n = F^n x_0,\ n \in \mathbb{N}\}$ *is convergent to* $z$.

◁ Let us consider the functional $\varphi(x) = \rho(x, Fx)$, which is continuous on $\mathfrak{D}$ because $F$ is a continuous operator. On the compact set $\overline{\mathfrak{F}}_{x_0}$ the functional $\varphi$ attains its minimal value, that is, there exists $z \in \overline{\mathfrak{F}}_{x_0}$ satisfying

$$\varphi(z) = \rho(z, Fz) = \min_{x \in \overline{\mathfrak{F}}_{x_0}} \varphi(x) = \min \rho(x, Fx).$$

Next, let us notice that the operator $F$ maps $\overline{\mathfrak{F}}_{x_0}$ into $\overline{\mathfrak{F}}_{x_0}$. Indeed, for any $y \in \mathfrak{F}_{x_0}$ there exists $n \in \mathbb{N}$ such that $y = F^n x_0$, so $Fy = F^{n+1} x_0$, hence $F(\mathfrak{F}_{x_0}) \subset \mathfrak{F}_{x_0}$. Since $F$ is continuous, then $E(\mathfrak{F}_{x_0}) \subset \overline{\mathfrak{F}}_{x_0}$, too. This implies that $\varphi(z) = 0$ because, otherwise, we obtain

$$\varphi(Fz) = \rho(Fz, F^2 z) < \rho(z, Fz) = \varphi(z) = \min_{x \in \overline{\mathfrak{F}}_{x_0}} \varphi(x),$$

a contradiction. The equality $\varphi(z) = 0$ is equivalent with $Fz = z$, i.e., $z$ is a fixed point. Its uniqueness follows from (2.2). It remains to show that $z$ is the unique limit point of the set $\overline{\mathfrak{F}}_{x_0}$. Assume, on the contrary, that there exists a subsequence $\{x_{n_k} \,:\, k \in \mathbb{N}\} \subseteq \overline{\mathfrak{F}}_{x_0}$ convergent to a point $y \in \mathfrak{D}$, $y \neq z$. Then $\rho(z, Fy) = \rho(Fz, Fy) < \rho(z, y)$. Consequently, the number $2\delta = \rho(z, y) - \rho(z, Fy)$ is positive. Choose $i \in \mathbb{N}$ such that $\rho(x_{n_i}, y) < \delta$. Then

$$\begin{aligned} \rho(z, x_{n_i+1}) = \rho(z, Fx_{n_i}) \leqslant \rho(z, Fy) + \rho(Fy, Fx_{n_i}) < \\ < \rho(z, Fy) + \rho(y, x_{n_i}) < \rho(z, y) - \delta. \end{aligned} \tag{2.3}$$

For any $m \in \mathbb{N}$, using (2.3) we obtain the inequalities

$$\begin{aligned} \rho(z, x_{n_i+m}) = \rho(Fz, Fx_{n_i+m-1}) < \\ < \rho(z, x_{n_i+m-1}) < \cdots < \rho(z, y) - \delta, \end{aligned} \tag{2.4}$$

which imply that $y$ ($\neq z$) is not a limit point of the sequence $\{x_{n_k} \,:\, k \in \mathbb{N}\}$, a contradiction. ▶

**COROLLARY 2.1.** *If a strictly non-expansive operator* $F$ *maps* $\mathfrak{D}$ *onto a precompact subset of* $\mathfrak{D}$, *then* $F$ *has an attractive fixed point.*

◁ In this case the orbit $\mathfrak{F}_x$ of any point $x \in \mathfrak{D}$ is a precompact subset of $\mathfrak{D}$. ▶

Let us notice the slight difference between conditions (2.2) or (2.1), and the contraction condition (1.2). In spite of this, the existence and the uniqueness of a fixed point do not necessarily imply its atractiveness. Moreover, a unique fixed point may be even repulsive.

**EXAMPLE 2.4.** Let $\ell^2$ be the Hilbert space of square sumable sequences over the field $\mathbb{K}$ of real or complex numbers. Define the linear operator $A : \ell^2 \to \ell^2$ by the equality $Ae_i = (1 - \alpha_i)e_{i+1}$, $i = 1, 2, \dots$, where $\{e_i : i \in \mathbb{N}\}$ is an orthonormal basis in $\ell^2$, and $\{\alpha_i : i \in \mathbb{N}\}$ is an arbitrary collection of real numbers satisfying the conditions $0 < \alpha_i < 1$ and $\prod_{i=1}^{\infty}(1 - \alpha_i) = \alpha > 0$.

Let $\mathfrak{X}$ be the space $L(\mathbb{K}, \ell^2)$ of all linear continuous operators from $\mathbb{K}$ into $\ell^2$. Set $Fx = Ax$. Clearly $F : \mathfrak{X} \to \mathfrak{X}$. For $h \in \mathbb{K}$ we have $xh = \sum_{i=1}^{n} \lambda_i(h)e_i$, where $\lambda_i(h) \in \mathbb{K}$. Therefore,

$$\|F^n x\|^2 = \|A^n x\|^2 = \sup_{|h|=1} \|A^n xh\| =$$

$$\sup_{|h|=1} \sum_{i=1}^{\infty} |\lambda_i(h)|^2 \prod_{j=1}^{i+n-1} (1 - \alpha_j)^2 < \sup_{|h|=1} \sum_{i=1}^{\infty} |\lambda_i(h)|^2 = \|x\|^2,$$

for all $n \in \mathbb{N}$, i.e., $\|Fx\| < \|x\|$, for $x \neq 0$. Since $F$ is a linear operator, it satisfies condition (2.2) and has a fixed point $z = 0$. On the other hand, from the inequalities

$$\|F^n x\|^2 \geqslant \sup_{|h|=1} \sum_{i=1}^{\infty} |\lambda_i(h)|^2 \prod_{j=1}^{\infty} (1 - \alpha_j)^2 = \alpha^2 \|x\|^2$$

it follows that the sequence $\{F^n x : n \in \mathbb{N}\}$ converges to $z = 0$ if and only if $x = 0$.

The examples above show that the claim on precompactness in Theorem 2.1 and Corollary 2.1 is essential.

If the considered space $\mathfrak{X}$ is a Banach space and the set $\mathfrak{D}$ is convex and bounded, then criterions for the existence of fixed points can be reformulated by slightly weakening the assumptions on $F$ in Theorem 2.1.

Let us recall (see §6 of Chapter 0) that an operator $\Phi : \mathfrak{D} \to \mathfrak{X}$ is called proper on $\mathfrak{D}$ if the preimage of any precompact subset of $\mathfrak{X}$ is a precompact subset of $\mathfrak{D}$.

**THEOREM 2.2.** *Let $\mathfrak{X}$ be a Banach space, $\mathfrak{D}$ a closed bounded convex subset of $\mathfrak{X}$, and $F : \mathfrak{D} \to \mathfrak{D}$ a non-expansive operator. Assume that the operator $\Phi = I - F$ (where $I$ is the identity mapping on $\mathfrak{X}$) is proper on $\mathfrak{D}$. Then the operator $F$ has at least one fixed point $z$ in $\mathfrak{D}$.*

◁ Choose an arbitrary element $y \in \mathfrak{D}$ and consider the operators $F_k$, $k = 1, 2, \dots$, defined by the equalities

$$F_k x = \left(1 - \frac{1}{k}\right) Fx + \frac{1}{k} y. \tag{2.5}$$

Since $\mathfrak{D}$ is convex, we obtain $F_k(\mathfrak{D}) \subseteq \mathfrak{D}$. Moreover, by (2.5) we have

$$\|F_k x - F_k x'\| = \left(1 - \frac{1}{k}\right) \|Fx - Fx'\| \leqslant \left(1 - \frac{1}{k}\right) \|x - x'\|,$$

for any $x, x' \in \mathfrak{D}$ and $k \in \mathbb{N}$, i.e., the operator $F_k$ is a contraction on $\mathfrak{D}$ for any fixed $k \in \mathbb{N}$. By Theorem 1.1 there exists a fixed point $x(k)$ in $\mathfrak{D}$ for the operator $F_k$, i.e., $F_k x(k) = x(k)$. Consider the sequence

$$\Phi x(k) = x(k) - Fx(k) = \left(1 - \frac{1}{k}\right) Fx(k) + \frac{1}{k} y - Fx(k) = \frac{1}{k}(y - Fx(k)), \quad k \in \mathbb{N}.$$

Since $\mathfrak{D}$ is bounded, we obtain that $\Phi x(k) \to 0$ as $k \to \infty$, whence, by our assumptions, it follows that the sequence $\{x(k) \ : \ k \in \mathbb{N}\}$ is precompact in $\mathfrak{D}$. This means that there exists a subsequence $\{x(k_m) \ : \ m \in \mathbb{N}\}$ convergent to a point $z \in \mathfrak{D}$. Then $\Phi z = \lim_{m \to \infty} \Phi x(k_m) = 0$, which is equivalent with the equality $z = Fz$. ▶

**COROLLARY 2.2.** *Let $\mathfrak{X}$ and $\mathfrak{D}$ be such as in Theorem 2.2 above. If a non-expansive operator $F$ maps $\mathfrak{D}$ onto a precompact subset of $\mathfrak{D}$, then there exists at least one fixed point for $F$. If, in addition, we assume that the space $\mathfrak{X}$ is strongly convex (see §9 of Chapter 0) then, for any $\theta$ with $0 < \theta < 1$, and any $x_0 \in \mathfrak{D}$, the sequence*

$$x_n = \theta x_{n-1} + (1 - \theta) F x_{n-1}, \quad n \in \mathbb{N}, \tag{2.6}$$

*converges to a fixed point for the operator $F$.*

◁ The first assertion follows directly from Theorem 2.2 if we notice that the operator $\Phi = I - F$ is proper. Indeed, let $\{y_n \ : \ n \in \mathbb{N}\}$ be a convergent sequence in $\Phi(\mathfrak{D})$. Then its preimages have the following form: $x_n = Fx_n + y_n$. Since $\{Fx_n \ : \ n \in \mathbb{N}\}$ is precompact it follows that $\{x_n \ : \ n \in \mathbb{N}\}$ is precompact, too, as requested.

Let us prove now the second assertion. First of all observe that, by the well known Mazur Lemma (see, for instance N. Dunford and J. T. Schwartz [1]), the sequence (2.6) is precompact, since it lies in the closed convex hull of the union of the precompact set $F(\mathfrak{D})$ and the point $x_0$. Denote by $y$ a certain limit point of this sequence and assume that $y$ is not a fixed point for the operator $F$. Let $z$ be a fixed point for the operator

$$F(\theta)x = \theta x + (1-\theta)Fx, \quad 0 < \theta < 1. \tag{2.7}$$

Suppose now that $\mathfrak{X}$ is strongly convex. Then

$$\begin{aligned} \|z - F(\theta)y\| = \|z - \theta y - (1-\theta)Fy\| = \|\theta z - \theta y + (1-\theta)z - (1-\theta)Fy\| < \\ < \theta\|z-y\| + (1-\theta)\|Fz - Fy\| \leqslant \|z-y\|. \end{aligned} \tag{2.8}$$

In the proof of (2.8) we supposed that the vectors $Fz - Fy$ and $z - y$ are not collinear. If it is not so, i.e., $Fz - Fy = \lambda(z-y)$, $\lambda > 0$, then, since $F$ is non-expansive, there are two possibilities: either $\lambda = 1$ and then $y$ is a fixed point for $F$, or $\lambda < 1$ and then inequality (2.8) is obvious.

It follows that $2\delta = \|z-y\| - \|z - F(\theta)y\| > 0$. Further, following verbatim the computations in (2.3), (2.4), we obtain that $y$ is not a limit point of the sequence (2.6) and, consequently, it is not a fixed point for the operator $F(\theta)$. However, if $y$ is a limit point of the sequence (2.6), i.e., there exists a subsequence $\{x_{n_k} : k \in \mathbb{N}\}$ convergent to $y$, from (2.6) we obtain

$$\|x_{n_k+1} - y\| = \|F(\theta)x_{n_k} - F(\theta)y\| \leqslant \|x_{n_k} - y\|,$$

and, by induction, we get

$$\|x_{n_k+m} - y\| \leqslant \|x_{n_k} - y\|,$$

for any $m \in \mathbb{N}$. This implies the convergence of the sequence of iterations

$$x_n = F(\theta)x_{n-1}$$

which, actually, coincides with sequence (2.6). ▶

**REMARK 2.1.** Let us emphasize that the second assertion in Corollary 2.2 does not at all mean that the fixed point $z$ for $F$ is an attractive fixed point for $F(\theta)$. More precisely, if $z$ is the unique fixed point for $F$, then it is an attractive fixed point for the operator $F(\theta)$. Otherwise $z$ can be a neutral point only.

Moreover, it can be shown that the set of all fixed points for a non-expansive operator on a strongly convex Banach space is a convex set. Consequently, if there exists, more than one fixed point, then there are no isolated points among them.

**REMARK 2.2.** The requirement in Theorem 2.2 for $\mathfrak{D}$ to be a convex set can be removed and replaced by the weaker condition on $\mathfrak{D}$ to be a star-shaped set (for the definition of the star-shaped set see Chapter III). In this case we have to set $y = 0$ in (2.5).

**REMARK 2.3.** Notice that the first assertion in Corollary 2.2 follows, as a matter of fact, from Schauder principle. However, we preferred a straightforward proof which enables us to consider not only convex sets (see the previous remark).

## §3. Fixed points for differentiable operators

**1.** The next result follows from Theorem I.2.1 and the Banach principle (Theorem 1.1).

**THEOREM 3.1.** *Let $\mathfrak{D}$ be a closed convex set in a Banach space $\mathfrak{X}$ and let $F$ be an operator on $\mathfrak{D}$ which is Gâteaux differentiable at any point $x \in \mathfrak{D}$ (i.e., for any point $x \in \mathfrak{D}$ and any direction $h$ there exists a neighborhood of $0 \in \mathbb{K}$ in which the vector-function $F(x + th)$ is defined and differentiable). Assume that the derivative of $F$ satisfies the condition*

$$l = \sup_{x \in \mathfrak{D}} \|F'(x)\| < 1. \tag{3.1}$$

*If $F(\mathfrak{D}) \subseteq \mathfrak{D}$, then there exists an s-fixed point $z \in \mathfrak{D}$, for $F$. Moreover, for any $\theta \in [0, 1]$ the sequence*

$$x_n = \theta x_{n-1} + (1 - \theta)Fx_{n-1}, \quad n = 1, 2, \dots, \; x_0 \in \mathfrak{D}, \tag{3.2}$$

*converges to $z$, with the rate of a geometric progression*

$$\|x_n - z\| \leqslant \frac{q^n}{1-q}\|x_0 - Fx_0\|, \tag{3.3}$$

*where $q = \theta + (1 - \theta)l < 1$.*

◁ It is enough to observe that the operator $F(\theta)$ defined by $F(\theta)x = \theta + (1 - -\theta)Fx$ satisfies all the conditions in the Banach principle, with the Lipschitz constant $q$. The sequence (3.2) is the sequence of iterations of this operator. Formula (3.3) follows straightforwardly from (1.3), in which we replace the operator by $F(\theta)$ and then take the limit. ▶

**COROLLARY 3.1.** *Let $\mathfrak{D}$ be a closed convex set in $\mathfrak{X}$ and $F : \mathfrak{D} \to \mathfrak{D}$ an operator that is Gâteaux differentiable at any point $x \in \mathfrak{D}$, such that*

$$\sup_{x \in \mathfrak{D}} \|(F^p)'x\| < 1, \tag{3.4}$$

*for a fixed $p \in \mathbb{N}$. Then the operator $F$ has a unique attractive fixed point $z \in \mathfrak{D}$.*

◁ The differentiability of the operator $F^p$ follows from Lemma I.4.2. In fact,

$$(F^p)'x = (F')^p x, \tag{3.5}$$

where the right hand side of (3.5) must be understood as the $p$-th power of the linear operator $F'$. Now our conclusion follows from Theorem I.2.1 and Corollary 1.1. ▶

**REMARK 3.1.** We will show below that for a Fréchet differentiable operator condition (3.4) actually implies the existence of an s-fixed point.

**REMARK 3.2.** Since for a Fréchet differentiable operator the constant $l$ in condition (3.1) is the smallest Lipschitz constant in inequality (I.2.1) then, by Remark 1.3, it follows that this condition is not necessary for the existence of s-fixed points.

◁ Namely, let $z$ be a certain fixed point for a Fréchet differentiable operator $F$, lying strictly inside $\mathfrak{D}$. Assume that $\|(F^p)'z\| < 1$ for some $p \in \mathbb{N}$. Then, by the continuity of $F'(x)$ in the uniform operator topology, there exists a closed ball $\overline{\mathfrak{U}}$ centered at $z$ and completely lying inside $\mathfrak{D}$, such that $\sup\limits_{x\in\overline{\mathfrak{U}}} \|(F^p)'x\| < 1$. Moreover,

$$\|F^p x - z\| = \|F^p x - F^p z\| \leqslant \sup_{y\in\overline{\mathfrak{U}}} \|(F^p)'y\| \cdot \|z - x\| < \|x - z\|,$$

for all $x \in \overline{\mathfrak{U}}$. Thus $F^p(\overline{\mathfrak{U}}) \subseteq \overline{\mathfrak{U}}$. By the continuity of $F$, there exists a closed neighborhood $\mathfrak{V} \subseteq \overline{\mathfrak{U}}$ of the point $z$ such that $F(\mathfrak{V}) \subseteq \mathfrak{V}$. Then, according to Corollary 3.1, the point $z$ is attractive in $\mathfrak{V}$. Moreover, if $z$ is globally attractive in $\mathfrak{D}$ then, by Meyers Theorem (Theorem 1.2) $z$ is s-fixed on the whole $\mathfrak{D}$, regardless of how large the value of the norm of $F'(x)$ on $\mathfrak{D}$ is. In particular, this explains why the point 0 is an s-fixed point for the operator $F$ in Example 1.3, where $F'(0) = 0$. ▶

In Section 2 (see Corolary 2.1) we established the existence of an attractive fixed point $y$ for a non-expansive operator mapping a closed set $\mathfrak{D}$ onto a precompact subset of it. For a differentiable operator the following version of Corollary 2.1 is true.

**THEOREM 3.2.** *Let $\mathfrak{D}$ be a closed convex set in the Banach space $\mathfrak{X}$ and let $F : \mathfrak{D} \to \mathfrak{D}$ be an operator which is indefinitely Fréchet differentiable in a neighborhood of each point $x \in \mathfrak{D}$. Assume that for some $p, q \in \mathbb{N}$ the following conditions are fulfilled:*

1) *the operator $F^p$ maps $\mathfrak{D}$ onto a precompact subset of $\mathfrak{D}$, and*
2) *$\|(F^q)'(x)\| < 1$ for all $x \in \mathfrak{D}$.*

*Then the operator $F$ has an attractive fixed point.*

◁ The operator $F^{pq}$ satisfies all the conditions in Corollary 2.1. Indeed, on one hand, $F^{pq} = (F^q)^p$ and, consequently, $\|(F^{pq})'(x)\| \leqslant \|(F^q)'(x)\|^p < 1$, for all $x \in \mathfrak{D}$, i.e., the operator $F^{pq}$ is strictly non-expansive. On the other hand, $F^{pq} = (F^p)^q$ and, consequently, the operator $F^{pq}$ maps $\mathfrak{D}$ onto a precompact subset of $\mathfrak{D}$. Thus $F^{pq}$ has an attractive fixed point $z \in \mathfrak{D}$ which, obviously, is an attractive fixed point for the operator $F$, too. ▶

**2.** A sufficient condition for the existence of an s-fixed point for an operator $F$ is, for instance, condition (3.1). At the same time, if the Fréchet derivative $F'(z)$ of $F$ at the fixed point $z$ has norm 1, then it does not follow that $z$ fails to be an s-fixed point. For example, on the two-dimensional space $\mathfrak{X}$, the linear operator $A$ defined, relatively to the canonical basis, by the matrix

$$\begin{pmatrix} 0 & 1 \\ 0 & 0 \end{pmatrix}$$

has norm 1 (recall that $A' = A$ — see Lemma I.4.1), yet the point $z = 0$ is an s-fixed point for $A$. Therefore, we would like to point out some criterions that characterize, at least locally, the uniformly attractive fixed points. The strongest sufficient — and, in the case of complex spaces, also necessary — condition which accomplishes this goal is given by

$$r((F^p)'(z)) < 1, \tag{3.6}$$

for some $p \in \mathbb{N}$, where $r(A)$ denotes the spectral radius of an operator $A$ (see §2 of Chapter IV).

◁ If (3.6) is fulfilled then, according to Theorem IV.2.2, we can introduce on the space $\mathfrak{X}$ a new norm $\| \ \|_*$, equivalent with the original one, such that $\|(F^p)'(z)\|_* < 1$. By Remarks 3.1 and 3.2 above the next result follows.

**THEOREM 3.3** *Assume that $F$ is a Fréchet differentiable operator on an open set $\mathfrak{D}$ and $z = F(z) \in \mathfrak{D}$. If $F$ satisfies condition (3.6), then there exists a neighborhood $\mathfrak{U}$ of the point $z$ such that $F(\overline{\mathfrak{U}}) \subseteq \overline{\mathfrak{U}}$ and $z$ is an s-fixed point for the operator $F$ on $\overline{\mathfrak{U}}$. If, in addition, the operator $F$ is continuous on the closure $\overline{\mathfrak{D}}$ of $\mathfrak{D}$, $F(\overline{\mathfrak{D}}) \subseteq \overline{\mathfrak{D}}$, and $z$ is an attractive fixed point in $\mathfrak{D}$, then $z$ is s-fixed.* ▶

**COROLLARY 3.2.** *Let $\mathfrak{D}$ be a closed convex set in $\mathfrak{X}$ and let $F : \mathfrak{D} \to \mathfrak{D}$ be an operator that is differentiable in $\mathfrak{D}$ and satisfies condition (3.4). Then $F$ has an s-fixed point in $\mathfrak{D}$.*

◁ The existence of a unique attractive fixed point $z$ is a consequence of Corollary 3.1. By equalities (3.5) and (IV.2.2), it follows that the spectral radius of the Fréchet derivative of $F$ at the point $z$ is less than 1. In order to conclude the proof, it remains to use Theorem 3.2. ▶

**REMARK 3.3.** Let us consider the linear equation

$$x = Ax + y, \quad y \in \mathfrak{X},\ A \in L(\mathfrak{X}, \mathfrak{X}), \tag{3.7}$$

where $r(A) < 1$. By the arguments above we can construct a new equivalent norm on $\mathfrak{X}$, relatively to which $A$ becomes a contraction. Consequently, for any $y \in \mathfrak{X}$ the operator $F$ defined by $F(x) = Ax + y$ is a contraction, too. Therefore, using inequalities (3.3) we get an estimate, with respect to this new norm, for the rate of convergence of the succesive approximations to the solution of equation (3.7). Unfortunately, this estimate does not hold for nonlinear equations, in general. Estimates (3.3), under the conditions in Theorem 3.2 or Corollary I.3.2 and with respect to a suitable norm, are true only when $x_0$ is sufficiently closed to $z$. If, in addition, the operator $F^p$ subject to the conditions in Corollary 3.2 can be computed explicitly, then it is convenient to use its iterations to approximate the fixed point $z$.

However, let us also point out the fact that if the Fréchet derivative of the operator $F$ at the fixed point $z$ is zero, then the iterations of $F$ on a certain neighborhood of the point $z$ converge to $z$, with the rate of a geometric progression of an as small as we please ratio. This remark can be strengthened if we assume the existence of higher order derivatives for the operator $F$.

◁ Namely, let us assume that the operator $F$ is $m$-times continouously differentiable on the closure of $\mathfrak{D}$, and $F^{(k)}(z) = 0$ for all $k = 1, 2, \ldots, m-1$. Then, using one of the Taylor formulas I.7.5 or II.4.4, we obtain the inequality

$$\|Fx - z\| \leqslant \frac{1}{m!} M \|x - z\|^m,$$

where $M = \sup\limits_{x \in \overline{\mathfrak{D}}} \|F^{(m)}(x)\|$, which is true for all $x$ sufficiently close to $z$, let us say, for those $x$ lying in the ball centered at $z$ and with radius $\delta$ ($\delta < \operatorname{dist}(z, \partial\mathfrak{D})$). Assume now that $z$ is an attractive point for $F$. Then there exists an integer $n_0$ such that $\|x_{n_0} - z\| < \delta$, where $x_n = Fx_{n-1}$, $n = 1, 2, \ldots$. Setting $L = M/m!$, for any $n \geqslant n_0$ we get

$$\|x_n - z\| \leqslant L\|x_{n-1} - z\|^m \leqslant L^{m+1}\|x_{n-2} - z\|^{m^2} \leqslant$$
$$\leqslant L^{m^2+m+1}\|x_{n-3} - z\|^{m^3} \leqslant \cdots \leqslant L^{1+m+m^2+\cdots+m^{n-n_0-1}}\|x_{n_0} - z\|^{m^{n-n_0}},$$

or

$$\|x_n - z\| \leqslant \frac{1}{L_1} \left(L_1 \|x_{n_0} - z\|\right)^{m^{n-n_0}}, \tag{3.8}$$

where $L_1 = \sqrt[m-1]{L}$. ▶

The estimate (3.8) shows that, starting with $n_0$, the iterations converge with a rate of convergence better even than the factorial rate.

Notice that, under these conditions, the sequence (3.2), for $\theta > 0$, converges more slowly than for $\theta = 0$.

## §4. Some applications of fixed point principles

The fixed point principles offer the possibility to apply a lot of methods from functional analysis in different other mathematical fields.

**1.** EQUATIONS WITH COMPOSITION OPERATORS. Let $\mathfrak{E}$ be a fixed Banach space over the field $\mathbb{K}$. Choose and fix a subset $\Omega$ of $\mathbb{K}$ and consider a set $\mathfrak{X}$ of vector-functions defined on $\Omega$ with values in $\mathfrak{E}$, such that $\mathfrak{X}$ has the structure of a Banach space.

For example, if $\Omega$ is a compact set and $\mathfrak{X}$ is the space of all continuous vector--functions from $\Omega$ into $\mathfrak{E}$, then we can set

$$\|x\| = \sup_{\sigma \in \Omega} \|x(\sigma)\|_{\mathfrak{E}},$$

i.e., $\mathfrak{X} = C(\Omega, \mathfrak{E})$.

If $\mathfrak{X}$ is the set of all vector-functions measurable in the sense of §1 of Chapter II and, at the same time, square summable on $\Omega$ in the usual sense, then $\mathfrak{X}$ becomes a Banach space, with the norm

$$\|x\|^2 = \int_{\Omega} \|x(\sigma)\|^2 \mathrm{d}\sigma.$$

Following the usual notation, we have $\mathfrak{X} = L_2(\Omega, \mathfrak{E})$. If $\mathfrak{E}$ is a Banach algebra, then it can be shown that $\mathfrak{X}$ is a Hilbert space.

Analogously we can introduce the spaces $L_p(\Omega, \mathfrak{E})$, and some other structures of Banach spaces.

Let us return to the general case and let $\mathfrak{X}$ be a given space of vector-functions defined on the set $\Omega$, with values in the Banach space $\mathfrak{E}$. Let $f(\cdot, \cdot)$ be a fixed operator defined on the topological product $\Omega \times \mathfrak{E}$ with values in $\mathfrak{E}$. If $f(\sigma, x(\sigma)) \in \mathfrak{X}$ then the operator $f(\cdot, \cdot)$ defines a composition operator $F : \mathfrak{X} \to \mathfrak{X}$ by the equality

$$Fx(\cdot) = f(\cdot, x(\cdot)). \tag{4.1}$$

The operator (4.1) is an analogue of the Nemytzki operator (see Example I.3.3). As in the case of the Nemytzki operator, set $\mathfrak{H} = \{x \in \mathfrak{E} : \|x\| \leqslant R\}$ and assume that $f : \Omega \times \mathfrak{H} \to \mathfrak{E}$ is a jointly continuous operator, such that the operator $f'_x(\sigma, x)$ exists and is jointly continuous on $\Omega \times \mathfrak{H}$, too. Suppose next that $\mathfrak{X} = C(\Omega, \mathfrak{E})$. Then the operator $F$ is Fréchet differentiable on the ball $\mathfrak{D}$ of radius $R$ centered at the origin and

$$F'(x(\cdot))h(\cdot) = f'_x(\sigma, x(\cdot))h(\cdot).$$

Further, if $\|f(\sigma, x)\| \leqslant R$ for all $(\sigma, x) \in \Omega \times \mathfrak{H}$, and

$$\sup_{(\sigma,x)\in\Omega\times\mathfrak{H}} \|f'_x(\sigma, x)\|_{\mathfrak{E}} < 1, \tag{4.2}$$

then, obviously, $F(\mathfrak{D}) \subseteq \mathfrak{D}$ and

$$\sup_{\|x\| \leqslant R} \|F'(x)\| < 1.$$

Hence, by Theorem 1.1, the operator $F$ has an s-fixed point. Thus we proved the next result.

**PROPOSITION 4.1.** *Let $f(\cdot, \cdot)$ be an operator defined on the topological product $\Omega \times \mathfrak{H}$ and with values in $\mathfrak{H}$ (with $\Omega$ and $\mathfrak{H}$ as above). Assume that both $f$ and its partial derivative $f'_x$, $x \in \mathfrak{H}$, are continuous on $\Omega \times \mathfrak{H}$, and that condition (4.2) is fulfilled. Then the equation*

$$x(\sigma) = f(\sigma, x(\sigma)) \tag{4.3}$$

*has a unique solution in the set of all continuous vector-functions from $\Omega$ into $\mathfrak{H}$. Moreover, the sequence*

$$x_n(\sigma) = f(\sigma, x_{n-1}(\sigma)), \quad n = 1, 2, \ldots, \ x_0 \in \mathfrak{H},$$

*converges to $x(\sigma)$ uniformly relatively to $x_0$, and, for a fixed $x_0$, uniformly on each compact subset of $\Omega$.*

By replacing condition (4.2) with the slightly weaker conditions

$$\sup_{(\sigma,x)\in\Omega\times\mathfrak{H}} \|f'_x(\sigma, x)\| \leqslant 1$$

or

$$\|f'_x(\sigma, x)\| < 1 \quad \text{for all } (\sigma, x) \in \Omega \times \mathfrak{H},$$

we can easily reformulate, in this new setting, counterparts of the results in §§2,3.

**2.** EQUATIONS WITH HAMMERSTEIN TYPE OPERATORS. Keeping the notations introduced in Subsection 1 above, let us set $\Omega = [a, b]$ and $\mathfrak{X} = C(\Omega, \mathfrak{E})$. Consider the operator $\Phi : \mathfrak{X} \to \mathfrak{X}$ defined by

$$\Phi x(t) = \int_a^b K(\sigma, t) f(\sigma, x(\sigma)) \mathrm{d}\sigma, \tag{4.4}$$

where $K(\sigma, t)$ — the kernel of the operator $\Phi$ — is a scalar-function defined for $a \leqslant \sigma, t \leqslant b$ and separately continuous on $[a, b] \times [a, b]$. Since the operator $f$ is jointly continuous, the intergal in (4.4) of the vector-function $K(\sigma, t) f(\sigma, x(\sigma))$ exists for any continuous function $x(\sigma)$, and depends continuously on the variable $t \in [a, b]$. Further, the equality

$$\Phi'(x(\cdot))h(\cdot) = \int_a^b K(\sigma, \cdot) f'_x(\sigma, x(\sigma)) h(\sigma) \mathrm{d}\sigma$$

holds.

**PROPOSITION 4.2.** *Assume that for any $\sigma \in [a, b]$ the operator $f(\sigma, \cdot)$ maps the ball $\mathfrak{H} = \{x \in \mathfrak{E} \,:\, \|x\| \leqslant R\}$ into the ball of radius $R_1$ with center at the zero element of the space $\mathfrak{E}$. Let*

$$L = \sup_{(\sigma,x)\in\Omega\times\mathfrak{H}} \|f'_x(\sigma, x)\|\,, \quad \kappa = \max_{\substack{a \leqslant \sigma \leqslant b \\ a \leqslant t \leqslant b}} |K(\sigma, t)|.$$

*If*

$$\begin{cases} R_1 \kappa (b - a) \leqslant R \\ \text{and} \\ L\kappa(b - a) < 1 \end{cases} \tag{4.5}$$

*then the equation*

$$x(t) = \int_a^b K(t, \sigma) f(\sigma, x(\sigma)) \mathrm{d}\sigma \tag{4.6}$$

*has a unique solution in $C(\Omega, \mathfrak{E})$, which is the uniform limit of the sequence*

$$x_n(t) = \int_a^b K(t, \sigma) f(\sigma, x_{n-1}(\sigma)) \mathrm{d}\sigma, \quad n = 1, 2, \ldots\,, \tag{4.7}$$

*where $x_0(t)$ is an arbitrary continuous vector-function from $[a,b]$ into $\mathfrak{H}$, i.e., $\|x_0(t)\| \leqslant R$.*

◁ The first inequality in (4.5) guarantees that $\|\Phi x(\cdot)\| \leqslant R$, for any $x$ with $\|x\| \leqslant R$ whereas the second one implies $\sup\limits_{\|x\| \leqslant R} \|\Phi'(x(\cdot))\| < 1$. It follows that the operator $\Phi$ is a contraction on the ball of radius $R$ with center at the zero element of the space $\mathfrak{X} = C(\Omega, \mathfrak{E})$. ▶

Further, let us remark that the operator $\Phi$ is the composition of the operator $F$ associated to $f$ as in Subsection 1, with a linear operator $A$, that is $\Phi = AF$, where

$$Ay(t) = \int_a^b K(t,\sigma)y(\sigma)\mathrm{d}\sigma.$$

We have

$$\|A\| = \sup_{\|y(\cdot)\| \leqslant 1} \left\| \int_a^b K(t,\sigma)y(\sigma) \right\| \leqslant \kappa(b-a).$$

This remark enables us to weaken slightly the second inequality in (4.5). More precisely, the next result is true.

**PROPOSITION 4.3.** *Assume that*

$$\begin{cases} R_1\kappa(b-a) \leqslant R, \\ Lr(A)(b-a) < b-a, \end{cases} \tag{4.8}$$

*where $r(A)$ is the spectral radius of the operator $A$. Then, equation (4.6) has a unique solution in $C(\Omega, \mathfrak{H})$, which is the uniform limit of sequence (4.7) for any $x_0(\cdot) \in C(\Omega, \mathfrak{H})$.*

◁ Indeed, from $\Phi = AF$ and by Corollary I.4.1, it follows that $\Phi' = AF'$ and $(\Phi^n)' = A^n(F^n)'$, where $\Phi^n$ and $F^n$ are the $n$-th iterations of the operator $\Phi$ and $F$, respectively. If the second condition in (4.8) is fulfilled, then, for any $\varepsilon > 0$, we can choose $p \in \mathbb{N}$ such that $\|A^p\| \leqslant (r(A)+\varepsilon)^p$. For a sufficiently small $\varepsilon$ we have $L(r(A)+\varepsilon)) < 1$ and, consequently

$$\sup_{\|x(\cdot)\| \leqslant R} \|(\Phi^p)'(x(\cdot))\| \leqslant \sup_{\|x(\cdot)\| \leqslant R} \|A^p\| \, \|(F^p)'(x(\cdot))\| < 1.$$

To conclude the proof it remains to use Corollary 3.1. ▶

Although it is practically easier to check condition (4.5), however condition (4.8) is a matter of principle. For example, it shows that if the spectral radius of $A$ is zero, then it is enough to check only the first inequality in condition (4.5).

Let us consider a rather general situation. Assume that the equation

$$x = AFx + y \tag{4.9}$$

is given on a Banach space $\mathfrak{X}$, where $F$ is a continuously differentiable operator on the closed ball $\mathfrak{D}$ of radius $R$ and centered at the zero element of the space $\mathfrak{X}$, $A$ is a linear continuous operator on $\mathfrak{X}$, and $y \in \mathfrak{X}$. (Such equations are usually called equations of the second genre.)

◁ By Theorem 3.1, if conditions

$$\begin{cases} \|A\| \cdot M + \|y\| \leqslant R, \\ \|A\| \cdot L < 1 \end{cases} \tag{4.10}$$

are fulfilled, where

$$M = \sup_{\|x\| \leqslant R} \|Fx\|, \quad L = \sup_{\|x\| \leqslant R} \|F'(x)\|,$$

then equation (4.9) has a unique solution in $\mathfrak{D}$, which is the limit, in the strong sense, of the sequence

$$x_n = AFx_{n-1} + y, \quad n = 1, 2, \dots , \tag{4.11}$$

with $x_0$ an arbitrary element of $\mathfrak{D}$.

Conditions (4.10) are straightforward generalizations of conditions (4.5). Setting $G = AF$ we obtain

$$G^2(\cdot) = AF(AF(\cdot))$$

and

$$(G^2)'(x) = AF'(AF(x))AF'(x).$$

Therefore

$$\sup_{\|x\| \leqslant R} \|(G^2(x))'\| \leqslant L^2\|A\|^2.$$

More generally, if $Lr(A) < 1$ then, using (IV.2.2), we obtain

$$\sup_{\|x\| \leqslant R} \|(G^n)'(x)\| \leqslant L^n\|A^n\| \leqslant L^n(r(A) + \varepsilon)^n < 1,$$

for a sufficiently small $\varepsilon > 0$ and a sufficiently large $n \in \mathbb{N}$. Thus, according to Corollary 3.1, conditions

$$\begin{cases} \|A\|M + \|y\| \leqslant R, \\ r(A)L < 1 \end{cases} \tag{4.12}$$

are sufficient for the existence of the solution of equation (4.9), and for the sequence (4.11) to converge to that solution. ▶

**REMARK 4.1.** Let us emphasize on the fact that the first inequality in (4.12), which means the invariance of the ball $\mathfrak{D}$ relatively to $G + y$, is a rather rough estimate, and for certain $y$ and $M$ it may not be fulfilled. Yet, taking into account that $M$ is a function of $R$, if the second inequality in (4.12) is fulfilled, then we can select $R$ such that, for any $y$ of a sufficiently small norm, the operator $G + y$ has an invariant domain.

◁ Indeed, we may assume that $F(0) = 0$ (otherwise we can consider the equation $x = AF_1(x) + y$, where $F_1(x) = F(x) - F(0)$, which is equivalent with equation (4.9)). Then $0 \in \mathfrak{X}$ is a fixed point for the operator $AF$. If the second inequality in (4.12) is fulfilled, then, for any sufficiently small $\varepsilon > 0$ there exist a norm $\|\cdot\|_*$ on $\mathfrak{X}$ equivalent to the original one, and a number $r > 0$ such that the operator $AF$ is a contraction on the ball $\{x : \|x\|_* < r\}$, with the constant $q = R(A)L + \varepsilon < 1$. If $y$ satisfies the condition

$$\|y\|_* < (1 - q)r,$$

then

$$\|G(x)\|_* \leqslant \|AF(x)\|_* + \|y\|_* =$$
$$= \|AF(x) - AF(0)\|_* + \|y\|_* < q\|x\|_* + (1 - q)r = r. \qquad ▶$$

Thus, the condition $r(A)L < 1$ is sufficient for the local solvability of equation (4.9). However, it is not necessary. (By "local solvability" we mean the existence of the solution for all $y$ in a certain neighborhood of zero, and for a sufficiently small value of $\|F(0)\|$.)

With the same notations as above, assume that

$$\begin{cases} \|A\|M + \|y\| \leqslant r, \\ r(A)L \leqslant 1. \end{cases}$$

Then we can try to construct a norm relatively to which the operator $G$ is non--expansive. In what follows we restrict ourselves to the case when $\|A\| = r(A)$ (this is the case, for instance, if $A$ is a self-adjoint operator on a Hilbert space). If at least one of the operators $A$ or $F$ maps $\mathfrak{D}$ into a compact set, then $G = AF$ will have the same property. In this case we can use the results in §2. In particular, from Corollary 2.2 we have:

**PROPOSITION 4.4.** *Let $\mathfrak{E} = \mathbb{K}^n$, for some $n \in \mathbb{N}$ ($\mathbb{K}$ is the field of real or complex numbers), and let $f(\sigma, u)$ be a vector-function defined for $\sigma \in [a, b]$ and $u \in \mathbb{K}^n$,*

*$\|u\| \leqslant R$, with values in $\mathbb{K}^n$. Assume that $f$ and its partial derivative $f'_u(\sigma, u)$ are continuous on the whole domain of definition. Further, let $K(\sigma, t)$ be a symmetric scalar-function which defines a linear completely continuous integral operator $A$ on the space $L_2(\Omega, \mathbb{K}^n)$, by the formula*

$$Ax(t) = (\mathrm{P}) \int_a^b K(t, \sigma) x(\sigma) \mathrm{d}\sigma, \quad x \in L_2(\Omega, \mathbb{K}^n).$$

*If the conditions*

$$\begin{cases} \kappa M(b-a) + \max\limits_{t \in [a,\, b]} \|y(t)\|_{\mathbb{K}^n} \leqslant R, \\ \kappa L(b-a) \leqslant 1 \end{cases}$$

*are fulfilled, then there exists at least one solution of the Hammerstein type equation of the second genre*

$$x(t) = (\mathrm{P}) \int_a^b K(t, \sigma) f(\sigma, x(\sigma)) \mathrm{d}\sigma + y(t).$$

*Moreover, the sequence of vector-functions*

$$x_n(t) = \theta x_{n-1}(t) + (1-\theta)\,(\mathrm{P}) \int_a^b K(t, \sigma) f(\sigma, x_{n-1}(\sigma)) \mathrm{d}\sigma + y(t), \quad n = 1, 2, \dots ,$$

*converges to a solution of this equation, for any $x_0$ with $\|x_0(t)\| \leqslant R$, $t \in [a,\, b]$, and any $\theta \in (0,\, 1)$.*

**3.** NONLINEAR VOLTERRA TYPE EQUATIONS. As a particular case of equation (4.9) let us consider an equation of the form

$$x(t) = \int_a^t K(t, \sigma) f(\sigma, x(\sigma)) \mathrm{d}\sigma + y(t), \tag{4.13}$$

where $K$ and $f$ satisfy the conditions in Proposition 4.2.

The spectral radius of the linear operator $A$ — the Volterra operator — defined by

$$Ax(\cdot) = \int_a^t K(\cdot, \sigma) x(\sigma) \mathrm{d}\sigma,$$

is equal to zero. Indeed, for any function $x(\cdot) \in C(\Omega, \mathfrak{E})$ we have

$$\|A^2 x(\cdot)\| = \left\| \int_a^t K(t,\sigma) A x(\sigma) d\sigma \right\| \leqslant \kappa^2 \|x(\cdot)\| \frac{(t-a)^2}{2}.$$

More generally, for any $n \in \mathbb{N}$ we obtain

$$\|A^n x(\cdot)\| \leqslant \kappa^n \|x(\cdot)\| \frac{(t-a)^n}{n!},$$

or,

$$\|A^n\| \leqslant \kappa^n \frac{(b-a)^n}{n!}. \tag{4.14}$$

Since the right hand side of the last inequality approaches zero as $n \to \infty$, we get

$$r(A) = \lim_{n \to \infty} \sqrt[n]{\|A^n\|} = 0.$$

From condition (4.10) we have:

**PROPOSITION 4.5.** *If the vector-functions $K$, $f$ and $y$ satisfy the condition*

$$\kappa M(b-a) + \max_{a \leqslant t \leqslant b} \|y(t)\| \leqslant R, \tag{4.15}$$

*then equation (4.13) has a unique solution, and the sequence*

$$x_n(t) = \int_a^t K(t,\sigma) f(\sigma, x_{n-1}(\sigma)) d\sigma + y(t), \quad n = 1, 2, \dots, \tag{4.16}$$

*where $\|x_0(t)\| \leqslant R$, converges to that solution. Moreover, the next estimate*

$$\sup_{\|x\| \leqslant R} \|(G^n)'(x)\| \leqslant \|A^n\| L^n \leqslant \frac{1}{n!} (\kappa L(b-a))^n$$

*can be established analogously to estimates (4.14). These estimates imply that the sequence (4.16) converges to the solution of equation (4.13) with a factorial type rate, i.e.,*

$$\|x_n(t) - x(t)\| \leqslant \frac{1}{n!} (\kappa L(b-a))^n \|x_1(t) - x_0(t)\|.$$

**REMARK 4.2.** As we have already noticed, condition (4.15) requires some rather strong restrictions on vector-functions $K$, $f$ and $y$. However, according to Subsection 2 above, we can assert that equation (4.13) has a unique solution, for any $\kappa$ and $M$, if

$y$ is sufficiently close to the element $f(\sigma, 0)$. Moreover, taking into account the specific form of the operator $AF$ (it involves an integral over an interval with a variable upper limit), we can assert the following: for any $y \in C(\Omega, \mathfrak{E})$, with $\|y\| < R$, there exists a number $\tau \in [a, b]$ for which equation (4.13) has a unique solution $x \in C(\Omega_\tau, \mathfrak{E})$, where $\Omega_\tau = [a, \tau]$. Indeed, condition (4.15) can be replaced, in this case, by the following condition:

$$\kappa M(\tau - a) + \max_{a \leqslant t \leqslant \tau} \|y(t)\| \leqslant R, \tag{4.17}$$

which is obviously fulfilled for

$$\tau \leqslant a + \frac{R - \|y(t)\|}{\kappa M}.$$

In the case when $R = \infty$, i.e., $f(\sigma, x)$ is defined on the whole $\Omega \times \mathfrak{E}$ and satisfies all the conditions listed above, then equation (4.13) has a unique solution for any $\kappa, M < \infty$ and any $y \in C(\Omega, \mathfrak{E})$.

**REMARK 4.3.** Let us note also the following possible approach. Consider the operator $G$ defined by the right hand side of equation (4.13). Then, it is easy to construct a norm on $\mathfrak{X}$, equivalent with the original norm, with respect to which $G$ becomes a contraction. Indeed, for any $\tau \in [a, b]$ set

$$\|x\|_\tau^* = \max_{a \leqslant t \leqslant \tau} \mathrm{e}^{-mt} \|x(t)\|_{\mathfrak{E}},$$

where $m > 0$ will be specified below. Clearly,

$$\|x\|_\tau \mathrm{e}^{-m\tau} \leqslant \|x\|_\tau^* \leqslant \|x\|_\tau,$$

where $\|x\|_\tau$ is the usual norm of an element $x \in C(\Omega_\tau, \mathfrak{E})$, i.e.,

$$\|x\|_\tau = \max_{a \leqslant t \leqslant \tau} \|x(t)\|_{\mathfrak{E}}.$$

If $\|x\|_\tau^* \leqslant R\mathrm{e}^{-m\tau}$ and $\|z\|_\tau^* \leqslant R\mathrm{e}^{-m\tau}$, then $x, z \in \mathfrak{H}$ and $\|F(x) - F(z)\|_\tau \leqslant L\|x - z\|_\tau$. Consequently,

$$\|G(x) - G(z)\|_\tau = \max_{a \leqslant t \leqslant \tau} \left\| \mathrm{e}^{-mt} \int_a^t K(t, \sigma)\{f(\sigma, x(\sigma)) - f(\sigma, z(\sigma))\} \mathrm{d}\sigma \right\|_{\mathfrak{E}} \leqslant$$

$$\leqslant \kappa L \max_{a \leqslant t \leqslant \tau} \left| \mathrm{e}^{-mt} \int_a^t \|x(\sigma) - z(\sigma)\|_{\mathfrak{E}} \mathrm{d}\sigma \right| \leqslant$$

$$\leqslant \kappa L \max_{a \leqslant t \leqslant \tau} \int_a^t e^{m(\sigma-t)} e^{-m\sigma} \|x(\sigma) - z(\sigma)\|_{\mathfrak{E}} d\sigma \leqslant$$

$$\leqslant \kappa L \|x - z\|_\tau^* \max_{a \leqslant t \leqslant \tau} \int_a^t e^{m(\sigma-t)} d\sigma =$$

$$= \kappa L \frac{1}{m}(1 - e^{-m\tau}) \|x - z\|_\tau^*.$$

Therefore, setting $m \geqslant \kappa L$, we obtain that $G$ is a non-expansive operator relatively to the norm $\| \ \|_\tau^*$. This remark enables us to specify not only a new estimate for the convergence rate of sequence (4.16), but also a new estimate for $y$, which guarantees that equation (4.13) is solvable. To this end, it is sufficient to repeat the arguments in Remark 4.1 for $q = \kappa L m^{-1}(1 - e^{m\tau})$ and $m \geqslant \kappa L$. Then we get the following condition

$$\|y\|_\tau^* \leqslant (1 - q)r, \quad r = Re^{-m\tau}.$$

In particular, for $m = \kappa L$ we obtain

$$\|y\|_\tau^* \leqslant Re^{-2m\tau}.$$

**4.** THE CAUCHY PROBLEM FOR DIFFERENTIAL EQUATIONS IN A BANACH SPACE. The next result is a straightforward consequence of the results in the previous subsection.

**PROPOSITION 4.6.** *As above, let $\Omega = [a, b]$, let $\mathfrak{E}$ be a Banach space, and $f(\sigma, u)$ an operator from $\Omega \times \mathfrak{H}$ into $\mathfrak{E}$, where $\mathfrak{H}$ is the open ball of radius $R$ in $\mathfrak{E}$. Assume that both $f(\sigma, u)$ and $f'_u(\sigma, u)$ are continuous on $\Omega \times \mathfrak{H}$. Then, for any element $y_0 \in \mathfrak{H}$, there exists a unique solution of the differential equation*

$$x'(\sigma) = f(\sigma, x(\sigma)), \tag{4.18}$$

*on a certain interval $[a, \tau]$, which satisfies the initial condition*

$$x(a) = y_0. \tag{4.19}$$

*In addition, if the norm $\|y_0\|$ is sufficiently small, or $\mathfrak{H} = \mathfrak{E}$, then the solution of the problem (4.18)–(4.19) can be continuously extended in a unique way to a solution on the whole interval $[a, b]$.*

◁ For the proof it is sufficient to replace problem (4.18)–(4.19) by the equivalent integral equation

$$x(t) = \int_a^t f(\sigma, x(\sigma)) d\sigma + y_0,$$

which is a particular case of equation (4.13), at least for a finite-dimensional space $\mathfrak{E}$. ▶

**REMARK 4.4.** The integral operators appearing in the problems considered above are completely continuous. Therefore, the existence of the solutions of these problems follows from Schauder principle, under some less restrictive conditions: it is not necessary to require the differentiability of the operator; it is sufficient to have only the invariance of a suitable domain relatively to that operator. (As we have already observed, in this case, we can not guarantee the uniqueness of the solution.) However, in many practical cases, this does not lead to significant improvements.

**5.** EQUATIONS WITH ANALYTIC INTEGRAL OPERATORS. In connection with some problems in nonlinear mechanics, many authors, such as, for example, A. M. Liapunov, Lichtenstein, E. Schmidt, A. N. Nekrasov, H. H. Nazarov, and others (see M. M. Vainberg and V. A. Trenogin [1]) studied the Hammerstein equation in which the function $f(\sigma, u)$ is analytic in $u$.

First, let us consider the simplest nonlinear equation that often occurs in a series of problems in electrotechnics:

$$x(t) = \int_a^b f(\sigma, t) x^m(\sigma) \mathrm{d}\sigma + y(t), \tag{4.20}$$

where the complex-valued functions $f(\sigma, t)$ and $y(t)$ are continuous for $a \leqslant t, \sigma \leqslant b$, and $m > 1$.

Since in many concrete problems the function $y(t)$ has a well-determined meaning, it is important to find an estimate for the absolute value of $y(t)$, which guarantees that equation (4.20) has solutions. In order to accomplish this we will consider the auxiliary, so-called, majorant numerical equation:

$$r = N r^m + \|y\|, \tag{4.21}$$

where $N$ is a positive number. Since the function in the right hand side of (4.21) is concave upward for $r > 0$, equation (4.21) has only two positive solutions $r_*$ and $r^*$, $r_* < r^*$, for

$$\|y\| < l = \frac{m-1}{m \sqrt[m-1]{mN}}; \tag{4.22}$$

for $\|y\| = l$ it has the unique positive solution $r_0 = r_* = r^* = (mN)^{-\frac{1}{m-1}}$ (see Figure 1 below).

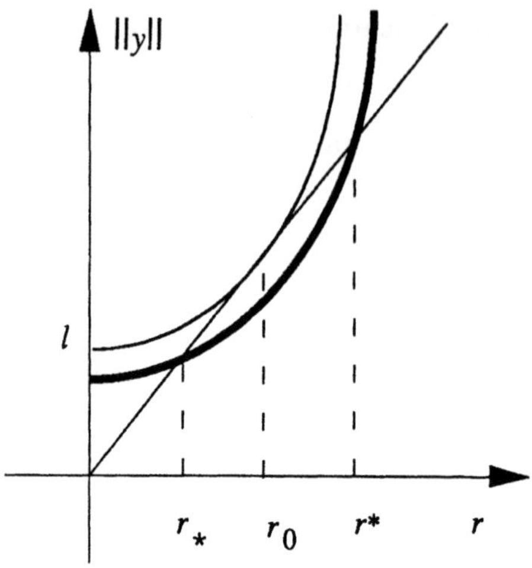


Figure 1

Thus, if we set $N = \kappa(b-a)$, where $\kappa = \max\limits_{a \leqslant \sigma, t \leqslant b} |f(\sigma, t)|$, then for

$$\|y\| \leqslant \frac{m-1}{m \sqrt[m-1]{m\kappa(b-a)}} \tag{4.23}$$

the operator $F$ defined by the right hand side of (4.20) satisfies the following condition

$$\|F(x)\| \leqslant N r^{*m} + \|y\| = r^*,$$

for $\|x\| \leqslant r^*$. According to Schauder principle, there exists at least a solution of equation (4.20) satisfying the estimate

$$\max_{a \leqslant t \leqslant b} |x(t)| \leqslant r^*. \tag{4.24}$$

Whenever $y$ satisfies the stronger inequality (4.22), with $N = \kappa(b-a)$, then the operator $F$ is a contraction on the ball $\|x\| \leqslant r_*$.

$\triangleleft$ Indeed, under these conditions we have $r_* < r_0$, hence

$$\|F(x_1) - F(x_2)\| \leqslant \kappa(b-a)\, \|x_1^m(t) - x_2^m(t)\| \leqslant$$
$$\leqslant m\kappa(b-a) r_*^{m-1} \|x_1 - x_2\| = q\|x_1 - x_2\|,$$

where $q = m\kappa(b-a)r_*^{m-1} < 1$. Moreover, $\|F(x)\| < r_*$ for $\|x\| < r_*$. Therefore, under condition (4.22), equation (4.20) has a unique solution $x(t)$, satisfying the estimate

$$\max_{a \leqslant t \leqslant b} |x(t)| \leqslant r_*. \tag{4.25}$$

In addition, the sequence $\{x_n(t)\}_{n\in\mathbb{N}}$ given by

$$x_n(t) = \int_a^b K(\sigma, t) x_{n-1}^m(t) \mathrm{d}\sigma + y(t), \qquad n = 1, 2, \dots , \tag{4.26}$$

converges uniformly on the interval $[a, b]$ to $x(t)$, for any continuous function $x_0(t)$ which satisfies the condition

$$\max_{a \leqslant t \leqslant b} |x_0(t)| \leqslant r_*.$$ ▶

Let us point out the next important remark.

**REMARK 4.5.** If condition (4.22) is fulfilled, then between the numbers $r_*$ and $r^*$ there exists a "clearance" (see Figure 1), which guarantees the existence of at least one solution for equation (4.20) continuous on $[a, b]$. However, in this case, following the same arguments, we can not infer from condition (4.27) the existence of a unique solution for equation (4.20) satisfying the above mentioned estimate. To achieve this goal we have to use essentially the properties of analytic operators.

**REMARK 4.6.** An upper-bound of the norm of $y$ that guarantees the existence of a unique continuous solution can be obtained using the classical Cauchy-Goursat majorant series method (see V. A. Trenogin [1]).

Using Cauchy inequalities (see §3 of Chapter III) for the coefficients $a_k(\sigma)$, we can estimate the Lipschitz constant $L$ for the function $f(\sigma, u)$ on a certain disk of radius $\rho_1 < \rho$ and centered at zero, as follows:

$$\begin{aligned} L = \sup_{\substack{|u| \leqslant \rho_1 \\ a \leqslant \sigma \leqslant b}} |f'_u(\sigma, u)| &= \sup_{\substack{|u| \leqslant \rho_1 \\ a \leqslant \sigma \leqslant b}} \left| \sum_{k=m}^{\infty} k a_k(\sigma) u^{k-1} \right| \leqslant \\ &\leqslant \sum_{k=m}^{\infty} k M \rho^{-k} |u|^{k-1} \leqslant \frac{M}{\rho} \sum_{k=m}^{\infty} k \left(\frac{\rho_1}{\rho}\right)^{k-1} = \\ &= \frac{M}{\rho} \cdot \frac{\left(\frac{\rho_1}{\rho}\right)^{m-1} \left(m - \frac{\rho_1}{\rho}(m-1)\right)}{\left(1 - \frac{\rho_1}{\rho}\right)^2}. \end{aligned}$$

Obviously $\rho_1$ can be chosen such that $L\kappa(b-a) < 1$. Now, assuming that $\rho_1$ is chosen and fixed, we can find a number $l_1 \leqslant l$ such that, for

$$\|y\| \leqslant l_1,$$

equation (4.22) has two solutions $r_*$ and $r^*$ satisfying the relations

$$0 < r_* \leqslant \rho_1 \leqslant r^*.$$

Then, the operator $F$, defined by the right hand side of (4.22), maps the ball $\|x\| \leqslant r_*$ into itself and is a contraction on it.

For example, assume that $m = 2$. Then the number $\rho_1$ must satisfy the quadratic inequality

$$\frac{M}{\rho} \frac{2\left(\frac{\rho_1}{\rho}\right) - \left(\frac{\rho_1}{\rho}\right)^2}{\left(1 - \frac{\rho_1}{\rho}\right)^2} < \frac{1}{\kappa(b-a)}.$$

In the general case the numbers $\rho_1$ and $l_1$ are rather difficult to be found.

## §5. Implicit and invertible operators. Connection with fixed points

**1.** The arguments presented in this subsection may be highly useful in finding a way to solve some concrete problems.

Let $\mathfrak{X}$, $\mathfrak{Y}$ and $\Lambda$ be Banach spaces over the field $\mathbb{K}$. Consider an operator $F : \mathfrak{X} \times \Lambda \to \mathfrak{Y}$ defined on a domain $\mathfrak{D} \times \Omega \subseteq \mathfrak{X} \times \Lambda$, with values in $\mathfrak{Y}$. An implicit operator $G : \mathfrak{U} \to \mathfrak{D}$, $\mathfrak{U} \subset \mathfrak{D}$, is defined as a solution for the equation

$$F(x, \lambda) = 0, \tag{5.1}$$

i.e., for $x = G(\lambda)$, the equality

$$F(G(\lambda), \lambda) \equiv 0 \tag{5.2}$$

holds for all $\lambda \in \mathfrak{U}$. (The operator $G$ defined in this way is not necessarily single-valued.)

If the operator $G$ is defined on the whole domain $\Omega$ and satisfies equality (5.2) we say that the implicit function problem is *globally solvable*, or that the equation (5.1) is globally solvable. However, the global solvability does not always take place. For instance, if $\mathfrak{X} = \mathfrak{Y} = \Lambda = \mathbb{K}$ then the equation $x - \lambda^3 = 0$ defines a global implicit operator $G(\lambda) = \lambda^3$, which is single-valued on the whole $\Lambda$, while the equation $x^2 + \lambda^2 = 0$ defines a single-valued operator $G(\lambda) = 0$ for $\lambda = 0$ only. We say that the implicit function problem is *locally solvable*, if $\mathfrak{U}$ is an open subset on $\Omega$. Quite often the local problem is formulated in the following way. If for a certain point $(x_0, \lambda_0) \in \mathfrak{D} \times \Omega$ we have

$$F(x_0, \lambda_0) = 0, \tag{5.3}$$

then we are interested in the existence of an implicit operator $G$, defined on a neighborhood $\mathfrak{U}$ of the point $\lambda_0$, and satisfying the condition

$$G(\lambda_0) = x_0. \tag{5.4}$$

If, in addition, such an operator $G$ is continuous on $\mathfrak{U}$, then it is called a *small solution* of equation (5.1) subject to conditions (5.3), (5.4).

Notice that in some certain cases the problem defined by (5.1)–(5.3), (5.4) may have a single-valued solution, while the more general problem (5.1)–(5.2) has a multi-valued solution.

For instance, when $\mathfrak{X} = \mathfrak{Y} = \Lambda = \mathbb{R}$, the equation $x^2 + \lambda^2 = 1$ has on the interval $[-1, 1]$ two continuous solutions: $x_1(\lambda) = \sqrt{1-\lambda^2}$ and $x_2(\lambda) = -\sqrt{1-\lambda^2}$. Considering the same equation with $\lambda_0 = \frac{1}{2}$, $x_0 = -\frac{\sqrt{3}}{2}$, then problem (5.1)–(5.3), (5.4) has the unique solution $G(\lambda) = x_2(\lambda) = -\sqrt{1-\lambda^2}$.

At the same time the additional conditions (5.3)–(5.4) do not always provide a single-valued branch of the implicit operator. For example, the equation $x^2 - \lambda = 0$ with the initial condition $x(0) = 0$ has two solutions:

$$x_1(\lambda) = +\sqrt{\lambda}, \quad x_2(\lambda) = -\sqrt{\lambda},$$

i.e., the operator $G(\lambda)$ is multi-valued.

Therefore, in each particular case, we have to point out exactly the type of the implicit operator problem, and what we mean by a single-valued implicit operator.

Consider now an operator $\Phi : \mathfrak{X} \to \mathfrak{Y}$ defined on a domain $\mathfrak{D} \subseteq \mathfrak{X}$. Assume that on a certain domain $\mathfrak{V} \subseteq \Phi(\mathfrak{D})$ there exists an operator $\Psi : \mathfrak{V} \to \mathfrak{D}$ such that $\Phi\Psi = I \mid \mathfrak{V}$, where $I$ is the identity operator on $\mathfrak{Y}$. Then the operator $\Psi$ is called an *inverse* of the operator $\Phi$ on $\mathfrak{V}$. Obviously, finding an inverse of the operator $\Phi$, is a particular case of an implicit operator problem. Indeed, the operator $\Psi$ is a solution for the equation

$$\Phi(x) = y,$$

which is a particular case of equation (5.1) if we set $\Lambda = \mathfrak{Y}$ and $F(x, y) = \Phi(x) - y$. Consequently, in this case, we can speak about a local or a global inverse, as well as about a single-valued inverse, or an inverse satisfying the initial condition

$$\Psi(y_0) = x_0, \tag{5.6}$$

if $\Phi(x_0) = y_0$.

The converse is also true. Any implicit operator problem can be reformulated as an inverse function problem. Indeed, let us assume that equation (5.1) is given. Consider the spaces $\mathfrak{E} = \mathfrak{X} \times \Lambda$, $\mathfrak{H} = \mathfrak{Y} \times \Lambda$, and the operator $\Phi : \mathfrak{E} \to \mathfrak{H}$ defined by the equality $\Phi = (F, I_\Lambda)$, where $I_\Lambda$ is the identity operator on the space $\Lambda$, i.e., the operator $\Phi$ maps each pair $(x, \lambda) \in \mathfrak{D} \times \Omega \subseteq \mathfrak{E}$ into $(y, \lambda)$, where $y = F(x, \lambda)$.

Suppose futher that the operator $\Phi$ has an inverse on a certain domain $\mathfrak{V} \subseteq \subseteq \Phi(\mathfrak{E} \times \Omega)$, i.e., there exists an operator $\Psi : \mathfrak{V} \to \mathfrak{E}$ such that

$$\Phi\Psi = I \mid \mathfrak{V}.$$

Since the equation $\Phi e = h$, where $e = (x, \lambda)$ and $h = (y, \lambda)$, is equivalent with the system

$$\begin{cases} F(x, \lambda) = y, \\ \lambda = \lambda \end{cases}$$

then, obviously, the operator $\Psi$ can be represented as $\Psi = (T, I_\Lambda)$, where $T : \mathfrak{Y} \to \mathfrak{E}$, i.e., the equality

$$F(T(y, \lambda), \lambda) \equiv y$$

is true for all $(y, \lambda) \in \mathfrak{V}$. But this means that the operator $G$, with $G(\lambda) = T(0, \lambda)$, satisfies equality (5.2) for all $\lambda \in \mathfrak{U} = \mathrm{Pr}_\Lambda \mathfrak{V}$.

Consider now the important particular case when $\mathfrak{X} = \mathfrak{Y}$. In this case equation (5.1) can be written as

$$x = P(x, \lambda). \tag{5.7}$$

Thus, for each given $\lambda$, the value $G(\lambda)$ of an implicit operator $G$ is a fixed point for the operator $P_\lambda = P(\cdot, \lambda)$ in the domain $\mathfrak{D}$.

Let us remark that there are many possibilities to rewrite equation (5.1) as in (5.7). For instance, consider the equation $x^2 + \lambda - 5x = 0$. By adding $x$ on both sides of this equation we get

$$x = x^2 + \lambda - 4x.$$

We can as well add $-x^2 - \lambda$ on both sides, and then divide by $-5$, to obtain

$$x = \frac{1}{5}(x^2 + \lambda).$$

Finally, since $x \neq 0$ for $\lambda \neq 0$, we can rewrite the equation as

$$x = \frac{5x - \lambda}{x},$$

and so on.

Apparently, it is impossible to give a general rule to find a suitable form (5.7) for equation (5.1). However, it obviously makes sense to consider the next two aspects:

1) the transformation of equation (5.1) into (5.7) has to be done such that the operator $P(\cdot,\cdot)$ to have "good" properties (continuity, differentiability, and so on);

2) equation (5.7) must satisfy one of the fixed point principles.

Let us point out also the following circumstance. In concrete problems the parameter $\lambda$ in equation (5.1), or the parameter $y$ in equation (5.5), has well-determined physical meanings. Therefore, it is important to obtain estimates which characterize the domain of existence $\mathfrak{U}$ of the implicit operator $G(\lambda)$, or the domain of existence $\mathfrak{V}$ of the inverse operator $\Psi(y)$ (as we have already noted, these problems are equivalent). It may also turn out that the operator $G = G(\lambda)$ is multi-valued on the whole domain of existence $\mathfrak{U}$, but it is single-valued on some subset $\tilde{\mathfrak{U}} \subseteq \mathfrak{U}$. The points lying in $\mathfrak{U} \setminus \tilde{\mathfrak{U}}$ at which the operator $G$ has a multi-value (a ramification) characterize, as a rule, the critical cases (for more details see our considerations below). Therefore, here too, an estimate, as accurate as it can be given, which characterize the set $\tilde{\mathfrak{U}}$, is sometimes more important than finding the operator $G$ itself.

The classical theorems on implicit operator-functions lack basically in providing such estimates, or give rather vague estimates. A typical method for obtaining estimates is to transform equation (5.1) into equation (5.7). Assume that this was done and, in addition, the operator $P$ appearing in equation (5.7) is completely continuous for each fixed $\lambda \in \Omega$. If we can find in $\Omega$ a ball $\mathfrak{U}$ centered at some point $\lambda_0 \in \Lambda$, such that $P(x,\lambda) \in \mathfrak{D}$ for all $(x,\lambda) \in \mathfrak{D} \times \mathfrak{U}$, then, according to Schauder principle, there exists an implicit operator $G(\lambda)$ on the ball $\mathfrak{U}$ satisfying equation (5.7) and, consequently, equation (5.1). The value of the radius of $\mathfrak{U}$ gives actually the desired estimate. Sometimes this estimate can be improved by choosing another set $\tilde{\mathfrak{D}} \subseteq \mathfrak{D}$, which is invariant relatively to the operator $P_\lambda$ for $(x,\lambda) \in \tilde{\mathfrak{D}} \times \mathfrak{U}_1$, where $\mathfrak{U}_1$ is a ball with a radius larger than that of $\mathfrak{U}$. We must point out that we have already followed this approach in Subsection 4 of §4. Indeed, let us return to equation (4.27): $x = Fx + y$, where $F$ is a Hammerstein integral operator. The solution $x = \Psi(y)$ of this equation can be considered as an inverse of the operator $\Phi = I - F$. Therefore, the estimates for $\|y\|$ obtained there are nothing else but estimates for the domain of existence of the inverse operator $\Psi$.

Finally, let us remark that the problem concerning the fixed points of a certain operator $S$ can be solved as well, by reducing it to an implicit operator problem. This can be done, as we will see below, by introducing a parameter in the right hand side of the equation $x = S(x)$. To be more specific, let us assume that we have already constructed an operator $P : \mathfrak{D} \times \Omega \to \mathfrak{X}$ which satisfies the equality $P(x, \lambda_0) = S(x)$ for a certain $\lambda_0 \in \Omega$. In the case when, for some reasons, there exists an implicit operator $G(x)$ defined on a certain set $\mathfrak{U} \ni \lambda_0$, which satisfies equation (5.1) for $F(x, \lambda) = x - P(x, \lambda)$, then the point $x^* = G(\lambda_0)$ is obviously a fixed point for the operator $S$.

If the set $\mathfrak{D}$ is a ball centered at the zero element of the space $\mathfrak{X}$, we can almost always choose the operator $P$ as $P(x, \lambda) = \lambda S(x)$, where $\lambda \in [0, 1]$ if $\mathfrak{X}$ is a real Banach space, or $\lambda \in \mathbb{C}$, $|\lambda| \leqslant 1$, if $\mathfrak{X}$ is a complex space. Taking into account the convexity of $\mathfrak{D}$ we can also set $P(x, \lambda) = \lambda S(x) + (1 - \lambda)y$, where $y$ is an arbitrary element of $\mathfrak{D}$ and $\lambda \in [0, 1]$. In this case, if the operator $G(\lambda)$ exists on a set containing the interval $[0, 1]$, then the point $x^* = G(1)$ is a fixed point for the operator $S$.

**2.** In the rest of this section we present some local theorems on implicit operators. A standard theorem which generalizes the classical case, can be formulated in the following way.

**THEOREM 5.1.** *Assume that the operators $F : \mathfrak{X} \times \Lambda \to \mathfrak{Y}$ and $F'_x : \mathfrak{X} \times \Lambda \to \mathfrak{Y}$ are defined and continuous on the set $\{(x, \lambda) \in \mathfrak{X} \times \Lambda \ : \ \|x - x_0\| \leqslant R, \|\lambda - \lambda_0\| \leqslant r\}$, for some $(x_0, \lambda_0) \in \mathfrak{X} \times \Lambda$ such that $F(x_0, \lambda_0) = 0$, and the linear operator $A = F'_x(x_0, \lambda_0)$ is continuously invertible. Then there exist the numbers $\rho, \delta > 0$, $\rho \leqslant R$, $\delta \leqslant r$, and a unique continuous implicit operator $x = G(\lambda)$ satisfying equation (5.1) and condition (5.4), which is defined on the ball $\{\lambda \in \Lambda \ : \ \|\lambda - \lambda_0\| < \delta\}$ and takes values in the ball $\{x \in \mathfrak{X} \ : \ \|x - x_0\| < \rho\}$. Moreover, if the operator $F'_\lambda(x_0, \lambda)$ exists, then the operator $G$ is differentiable at the point $\lambda_0$ and*

$$G'(\lambda_0) = -\left(F'_x(x_0, \lambda_0)\right)^{-1} \circ F'_\lambda(x_0, \lambda_0). \tag{5.8}$$

$\triangleleft$ Set $P = I - A^{-1}F$. Obviously, the operator $P$ acts from $\mathfrak{X} \times \Lambda$ into $\mathfrak{X}$ and equation (5.1) is equivalent to equation (5.7):

$$x = P(x, \lambda) = x - A^{-1}F(x, \lambda).$$

Then we have $x_0 = P(x_0, \lambda_0)$ and

$$P'_x(x, \lambda) = I - A^{-1}F'_x(x, \lambda) = A^{-1}\left(F'_x(x_0, \lambda) - F'_x(x, \lambda)\right).$$

In particular, $P'_x(x_0, \lambda_0) = 0$. Since the operator $F'_x(\cdot, \cdot)$ is continuous at the point $(x_0, \lambda_0)$, then for any $0 < q < 1$ we can find two numbers $\mu, \eta > 0$, with $\mu \leqslant r$, $\eta \leqslant R$, such that the inequality

$$\|P'_x(x, \lambda)\| \leqslant \|A^{-1}\| \ \|F'_x(x_0, \lambda_0) - F'_x(x, \lambda)\| \leqslant q < 1 \tag{5.9}$$

is fulfilled for $\|x - x_0\| \leqslant \eta$ and $\|\lambda - \lambda_0\| \leqslant \mu$. In addition, taking into account the continuity of the operator $F$, we can choose a number $0 < \nu < r$ such that

$$\|A^{-1}\| \, \|F(x_0, \lambda)\| \leqslant (1 - q)\eta$$

for $\|\lambda - \lambda_0\| < \nu$.

Let us denote $\min\{\mu, \nu\} = \tau$. Then

$$\begin{aligned} \|P(x,\lambda) - x_0\| &\leqslant \|P(x,\lambda) - P(x_0,\lambda)\| + \|P(x_0,\lambda) - x_0\| \leqslant \\ &\leqslant \sup_{\substack{\|x - x_0\| \leqslant \eta \\ \|\lambda - \lambda_0\| \leqslant \tau}} \|P'_x(x,\lambda)\| \, \|x - x_0\| + \|A^{-1}\| \, \|F(x_0,\lambda)\| \leqslant \\ &\leqslant q\eta + (1-q)\eta \leqslant \eta. \end{aligned} \tag{5.10}$$

The last inequality shows that for $\lambda$ with $\|\lambda - \lambda_0\| < \tau$ the operator $P(\cdot, \lambda)$ maps the ball $\{x \in \mathfrak{X} \, : \, \|x - x_0\| \leqslant \eta\}$ into itself, and inequality (5.9) guarantees that the operator $P(\cdot, \lambda)$ is a contraction on that ball. According to Banach principle, for any $\lambda$ with $\|\lambda - \lambda_0\| \leqslant \tau$ there exists a unique solution $x = G(\lambda)$ of equation (5.7), hence, of equation (5.1), which satisfies the condition $\|G(\lambda) - x_0\| \leqslant \eta$. To get the first assertion of the theorem it remains to put $\rho = \eta$ and $\delta = \tau$.

The continuity of operator $G$ follows by some standard arguments. The proof of this fact, as well as the proof of the second assertion of the theorem, are left to the reader. Notice however that (5.8) can be obtained directly differentiating the identity $F(G(\lambda), \lambda) = 0$ with respect to $\lambda$, for $\|\lambda - \lambda_0\| \leqslant \delta$. ▶

The numbers $\delta$ and $\rho$ play an essential role in different applications. Therefore, it is desirable to find some estimates for $\delta$ and $\rho$. To this end we need some additional assumptions. Suppose that the operators $F'_x(\cdot, \cdot) : \mathfrak{X} \times \Lambda \to \mathfrak{Y}$ and $F(x_0, \cdot) : \Lambda \to \mathfrak{Y}$ satisfy the Lipschitz conditions with constants $L$ and $N$, respectively, i.e.,

$$\|F'_x(x,\lambda) - F'_x(x_0,\lambda_0)\| \leqslant L(\|x - x_0\| + \|\lambda - \lambda_0\|) \tag{5.11}$$

(we considered here that the norm $\|(x, \lambda)\|$ is equivalent with the norm $\|x\| + \|\lambda\|$) and

$$\|F(x_0,\lambda) - F(x_0,\lambda_0)\| \leqslant N\|\lambda - \lambda_0\|, \tag{5.12}$$

for $(x, \lambda) \in \mathfrak{X} \times \Lambda$ with $\|x - x_0\| \leqslant R$, $\|\lambda - \lambda_0\| \leqslant r$. Then inequalities (5.9) and (5.10) are fulfilled for

$$\delta + \rho \leqslant q(Lm)^{-1}, \quad \tau \leqslant (1-q)\eta(Nm)^{-1},$$

where $m = \|A^{-1}\|$.

Solving this system of linear inequalities we obtain

$$\eta \leqslant \frac{1}{mL} \cdot \frac{qmN}{mN+1-q}, \quad \tau \leqslant \frac{1}{mL} \cdot \frac{q(1-q)}{mN+1-q}.$$

It is clear that the numbers $\eta$, $\tau$ are functions of the parameter $q$, where $q \in [0, 1]$. For $q = 0$ and $q = 1$ we have $\tau = 0$. Therefore, the largest value of $\tau$ is reached for some $q_* \in (0, 1)$. It is easy to verify that

$$q_* = \frac{\sqrt{1+mN}}{\sqrt{1+mN}+\sqrt{mN}} \tag{5.13}$$

and

$$\tau_* = \tau(q_*) = \max_{q \in (0,\,1)} \tau(q) = \frac{1}{mN(\sqrt{1+mN}+\sqrt{mN})^2}. \tag{5.14}$$

Set also

$$\eta_* = \eta(q_*) = \frac{\sqrt{mN}}{(\sqrt{1+mN}+\sqrt{mN})mL}. \tag{5.15}$$

Finally, we obtain the next result.

**THEOREM 5.2.** *Assume that the conditions in Theorem* 5.1 *and inequalities* (5.11), (5.12) *are fulfilled. Then, on the ball*

$$\|\lambda - \lambda_0\| \leqslant \delta = \min\{r, \tau_*\},$$

*there exists a unique continuous implicit operator* $G(\lambda)$ *defined by equation* (5.1) *and relation* (5.4), *which satisfies the condition*

$$\|G(\lambda) - x_0\| \leqslant \rho = \min\{R, \eta_*\},$$

*where the numbers* $\tau_*$, $\rho_*$ *are defined by formulas* (5.14), (5.15), *respectively.*

Taking into account the arguments developed in Subsection 1, from Theorem 5.2 we straightforwardly obtain the next result.

**COROLLARY 5.1.** *Let* $\mathfrak{D} = \{x \in \mathfrak{X} : \|x - x_0\| \leqslant R\}$ *and let* $\Phi : \mathfrak{D} \to \mathfrak{Y}$ *be an operator satisfying the condition* $\Phi(x_0) = y_0$. *Assume that*

1) $\Phi$ *is Fréchet differentiable on* $\mathfrak{D}$;
2) $A = \Phi'(x_0) : \mathfrak{X} \to \mathfrak{Y}$ *is continuously invertible;*
3) *there exists a number* $L > 0$ *such that*

$$\|\Phi'(x_1) - \Phi'(x_2)\| \leqslant L\|x_1 - x_2\|,$$

*for all* $x_1, x_2 \in \mathfrak{D}$.

*Then, on the ball* $\mathfrak{V} = \{y \in \mathfrak{Y} \ : \ \|y - y_0\| \leqslant \rho\}$, *where*

$$\rho = \min\left\{1, \frac{1}{mL(\sqrt{1+m}+\sqrt{m})^2}\right\}, \qquad m = \|A^{-1}\|,$$

*there exists a unique Fréchet differentiable operator* $\Phi^{-1} : \mathfrak{V} \to \mathfrak{X}$ *which satisfies the condition* $\Phi^{-1}(y_0) = x_0$ *and is an inverse of* $\Phi$.

*Moreover, we have*

$$\|\Phi^{-1}(y)\| \leqslant \min\{R, \sqrt{m}(mL(\sqrt{m+1}+\sqrt{m}))^{-1}\},$$

*for all* $y \in \mathfrak{V}$.

As we have already noticed, the transition from an inverse operator problem to an implicit operator problem is also possible. We illustrate this by an example involving analytic operators, for which we will indicate some other estimates. We will use the Cauchy-Goursat method of majorant series.

$\triangleleft$ Let $\overline{\mathfrak{D}} = \{x \in \mathfrak{X} \ : \ \|x - x_0\| \leqslant R\}$ and let $\Phi : \overline{\mathfrak{D}} \to \mathfrak{Y}$ be an operator analytic on $\overline{\mathfrak{D}}$ satisfying the condition $\Phi(x_0) = y_0$ and such that $A = \Phi'(x_0)$ is continuously invertible. Then the operator $\Psi = A^{-1}\Phi : \mathfrak{X} \to \mathfrak{X}$ is analytic on $\overline{\mathfrak{D}}$, too, and admits the representation

$$\Psi x = \Psi x_0 + \sum_{k=1}^{\infty} \delta_k \Psi(x_0, h),$$

where $h = x - x_0$, $\Psi x_0 = A^{-1}y_0$, $\delta_1\Psi(x_0, h) = A^{-1}\Phi'(x_0)h = h$, and $\delta_k\Psi(x_0, h) = A^{-1}\delta_k\Phi(x_0, h)$ are homogeneous forms of order $k$.

The equation $\Phi x = y$ is obviously equivalent to the next equation

$$h + \sum_{k=2}^{\infty} \delta_k \Psi(x_0, h) = z, \tag{5.16}$$

where $z = A^{-1}(y - y_0) \in \mathfrak{X}$ and $\|h\| \leqslant R$. We will try to find a solution for this equation of the form

$$h = P(z) = \sum_{k=0}^{\infty} \delta_k P(0, z), \tag{5.17}$$

where $\delta_k P(0, z)$ are homogeneous forms of order $k$. Substitute series (5.17) in equation (5.16) and equate the forms of the same order. As a result we obtain an infinite recurrent system from which we get all the forms $\delta_k P(0, z)$, succesively. (Explicit formulas for $\delta_k P(0,z)$ can be found, for example, in J. Orava and A. Halme [1] and A. P. Yuzhakov [2].) If we show that the formal series (5.17) converges normally on some neighborhood $\{z \in \mathfrak{X} \ : \ \|z\| < \delta\}$ of the point $z = 0$, then this will prove the existence of an analytic solution of equation (5.16), and the existence and the analyticity of the operator $\Phi^{-1} = \Psi^{-1}A^{-1} : \mathfrak{Y} \to \mathfrak{X}$ for all $y$ with $\|y - y_0\| < \rho = \delta m^{-1}$, where $m = \|a^{-1}\|$, as well.

For the proof let us consider the numerical series

$$\varphi(\zeta) = \sum_{k=2}^{\infty} \zeta_k \|\delta_k \Psi(x_0, h)\|$$

which converges for all $\zeta$ with $|\zeta| \leqslant R$, $\zeta \in \mathbb{K}$, $\|h\| = 1$. By Abell Theorem the series is convergent for all $\zeta$ lying in the disk of radius $R$ of the complex plane. Consequently, the function $\varphi(\zeta)$ is analytic on this disk. Let $\sup\limits_{|\zeta| \leqslant R,\ \zeta \in \mathbb{C}} |\varphi(\zeta)| \leqslant M < \infty$. Then, by Cauchy inequalities, we have

$$\|\delta_k \Psi(x_0, h)\| \leqslant M|\zeta|^k R^{-k} \tag{5.18}$$

and, consequently,

$$|\varphi(\zeta)| \leqslant M \sum_{k=2}^{\infty} |\zeta|^k R^{-k} \leqslant M \frac{|\zeta|^2}{R^2} \left(1 - \frac{|\zeta|}{R}\right)^{-1} = q(|\zeta|).$$

Consider the auxiliary numerical quadratic equation

$$t = \eta + q(t), \quad t = |\zeta|,$$

which, obviously, has the unique solution

$$t(\eta) = \left(\eta + R - \sqrt{(\eta + R)^2 - 4(M + R)\eta}\right) (2(M + R))^{-1},$$

satisfying the condition $t(0) = 0$. It is clear that for $\eta$ with

$$|\eta| \leqslant \delta = \frac{R^2}{4(M + R)}$$

the discriminant of this equation is greater than or equal to $|\eta| R \geqslant 0$ and, therefore, the function $t(\eta)$ is analytic on $\eta$ in the disk $|\eta| \leqslant \delta$. Hence it can be represented as a convergent power series

$$t(\eta) = \sum_{k=0}^{\infty} c_k \eta^k.$$

Since the coefficients $c_k$ can be also found from the recurrent relations obtained substituting the last series in equation (5.19), using relation (5.18) it is not difficult to get the estimates

$$\|\delta_k P(0,z)\| \leqslant c_k, \quad k \geqslant 0.$$

From these estimates it follows that the series (5.17) converges normally for all $z$ with $\|z\| < \delta$. This means that the operator $\Psi^{-1}$ is analytic on the ball $\|y - y_0\| < \rho = \delta m^{-1}$. Thus we proved the next result.

**THEOREM 5.3.** *Let $\mathfrak{X}$, $\mathfrak{Y}$ be arbitrary Banach spaces over the field $\mathbb{K}$ and $\Phi : \mathfrak{X} \to \mathfrak{Y}$ an operator analytic on the ball $\mathfrak{D} = \{x \in \mathfrak{X} \;:\; \|x - x_0\| \leqslant R\}$ for a certain $x_0 \in \mathfrak{X}$. Assume that the operator $A = \Phi'(x_0) : \mathfrak{X} \to \mathfrak{Y}$ is continuously invertible. Then, on the ball $\mathfrak{V} = \{y \in \mathfrak{Y} \;:\; \|y - y_0\| < \rho\}$, where $y_0 = \Phi(x_0)$ and*

$$\rho = R^2(4(M+R)\|A^{-1}\|)^{-1},$$

$$M = \|A^{-1}\| \sup_{\substack{\zeta \in \mathbb{C},\ |\zeta| \leqslant R \\ \|h\|=1}} \left\| \sum_{k=2}^{\infty} \zeta^k \delta_k \Phi(x_0, h) \right\|,$$

*there exists an analytic operator $\Phi^{-1}$ which is an inverse of the operator $\Phi$ on this ball.* ▶

Considering now the related implicit operator problem we obtain:

**COROLLARY 5.2.** *Let $\mathfrak{X}$, $\mathfrak{Y}$, $\Lambda$ be Banach spaces over the field $\mathbb{K}$ and let $F : \mathfrak{X} \times \Lambda \to \mathfrak{Y}$ be an operator analytic in the domain*

$$\{(x, \lambda) \;:\; \|(x, \lambda) - (x_0, \lambda_0)\| \leqslant R\}.$$

*Assume that the operator $A = F_x'(x_0, \lambda_0) : \mathfrak{X} \to \mathfrak{Y}$ is continuously invertible. Then on the ball $\{\lambda \in \Lambda \;:\; \|\lambda - \lambda_0\| \leqslant \rho\}$, where*

$$\rho = \|B^{-1}\|^{-1} R^2 (4(M+R))^{-1}, \tag{5.20}$$

$$B = \begin{pmatrix} A & F_\lambda'(x_0, \lambda_0) \\ 0 & I_\Lambda \end{pmatrix}, \quad M = \|B^{-1}\| \sup_{|\zeta| \leqslant R} \sum_{k=2}^{\infty} \|\zeta^k \delta_k F(u_0, h)\|,$$

$$u_0 = (x_0, \lambda_0), \quad h = (x - x_0, \lambda - \lambda_0),$$

*and $I_\Lambda$ is the identity operator on $\Lambda$, there exists a unique implicit operator $G(\lambda)$ satisfying the identity $F(G(\lambda), \lambda) \equiv 0$ and the condition $G(\lambda_0) = x_0$.*

**REMARK 5.1.** The method of constructing the inverse operator by formula (5.17), described in the proof of Theorem 5.3, is called the *method of undetermined coefficients.* It was used more than once by different authors, in order to solve some concrete integral equations (see, for example, N. N. Nazarov [1] and A. I. Nekrasov [1]). It seems that this method in its general form was used for the first time by K. T. Ahmedov [1] for an implicit operator depending on a numerical parameter, and by J. Orava and A. Halme [1] for the inverse operator problem.

Formulas (5.13)–(5.15) were established by V. A. Trenogin (see [1]). Using the same approach we can obtain some other, yet equivalent, estimates for $\tau$, $\eta$ if we set

$$\|(x, \lambda)\| = \max\{\|x\|, \|\lambda\|\}$$

or

$$\|(x, \lambda)\| = \sqrt{\|x\|^2 + \|\lambda\|^2}.$$

In §5 of Chapter VI below we will establish some more precise — and more convenient for computations — estimates for the numbers $\rho$, $\delta$, in the case of complex spaces.

# Chapter VI

# Nonlinear equations with holomorphic operators

Throughout this chapter we will consider complex Banach spaces only. In this case, many local features related to the solvability of equations with operators that are differentiable in the complex sense, turn out to be global.

## §1. s-fixed points for holomorphic operators. A converse of Banach principle

In this section we deal with differentiable operators and their fixed points that have the property of "succesful approximation", i.e., s-fixed points (see Chapter V).

Let $\mathfrak{D}$ be an arbitrary bounded domain in $\mathfrak{X}$ and $F$ an operator holomorphic in $\mathfrak{D}$ (i.e., $F$ is Fréchet differentiable in the complex sense on a neighborhood of each point of $\mathfrak{D}$). Assume that $F$ satisfies the invariance condition

$$F(\mathfrak{D}) \subseteq \mathfrak{D}. \tag{1.1}$$

Suppose further that $z \in \mathfrak{D}$ is a fixed point for $F$, that is, $Fz = z$.

As we have already noticed, if $F \in C(\overline{\mathfrak{D}}, \mathfrak{X})$, i.e., $F$ admits a continuous extension on $\partial\mathfrak{D}$ — the boundary of the domain $\mathfrak{D}$ —, and if the point $z$ is an attractive fixed point for the operator $F$ (see §1 of Chapter V), then, as soon as $z$ is a local s-fixed point, it is a global s-fixed point too. Indeed, by Meyers Theorem (Theorem V.1.2), the next assertion is a consequence of our assumptions:

(∗) *for any $q$ with $0 < q < 1$ there exists a metric $\rho$ on the set $\overline{\mathfrak{D}}$, equivalent to the metric given by the norm of $\mathfrak{X}$, relatively to which the operator $F$ satisfies the*

*Lipschitz condition with constant q, that is,*

$$\rho(Fx, Fy) \leqslant q\rho(x, y), \quad x, y \in \overline{\mathfrak{D}}. \tag{1.2}$$

Thus the assumptions in Theorem V.1.1 — the Banach principle — are fulfilled. If inequality (1.2) is satisfied on some closed subset $\mathfrak{U} \subseteq \mathfrak{D}$ and $F(\mathfrak{U}) \subseteq \mathfrak{U}$, then we will say that the operator $F$ is a $q$-contraction on $\mathfrak{U}$.

The following result holds.

**THEOREM 1.1.** *Let $F$ be a holomorphic operator on $\mathfrak{D}$, satisfying condition (1.1), let $z \in \mathfrak{D}$ be a fixed point for $F$, and let $A = F'(z)$ be the Fréchet derivative of $F$ at $z$.*

*a) The following assertions are equivalent:*

*1) $F^n x \to z$ for any $x \in \mathfrak{D}$ and the operator $e^{i\theta} I - A$ is normally solvable (see §5 of Chapter IV) for all $\theta \in [0, 2\pi]$, where $I$ is the identity operator on $\mathfrak{X}$;*

*2) $r(A) < 1$, where $r(A)$ denotes the spectral radius of the operator $A$;*

*3) there exist a number $r > 0$ and a norm $\|\cdot\|_*$ equivalent to the original norm of $\mathfrak{X}$, such that $F$ is a $q_1$-contraction on the ball $\overline{\mathfrak{U}}_r = \{x \in \mathfrak{D} : \|x - z\|_* \leqslant r\}$ for a certain $q_1$ with $0 \leqslant q_1 < 1$;*

*4) there exist a number $q_2$ with $0 \leqslant q_2 < 1$ and a metric equivalent to the metric given by the norm of $\mathfrak{X}$ relatively to which the operator $F$ is a $q_2$-contraction on some neighborhood of the point $z$.*

*b) If, in addition, $F \in C(\overline{\mathfrak{D}}, \mathfrak{X})$ and $F^n x \to z$ for any $x \in \partial\mathfrak{D}$ then conditions 1)–4) are equivalent to condition $(*)$.*

**EXPLANATIONS:** 1. Under condition b) the implication 1) $\Rightarrow$ $(*)$ is a global converse of the Banach principle with respect to a metric equivalent to the metric given by the norm of $\mathfrak{X}$. The implication 1) $\Rightarrow$ 3) is a local converse of that principle, but this time, with respect to a norm equivalent to the norm of $\mathfrak{X}$. In A. A. Ivanov's book [1] a construction of a metric such as in condition 4) above is given. In many concrete problems this construction turns out to be rather difficult. If we know already the spectral radius $r(A)$, then it is very simple to find the norm $\|\cdot\|_*$ and the number $r$ appearing in condition 3) (see Lemma V.3.1). This enables us to estimate the rate of convergence to $z$ of the iterations $\{F^n x\}_{n\in\mathbb{N}}$, starting with an arbitrary $x \in \overline{\mathfrak{U}}_r$.

2. The implication 2) $\Rightarrow$ 1) means that condition $r(A) < 1$ is, in fact, a global feature of the s-fixed point $z$ (see §3 of Chapter V) and, consequently, it guarantees the uniqueness of the fixed point $z$ on the whole $\mathfrak{D}$ and also the convergence of the iterations, for any $x \in \mathfrak{D}$.

$\lhd$ *Proof of Theorem 1.1.* The implication b)&1)&2)$\Rightarrow$ $(*)$ follows by the already mentioned Meyers Theorem. The implication $(*) \Rightarrow 4)$ is obvious. Therefore it is sufficient to prove only the equivalence of conditions 1)–4).

Let us prove that $1) \Rightarrow 2)$. Denote by $A_n$ the Fréchet derivative of the operator $F^n$ at the point $z$, i.e., $A_n = (F^n)'(z)$, $n \geqslant 1$ (then $A_1 = A$). From condition $F^n x - z \to 0$ and by Cauchy inequalities for the holomorphic operator $F^n x - z$ (see Chapter III) it follows that

$$\lim_{n\to\infty} A_n h = 0, \quad \text{for all } h \in \mathfrak{X} \tag{1.3}$$

(here the limit is considered in the strong sense). Moreover, $A_n = A^n$, where $A^n$ is the $n$-th iteration of the operator $A$ (see identity (3.5) in Chapter V). Further, it will be enough to show that the set $\Omega = \{\lambda \in \mathbb{C} : |\lambda| \geqslant 1\}$ is contained in the resolvent set of the operator $A$. This will clearly imply that $r(A) < 1$. By (1.6) the set $\Omega$ does not intersect the point spectrum $\sigma_1(A)$ of the operator $A$ (the set of eigenvalues of the operator $A$). Indeed, if $\lambda \in \Omega \cap \sigma_1(A)$ then there exists $y \in \mathfrak{X}$, $y \neq 0$, such that $Ay = \lambda y$, and consequently, $A^n y = \lambda^n y$, whence $\lim\limits_{n\to\infty} A^n y \neq 0$, which contradicts (1.3). Moreover, by the Uniform Boundedness Principle it follows that

$$\|A^n\| \leqslant N < \infty. \tag{1.4}$$

Therefore the sequence

$$S_m = \sum_{n=0}^{m} \lambda^{-(n+1)} A^n$$

converges uniformly to the operator $(\lambda I - A)^{-1}$ for all $\lambda$ with $|\lambda| > 1$. If $|\lambda| = 1$ we have

$$(\lambda I - A) S_m x = x - \lambda^{-(m+1)} AX, \quad x \in \mathfrak{X}.$$

It follows that the set $R(\lambda I - A)$ — the range of the operator $\lambda I - A$ — is dense in $\mathfrak{X}$. Taking into account condition 1) once again, we conclude that $R(\lambda I - A) = \mathfrak{X}$. Since, on the other hand, $\lambda \notin \sigma_1(A)$, it follows that the operator $\lambda I - A$ is continuously invertible. Thus $r(A) < 1$. The implication $1) \Rightarrow 2)$ is proved.

The implication $2) \Rightarrow 3)$ follows from Theorem V.3.3, and the implication $3) \Rightarrow 4)$ is trivial. It remains to prove that $4) \Rightarrow 1)$. By 4) it follows that the iterations $\{F^n x\}_{n\in\mathbb{N}}$ converge to $z$ uniformly on some neighborhood of the point $z$. Since the set $\mathfrak{D}$ is simply connected, then, by Vitali Theorem, the above mentioned sequence converges to $z$ for any $x \in \mathfrak{D}$. Let $\mathfrak{U}_{r_1} = \{x \in \mathfrak{D} : \|x - z\| < r_1\}$, where $r_1$ is sufficiently small. For any $\varepsilon$ with $0 < \varepsilon < r_1$ we can find $n$ such that $\|F^n x - z\| < r_1 - \varepsilon$ for all $x \in \mathfrak{U}_{r_1}$. Hence and by Cauchy inequalities it follows that

$$\|A^n h\| \leqslant r_1 - \varepsilon \quad \text{for all } h \in \mathfrak{X} \text{ with } \|h\| \leqslant r_1,$$

or, equivalently, $\|A^n\| < 1$, which means that the operator $e^{i\theta}I - A$ is continuously invertible for all $\theta \in [0, 2\pi]$. Thus condition 1) is fulfilled. ▶

**COROLLARY 1.1.** *If for some $p \geqslant 1$ the operator $F^p$ has the* s*-fixed point $z \in \mathfrak{D}$, then $z$ is an* s*-fixed point for $F$, too.*

**REMARK 1.1.** By Theorem III.8.4, in condition 1) above it is enough to require the convergence of the iterations $\{F^n x\}_{n\in\mathbb{N}}$ for $x$ in a dense subset of the domain $\mathfrak{D}$.

If for some $p \in \mathbb{N}$ the operator $F^p$ is completely continuous, then, by Theorem I.8.1, the operator $A_p = A^p$ is completely continuous, too. It follows that the spectrum of the operator $A$ coincides with the point spectrum $\sigma_1(A)$. Thus we get the next result.

**COROLLARY 1.2.** *Let $F \in \mathcal{H}(\mathfrak{D}, \mathfrak{X}) \cap C(\overline{\mathfrak{D}}, \mathfrak{X})$ be an operator satisfying condition* (1.1). *Assume that for some $p \in \mathbb{N}$ the operator $F^p$ is completely continuous. If the iterations $\{F^n x\}_{n\in\mathbb{N}}$ converge to a certain point $z \in \mathfrak{D}$ for all $x \in \mathfrak{D}' \cup \partial\mathfrak{D}$, where $\mathfrak{D}'$ is a dense subset of $\mathfrak{D}$, then $z$ is an* s*-fixed point for the operator $F$.*

Recall that there exist examples of operators that are not completely continuous but have completely continuous powers (see, for instance, L. V. Kantorovic and G. P. Akilov [1]).

**REMARK 1.2.** Actually, inequality (1.4) follows from the invariance condition (1.1), and the boundedness of the domain $\mathfrak{D}$ only. Indeed, fix a ball of radius $R$ and center $z$ completely contained in $\mathfrak{D}$, and a ball of radius $M$ with center at the zero element of $\mathfrak{X}$, completely containing $\mathfrak{D}$. Then for all $x$ with $\|x - z\| \leqslant R$ we have

$$\|F^n x\| \leqslant M,$$

and, by Cauchy inequalities, it follows that

$$\|A^n h\| \leqslant M\|h\|R^{-1},$$

for all $h$ with $\|h\| \leqslant R$. Therefore

$$\|A^n\| \leqslant MR^{-1} = N < \infty.$$

But this means (according to the proof of implication 1) $\Rightarrow$ 2)) that the spectrum of the operator $A$ lies completely in the unit disk. Thus we have proved the next important result.

**LEMMA 1.1.** *Let $F$ be an operator satifying condition (1.1) and let $z \in \mathfrak{D}$ be a fixed point for $F$. If $A = F'(z)$, then $r(A) \leqslant 1$, where $r(A)$ is the spectral radius of the operator $A$. The equality $r(A) = 1$ holds if and only if $z$ is not an s-fixed point.*

This assertion can be also considered as an analogue of Schwartz Lemma (see §5 of Chapter III).

Assume further that the operator $F$ is continuous on the closure of the domain $\mathfrak{D}$ and satisfies the stronger invariance condition

$$F(\overline{\mathfrak{D}}) \subseteq \mathfrak{D}. \tag{1.5}$$

In this case, as it will be shown below, the structure of the spectrum of the operator $A = F'(z)$ is relevant only on the boundary of the unit disk of the complex plane, and not inside this disk.

For simplicity we will consider that $\mathfrak{D}$ is a ball in $\mathfrak{X}$ centered at the point $z = 0$.

**THEOREM 1.2.** *Let us assume, in addition to condition (1.5), that $F(0) = 0$ and*

$$\sigma(A) \cap \{\lambda \in \mathbb{C} \;:\; |\lambda| = 1\} \subseteq \sigma_1(A),$$

*where $\sigma_1(A)$ is the point spectrum of the operator $A$ (i.e., the continuous and residual spectra of the operator $A$ lie completely inside the open disk $\Delta = \{\lambda \in \mathbb{C} \;:\; |\lambda| < 1\}$). Then $z = 0$ is an s-fixed point for the operator $F$.*

◁ Suppose, on the contrary, that $z = 0$ is not an s-fixed point for the operator $F$. Then, by Lemma 1.1, we get $r(A) = 1$, and, consequently, there exists a number $\theta \in [0, 2\pi]$ such that $e^{i\theta}$ is an eigenvalue of the operator $A$. It follows that we can find an element $x \in \partial\mathfrak{D}$ satifying $Ax = e^{i\theta}x$.

Choose a linear functional $f \in \mathfrak{X}^*$ such that $\langle x, f\rangle = 1$, $\|f\| = 1$, and consider the analytic function

$$\begin{aligned}\varphi(\lambda) &= \langle F(\lambda x), f\rangle = \langle A(\lambda x) + Q(\lambda x), f\rangle = \\ &= \lambda e^{i\theta}\langle x, f\rangle + \langle Q(\lambda x), f\rangle = \lambda e^{i\theta} + \langle Q(\lambda x), f\rangle,\end{aligned}$$

where $Q = F - A$ and $\lambda \in \mathbb{C}$ with $|\lambda| < 1$. From (1.5) it follows that $\|\varphi(\lambda)\| < 1$, for all $\lambda \in \mathbb{C}$ with $|\lambda| < 1$, and $\varphi'(0) = e^{i\theta}$. By Schwartz Lemma (see §5 of Chapter III) we find that $Q(\lambda x) = 0$ for all $\lambda \in \mathbb{C}$ with $|\lambda| < 1$. Hence $|\langle Fx, f\rangle| = 1$. But, on the other hand, by Cauchy-Bunyakowsky inequality and by (V.1.5) we have $|\langle Fx, f\rangle| \leqslant \|Fx\| \cdot \|f\| < 1$, a contradiction which proves the theorem. ▶

**COROLLARY 1.3.** *If the operator $F$ satisfies (1.5), $F(0) = 0$, and for a certain $p \in \mathbb{N}$, $p \geqslant 1$, the operator $F^p$ is completely continuous, then 0 is an s-fixed point for the operator $F$.*

◁ Indeed, the spectrum of the operator $A^p = (F^p)'(0)$ satisfies the condition

$$\sigma(A^p) \cap \{\lambda \in \mathbb{C} \,:\, |\lambda| \leqslant 1\} \setminus \{0\} \subseteq \sigma_1(A^p).$$

In particular we have

$$\sigma(A^p) \cap \{\lambda \in \mathbb{C} \,:\, |\lambda| = 1\} \subseteq \sigma_1(A^p).$$

Therefore 0 is an s-fixed point for the operator $F^p$. By Corollary 1.1 it follows that 0 is an s-fixed point for the operator $F$, too. ▶

Thus, we see that the geometric conditions of invariance and strong invariance enable us to infer from some local features of an operator at a fixed point, certain global properties of that point such as uniqueness, succesful approximation, and so on. In addition, the condition of strong invariance allows us to weaken slightly the requirements imposed on the spectrum of the operator $A$. Let us note here that the strong invariance condition alone is not enough to guarantee the attractiveness of a fixed point. Indeed, in Example V.2.4 we considered a linear, therefore holomorphic, operator $A$ satisfying the strong invariance condition

$$\|Ax\| < \|x\|, \quad x \neq 0,$$

and such that its iterations $A^n x$ converge to 0 if and only if $x = 0$.

A sufficient condition for the above mentioned matter is the so-called condition of *uniform* or *strict invariance*:

$$F(\mathfrak{D}) \subseteq \mathfrak{D}', \tag{1.6}$$

where $\overline{\mathfrak{D}'}$ is strictly contained in $\mathfrak{D}$. Moreover, as we will show in the sequel (Theorem 2.2), condition (1.6) is also sufficient for the existence of a fixed point (and, at the same time, for its uniqueness and for its uniform approximation by iterations of the operator $F$ at any element $x \in \mathfrak{D}$).

## §2. Criterions for the existence of an s-fixed point and its extension with respect to a parameter

Let $\mathfrak{X}$ and $\Lambda$ be complex Banach spaces. Assume that $\mathfrak{D}$ is an open bounded set in $\mathfrak{X}$ and $\Omega$ is a finitely connected $\rho$-open set in $\Lambda$ (see §2 of Chapter III). Consider an operator $\Phi : \mathfrak{D} \times \Omega \to \mathfrak{D}$ defined for any $(x, \lambda) \in \mathfrak{D} \times \Omega$, such that $\Phi(\cdot, \lambda) \in \mathcal{H}(\mathfrak{D}, \mathfrak{X})$ for all fixed $\lambda \in \Omega$, and $\Phi(x, \cdot) \in \mathcal{H}_\rho(\Omega, \mathfrak{X})$ for all fixed $x \in \mathfrak{D}$ (i.e., the operator $\Phi$ is holomorphic on $x$ and $\rho$-holomorphic on $\lambda$).

A solution of the equation

$$x(\lambda) = \Phi(x(\lambda), \lambda) \tag{2.1}$$

is called an s-*solution*, if for any operator $x_0 : \Omega \to \mathfrak{D}$ such $x_0$ $(= x_0(\lambda)) \in \mathcal{H}_\rho(\Omega, \mathfrak{D})$, the sequence $\{x_n(\lambda)\}_{n \in \mathbb{N}}$ defined by

$$x_{n+1}(\lambda) = \Phi(x_n(\lambda), \lambda), \quad n = 0, 1, 2, \dots , \tag{2.2}$$

converges in the strong topology of $\mathfrak{X}$ to $x(\lambda)$ uniformly on each compact subset of $\Omega$, and, for a fixed $\lambda \in \Omega$, uniformly relatively to those $x_0(\lambda)$ with values inside $\mathfrak{D}$; in other words, $x(\lambda)$ is an s-fixed point of the operator $\Phi(\cdot, \lambda)$, for each $\lambda \in \Omega$.

Now let us state the main result of this chapter.

**THEOREM 2.1.** *Let $\Phi : \mathfrak{D} \times \Omega \to \mathfrak{D}$ be an operator which satisfies the above mentioned conditions. If, for at least one value $\lambda_0 \in \Omega$ there exists an s-fixed point $x^*$ for the operator $\Phi(\cdot, \lambda_0)$, then there exists an s-solution for equation* (2.1). *Moreover, for any class of operators $\mathfrak{V} \subseteq \mathcal{H}_\rho(\Omega, \mathfrak{D})$ which is closed in the topology of uniform convergence on the compact subsets of $\Omega$, and contains the orbit $\mathfrak{F} = \{v_n : v_{n+1} = = \Phi(v_n, \lambda),\ n > N \geqslant 1,\ v_0 \in \mathcal{H}_\rho(\Omega, \mathfrak{D})\}$ of a certain element $v_0$, we have $x(\lambda) \in \mathfrak{V}$. In particular, it is always true that $x(\lambda) \in \mathcal{H}_\rho(\Omega, \mathfrak{D})$.*

◁ First of all let us show that for any operator $u$ $(= u(\cdot)) \in \mathcal{H}_\rho(\Omega, \mathfrak{D})$ the composed operator $\Phi(u(\cdot), \cdot)$ belongs to the class $\mathcal{H}_\rho(\Omega, \mathfrak{D})$, too. For any $x \in \mathfrak{D}$, $\lambda \in \Omega$, $\tau \in \Lambda$ and $\zeta \in \mathbb{C}$ with $|\zeta| < \rho(\lambda, \tau)$, the operator $\Phi$ admits the representation

$$\Phi(x, \lambda + \zeta\tau) = \sum_{n=0}^{\infty} a_n(x, \lambda, \tau)\zeta^n, \tag{2.3}$$

where, for any fixed $\lambda$ and $x$, $a_n$ is a homogeneous form in $\tau$ of order $n$, and the series in the right hand side of (2.3) converges normally for all $\zeta \in \mathbb{C}$ with $|\zeta| < \rho(\lambda, \tau)$. Explicitely we have

$$a_n = \frac{1}{2\pi \mathrm{i}} \,(\mathrm{P}) \int\limits_{|\zeta| = \rho < \rho(\lambda, \tau)} \Phi(x, \lambda + \zeta\tau)\zeta^{-(n+1)} \mathrm{d}\zeta. \tag{2.4}$$

On the other hand, since the operator $\Phi$ is holomorphic on $x$, it follows that for any fixed $\zeta$ with $|\zeta| = \rho$ and any element $h \in \mathfrak{X}$ with a sufficiently small norm, a representation of the following form

$$\Phi(x+h, \lambda+\zeta\tau) = \Phi(x, \lambda+\zeta\tau) + \Phi'_x(x, \lambda+\zeta\tau)h + Q(x, h, \lambda+\zeta\tau) \tag{2.5}$$

holds, where $\|Q(x, h, \lambda + \zeta\tau)\| = \mathrm{o}(\|h\|)$. Moreover, by the generalized Schwartz Lemma (see §5 of Chapter III) we get $\|Q(x, h, \lambda+\zeta\tau)\| \leqslant M\|h\|^2$, where, due to the boundedness of the set $\mathfrak{D}$, the constant $M$ can be chosen independently of $\lambda \in \Omega$ and $x \in \mathfrak{D}$. In its turn, the linear part in (2.5) can be represented in the integral form

$$\Phi'_x(x, \lambda+\zeta\tau)h = \frac{1}{2\pi \mathrm{i}}\,(\mathrm{P})\int\limits_{|t|=\delta} t^{-2}\Phi(x+th, \lambda+\zeta\tau)\mathrm{d}t,$$

where the number $\delta$ is small enough, and $t \in \mathbb{C}$. Hence it follows that the vector--function $\Phi'_x(x, \lambda+\zeta\tau)h$ is continuous on $\zeta$ with $|\zeta| = \rho$, for any fixed $\lambda \in \Omega$, $x \in \mathfrak{D}$, $h \in \mathfrak{X}$, $\tau \in \Lambda$. Consequently, from (2.4) and (2.5) we obtain

$$a_n(x+th, \lambda, \tau) = a_n(x, \lambda, \tau) + b_n(x, \lambda, \tau)h + c_n(x, h, \lambda, \tau),$$

where

$$b_n(x, \lambda, \tau)h = \frac{1}{2\pi \mathrm{i}}\,(\mathrm{P})\int\limits_{|\zeta|=\rho} \Phi'_x(x, \lambda+\zeta\tau)h\zeta^{-(n+1)}\mathrm{d}\zeta$$

is a linear operator on $h$ ($b_n \in L(\mathfrak{X})$), and

$$\|c_n(x, h, \lambda, \tau)\| = \left\| \frac{1}{2\pi \mathrm{i}}\,(\mathrm{P})\int\limits_{|\zeta|=\rho} \zeta^{-(n+1)}Q(x, h, \lambda+\zeta\tau)\mathrm{d}\zeta \right\| \leqslant$$
$$\leqslant \frac{1}{2\pi}\cdot M\|h\|^2 2\pi\rho^{-n} = \mathrm{o}(\|h\|),$$

which clearly means that the operator $a_n(\cdot, \lambda, \tau)$ is Fréchet differentiable on $x$ for any fixed $\lambda \in \Omega$, $\tau \in \Lambda$.

From the above proof it follows that the vector-functions $a_n(u(\lambda+\zeta\tau), \lambda, \tau)$ are holomorphic in the disk $|\zeta| < \rho(\lambda, \tau)$ for any $n = 0, 1, 2, \ldots$ . Then, by Weierstrass Theorem (see B. V. Shabat [1]), we obtain that for any functional $f \in \mathfrak{X}^*$ with $\|f\| = 1$, the vector-function $\varphi(\zeta) = \sum\limits_{n=0}^{\infty}\langle a_n(u(\lambda+\zeta\tau), \lambda, \tau), f\rangle$ is holomorphic in the same disk. Notice that the last series converges absolutely by virtue of the normal convergence of the series (2.3), and

$$\varphi(\zeta) = \langle \Phi(u(\lambda+\zeta\tau), \lambda+\zeta\tau)), f\rangle.$$

By Theorem III.2.1 the vector-function $\Phi(u(\lambda+\zeta\tau), \lambda+\zeta\tau))$ is also holomorphic in the disk $|\zeta| < \rho(\lambda,\tau)$. Hence it follows that the operators $\{x_n(\cdot)\}_{n\in\mathbb{N}}$ defined by (2.2) are $\rho$-holomorphic in $\Omega$, i.e., if $x_0 \in \mathcal{H}_\rho(\Omega,\mathfrak{D})$, then $x_n(\cdot) \in \mathcal{H}_\rho(\Omega,\mathfrak{D})$, for any $n = 1, 2, \ldots$ .

Let us show now that the sequence $\{x_n(\cdot)\}_{n\in\mathbb{N}}$ converges uniformly on any compact subset of $\Omega$. Taking into account the structure of the set $\Omega$ and by the generalized Vitali Theorem it is sufficient to prove the convergence on any disk $\Omega_\rho(\lambda_0,\tau) \subseteq \Omega$ centered at a point $\lambda_0$ and with a non-zero radius $\rho$. Since the point $x^*$ is an s-fixed point for the operator $\Phi(\cdot,\lambda_0) : \mathfrak{D} \to \mathfrak{D}$, then, by Theorem 1.1, there exists a neighborhood $\mathfrak{U}$ of the point $x^*$ on which the operator $\Phi(\cdot,\lambda_0)$ is a contraction with a certain constant $q$ $(0 \leqslant q < 1)$, relatively to a norm $\|\cdot\|_*$ equivalent to the original norm of the space $\mathfrak{X}$. Moreover, we can find a ball of radius $M$ with respect to the norm $\|\cdot\|_*$ which completely contains the domain $\mathfrak{D}$. Then, by (2.3) and the Cauchy inequalities (III.3.8), it follows that

$$\|a_n(x,\lambda_0,\tau)\|_* \leqslant M\rho^{-n}, \qquad n = 0, 1, 2, \ldots .$$

Since the operators $a_n(\cdot,\lambda_0,\tau)$ are holomorphic in $\mathfrak{D}$, using the estimates in Corollary III.6.1 we get

$$\|a_n(x,\lambda_0,\tau) - a_n(y,\lambda_0,\tau)\| \leqslant 2M\rho^{-n}r^{-1}\|x-y\|,$$

for all $x, y$ in the ball $\mathfrak{U}_r$ with a sufficiently small radius $r > 0$ and centered at the point $x^*$. Then

$$\begin{aligned}
&\|\Phi(x,\lambda_0+\zeta\tau) - \Phi(y,\lambda_0+\zeta\tau)\|_* \leqslant \\
&\leqslant \sum_{n=0}^{\infty} \|a_n(x,\lambda_0,\tau) - a_n(y,\lambda_0,\tau)\|_* |\zeta|^n \leqslant \\
&\leqslant q\|x-y\|_* + \frac{2Mr^{-1}\frac{|\zeta|}{\rho}}{1-\frac{|\zeta|}{\rho}}\|x-y\|_*.
\end{aligned}$$

Clearly, for any $q_1$ with $q < q_1 < 1$ we can choose a number $\rho_1$ with $0 < \rho_1 < \rho$, such that the right hand side of the last inequality does not exceed $q_1\|x-y\|_*$. Moreover, by Schwartz Lemma, taking into account that $\Phi(x^*,\lambda_0) = x^*$, we obtain

$$\|\Phi(x^*,\lambda_0+\zeta\tau) - \Phi(x^*,\lambda_0)\| \leqslant 2M|\zeta|\rho^{-1} \leqslant r(1-q_1),$$

for $|\zeta| \leqslant \rho_2 = \min\{\rho(2M)^{-1}, \rho_1\} \cdot r(1-q_1)$. Hence it follows that

$$\|\Phi(x,\lambda_0+\zeta\tau) - x^*\| \leqslant$$
$$\leqslant \|\Phi(x,\lambda_0+\zeta\tau) - \Phi(x^*,\lambda_0+\zeta\tau)\| + \|\Phi(x^*,\lambda_0+\zeta\tau) - x^*\| \leqslant$$
$$\leqslant q_1\|x - x^*\| + r(1-q_1) \leqslant r.$$

for any $\zeta$ with $|\zeta| \leqslant \rho_2$ and $x \in \mathfrak{U}_r$. Thus, for all $\lambda \in \Omega_{\rho_2}(\lambda_0, \tau)$, i.e., $\lambda = \lambda_0 + \zeta\tau$, where $\|\zeta\| < \rho_2$, the operator $\Phi(\cdot, \lambda)$ maps the ball $\mathfrak{U}_r$ centered at $x^*$ into itself, and $\Phi(\cdot, \lambda)$ is a $q_1$-contraction on $\mathfrak{U}_r$. Consequently, the sequence $\{x_n(\lambda)\}_{n\in\mathbb{N}}$ converges to a certain element $x(\lambda) \in \mathfrak{D}$, for all $\lambda \in \Omega_{\rho_2}(\lambda_0, \tau)$. Since $\|x_n(\lambda)\| \leqslant M$, for all $n = 0, 1, 2, \ldots$, it follows that the sequence converges uniformly, so our claim is proved. By Theorem III.6.4 we have $x_n(\cdot) \in \mathcal{H}_\rho(\Omega, \mathfrak{D})$. It is clear that $x(\cdot) \in \mathfrak{V}$ whenever $\mathfrak{V}$ is a subset of $\mathcal{H}_\rho(\Omega, \mathfrak{D})$ closed in the topology of uniform convergence on the compact subsets of $\Omega$, and containing the orbit $\mathfrak{F} = \{v_{n+1} = \Phi(v_n, \lambda) \; : \; n > N\}$ of a certain element $v_0 \in \mathcal{H}_\rho(\Omega, \mathfrak{D})$. ▶

**REMARK 2.1.** The last assertion of our theorem turns out to be interesting in the following situation. Consider an open set $\Omega_1 \supset \Omega$. The class $\mathfrak{V} = \mathcal{H}(\Omega_1, \mathfrak{D})$ is contained in $\mathcal{H}_\rho(\Omega, \mathfrak{D})$. Although the operator $\Phi(x, \cdot) \in \mathcal{H}_\rho(\Omega, \mathfrak{D})$ is not necessarily holomorphic in $\Omega_1$, it is possible for the orbit $\mathfrak{F}$ of an element $v_0 \in \mathcal{H}(\Omega_1, \mathfrak{D})$ to be contained in $\mathcal{H}(\Omega_1, \mathfrak{D})$. Then, taking into account that the class $\mathfrak{V} = \mathcal{H}(\Omega_1, \mathfrak{D})$ is closed in the topology of uniform convergence on compact subsets, we obtain that the solution $x(\lambda)$ of equation (1.1) is holomorphic in $\Omega_1$. Some related examples will be presented below (see §6).

As a consequence of Theorem 2.1 we obtain the next criterion for the existence of an s-fixed point for a holomorphic operator.

**THEOREM 2.2.** *Let $\mathfrak{D}$ be a bounded domain which is $\mathbb{C}$-star-shaped relatively to $0 \in \mathfrak{X}$. An operator $F \in \mathcal{H}(\mathfrak{D}, \mathfrak{X})$ satisfying the condition $F(\mathfrak{D}) \subseteq \mathfrak{D}$ has an s-fixed point $z \in \mathfrak{D}$ if and only if there exist a subset $\tilde{\mathfrak{D}} \subseteq \mathfrak{D}$, a number $p \in \mathbb{N}$, and a number $\varepsilon > 0$ such that*

$$\operatorname{dist}(F^p(\tilde{\mathfrak{D}}), \partial\tilde{\mathfrak{D}}) = \inf_{\substack{y \in F^p(\tilde{\mathfrak{D}}) \\ x \in \partial\tilde{\mathfrak{D}}}} \|x - y\| > \varepsilon. \tag{2.6}$$

◁ Necessity. Let $z \in \mathfrak{D}$ be an s-fixed point for $F$ in $\mathfrak{D}$ and set $r = \inf\limits_{x \in \partial\mathfrak{D}} \|x - y\|$. Then, by the uniform convergence of the iterations $F^n(x)$ to $z$ it follows that for any $\varepsilon$ with $0 < \varepsilon < r$ and any set $\tilde{\mathfrak{D}}$ completely contained in $\mathfrak{D}$ together with its boundary, there exists a number $p$ such that $\|F^p(x) - z\| < r - \varepsilon$, for all $x \in \tilde{\mathfrak{D}}$. Clearly (2.6) is fulfilled for such $\varepsilon$ and $p$.

Sufficiency. Consider the operator $\Phi(\cdot,\cdot) : \mathfrak{D}\times\Omega \to \mathfrak{X}$, defined by the equality

$$\Phi(x, \lambda) = \lambda F^p x,$$

where $\Omega = \{\lambda \in \mathbb{C} : |\lambda| < 1+\varepsilon\}$. By (2.6) we have $\Phi(x, \lambda) \in \mathfrak{D}$ for all $(x, \lambda) \in \mathfrak{D} \times \Omega$ and $\Phi(x, 0) = 0$. Therefore, the point $x^* = 0$ is an s-fixed point for the operator $\Phi(\cdot, 0)$, and, consequently, there exists an s-solution $x(\lambda)$ of equation (2.1), for all $\lambda \in \Omega$.

In particular, setting $z = x(1)$ we obtain $z = F^p(z)$. It follows that $z$ is an s-fixed point for the operator $F^p$, whence, as above, we conclude that $z$ is an s-fixed point for the operator $F$, too. ▶

If, in particular, $\mathfrak{D}$ is the open ball of radius $R$ centered at the zero element of $\mathfrak{X}$, then the following corollary provides a sufficient condition for the existence of an s-fixed point.

**COROLLARY 2.1.** *Let $F$ be an operator holomorphic in $\mathfrak{D}$ which satisfies the strict invariance condition, i.e., there exists $r < R$ such that*

$$\|Fx\| \leqslant r,$$

*for all $x$ with $\|x\| < R$. Then the operator $F$ has an s-fixed point $z \in \mathfrak{D}$.*

It seems that this result was obtained for the first time by M. Helve [1] in the finite-dimensional case and by C. Early and R. Hamilton [1] in the general case of a Banach space, using the generalized Poincaré metric (see also L. Harris [4], T. Hayden and T. Suffridge [1], and K. Goebel and S. Reich [1]).

The results in this section have various applications which will be listed in the sequel. Now we return to an example of an integral equation studied in the previous chapter.

Let us consider equation (V.4.20):

$$x(t) = \int_a^b K(\sigma, t)x^m(\sigma)\mathrm{d}\sigma + y(t).$$

As it was proved in §4 of Chapter III, for $y$ $(= y(t))$ with $\|y\| < \dfrac{m-1}{m\sqrt{m\kappa(b-a)}}$ the relation

$$\|F(x)\| \leqslant \kappa(b-a)r^m + \|y\| < r,$$

is fulfilled for all $x$ such that $\|x\| \leqslant r < r^*$, where $r^*$ is the largest rooth of the equation $r = Nr^m + \|y\|$, where $N = \kappa(b-a)$ and $\kappa = \max\limits_{\sigma,t\in[a,\,b]} |K(\sigma,t)|$.

By Corollary 2.1 the integral equation above has a unique solution $x(t)$, satisfying the condition

$$\max_{[a,\,b]} |x(t)| < r.$$

Moreover, for any function $x_0(t)$ continuous on $[a,\,b]$ and such that $\max\limits_{[a,\,b]} |x_0(t)| \leqslant r$, the sequence

$$x_{n+1}(t) = \int_a^b K(\sigma,t)x_n^m(\sigma)\mathrm{d}\sigma + y(t),$$

$n = 0, 1, 2, \ldots$ , converges to $x(t)$ uniformly on $[a,\,b]$.

Let us emphasize once more that for $r_* < \|x\|$, where $r_*$ is the smallest positive rooth of the equation $Nr^m + \|y\| = r$, the operator fails to be a contraction (see Subsection 4 in §4 of Chapter IV).

## §3. Regular fixed points. Geometric criterions

Let $\mathfrak{D}$ be a bounded domain in $\mathfrak{X}$. In this section we will assume that the operator $F$, holomorphic in $\mathfrak{D}$, admits a continuous extension on the boundary $\partial\mathfrak{D}$ of the domain $\mathfrak{D}$, i.e., $F \in \mathcal{H}(\mathfrak{D},\mathfrak{X}) \cap C(\overline{\mathfrak{D}},\mathfrak{X})$. The invariance condition (1.1) is in this case equivalent to condition

$$F(\overline{\mathfrak{D}}) \subseteq \overline{\mathfrak{D}}. \tag{3.1}$$

Example V.2.1 considered in the previous chapter shows that condition (3.1) above is not enough for the existence of a fixed point for the operator $F$ (the operator in Example V.2.1 is holomorphic and, moreover, non-expansive). Therefore in the subsequent considerations it will be necessary to impose supplementary restrictions.

**1.** Let us start with a simple assertion.

**THEOREM 3.1.** *Assume that the domain $\mathfrak{D}$ is $\mathbb{C}$-star-shaped relatively to the point $0 \in \mathfrak{X}$. Let $F \in \mathcal{H}(\mathfrak{D},\mathfrak{X}) \cap C(\overline{\mathfrak{D}},\mathfrak{X})$ be an operator that satisfies condition (3.1) and such that $I - F$ is a proper operator. Then there exists at least one fixed point for the operator $F$.*

◁ Consider the operators $F_t = tF$. By Theorem 2.2, the operator $F_t$ has an s-fixed point $x_t$ for each $t \in (0,\,1)$. Since

$$\|x_t - F(x_t)\| = \|tF(x_t) - F(x_t)\| \leqslant \|F(x_t)\|\ |t-1|,$$

by the boundedness of the operator $F$ on $\overline{\mathfrak{D}}$, we obtain that the sequence $\{y_t\} = \{(I - F)(x_t)\}$ converges for $t \to 1$. Then, under our assumptions, it follows that the set $\{x_t\}$ is compact. Since the operator $F$ is continuous on $\overline{\mathfrak{D}}$ we infer that any limit point of the set $\{x_t\}$ is a fixed point for the operator $F$. ▶

In particular, if the operator $F$ is completely continuous, then the operator $I - F$ is proper. Therefore, Theorem 3.1 represents an analogue of the Schauder principle for holomorphic operators acting on a $\mathbb{C}$-star-shaped domain (recall that the classical Schauder principle is stated for convex domains).

However, like the Schauder principle, Theorem 3.1 above does not guarantee the uniqueness of a fixed point. Moreover, neither the isolation, nor the succesful approximation of a fixed point follows.

Note also that, if we deal with assumptions weaker than the strict invariance condition (V.2.6), then, naturally, we have to replace the succesful approximation property with a weaker condition. A suitable substitute is the property of regularity.

**DEFINITION 3.1.** Let $z$ be a fixed point for an operator $F$ which is Fréchet differentiable in a neighborhood of $z$. The point $z$ is called a *regular point* (or a *proper point*) for the operator $F$ if 1 is not an eigenvalue of the operator $A = F'(z)$. Otherwise the point $z$ is called *singular*.

By Theorem 1.1 an s-fixed point for a differentiable operator on a complex Banach space is regular.

For a completely continuous differentiable operator the regularity of a fixed point implies its isolation. Indeed, the equation

$$x = F(x)$$

can be written as

$$z + h = z + Ah + Q(z, h),$$

where $h$ has a sufficiently small norm and $\|Q(z,h)\| = o(\|h\|)$. Since 1 is not an eigenvalue of the compact operator $A$, the operator $I - A$ is continuously invertible and the last equation is equivalent to

$$h = (I - A)^{-1}Q(z,h) = \Phi(h).$$

Taking into account the equality $\Phi'(0)=0$ we obtain that $h=0$ is an s-fixed point for the operator $\Phi$ on a certain ball with a sufficiently small radius. Hence the considered equation has the unique solution $x=z$ in some neighborhood of the point $z$.

We will show below that in some special cases a regular fixed point for a not necessarily completely continuous holomorphic operator may be isolated as well.

The next result introduces a geometric condition which guarantees not only the isolation of a fixed point for a non-compact operator, but also its uniqueness.

**THEOREM 3.2.** *Let $F \in \mathcal{H}(\mathfrak{D}, \mathfrak{X}) \cap C(\overline{\mathfrak{D}}, \mathfrak{X})$ be an operator which maps $\mathfrak{D}$ into itself. Assume that for some $p \geqslant 1$ the operator $F^p$ is compact and has no fixed points on the boundary $\partial\mathfrak{D}$. Then the operator $F$ has a unique fixed point inside $\mathfrak{D}$ and this point is regular.*

◁ Let $\mathfrak{J}$ be the set of all fixed points for the operator $F^p$. Under our assumptions there exists a domain $\mathfrak{D}_1$ such that $\overline{\mathfrak{D}}_1 \supset \mathfrak{J}$ and $\overline{\mathfrak{D}}_1 \subset \mathfrak{D}$. Consider on $\overline{\mathfrak{D}}_1$ the vector field $I - tF^n$ which, for $t \in (0, 1)$ and $t$ close to 1, is homotopic to the field $I - F^n$ and has no zeroes on $\partial\mathfrak{D}_1$ (see M. A. Krasnoselskii and P. P. Zabreiko [1]). By Theorem 2.2, for such $t$'s the operator $F_t = tF^p$ has an s-fixed point $x_t \in \mathfrak{D}$. It follows that these points are regular. Since the winding number $\gamma$ of two homotopic non-degenerate vector fields is the same, then

$$\gamma(I - tF^p, \partial\mathfrak{D}_1) = \gamma(I - F^p, \partial\mathfrak{D}_1) \neq 0.$$

Therefore $x_t \in \overline{\mathfrak{D}}_1$, for any $t \in (0, 1)$ close to 1, and, moreover

$$\gamma(I - tF^p, \partial\mathfrak{D}_1) = \operatorname{ind}(I - tF^p, x_t) = 1.$$

Hence the equality

$$\gamma(I - tF^p, \partial\mathfrak{D}_1) = 1$$

is equivalent to the fact that the set $\mathfrak{J}$ has but one point $z$ which is regular. By uniqueness, the point $z$ is a fixed point for the operator $F$, too (see §1 of Chapter V). Moreover, if we assume that $z$ is not a regular point for the operator $F$, then there exists an element $u \neq 0$, $u \in \mathfrak{X}$, such that $Au = u$, where $A = F'(z)$. But then $A^p u = (F^p)'(z)u = u$, a contradiction. ▶

Thus we obtained an analogue of the extension to the boundary principle for a non-discrete analytic set (see M. Herve [1]): *if the set of fixed points of a compact holomorphic operator has more than one point inside a domain, then this set contains points from the boundary of the domain, too.*

**2.** For operators that are holomorphic in domains of a strongly convex space we may give up the compactness condition. Moreover, for a holomorphic operator defined on a ball of a strongly convex space and mapping that ball into itself, the set of all its fixed points has an affine structure.

**THEOREM 3.3** (W. Rudin [1]). *Let $\mathfrak{D}$ be the unit ball — centered at 0 — of a strongly convex Banach space $\mathfrak{X}$, and $F \in \mathcal{H}(\mathfrak{D}, \mathfrak{X}) \cap C(\overline{\mathfrak{D}}, \mathfrak{X})$ an operator satisfying condition* (3.1). *Let $\mathfrak{F}$ be the set of all fixed points for the operator $F$. If $0 \in \mathfrak{F}$ then $\mathfrak{F} \cap \mathfrak{D} = \mathfrak{E} \cap \mathfrak{D}$, where $\mathfrak{E}$ is the set of all fixed points for the linear operator $A = F'(0)$.*

◁ Let $z \in \mathfrak{F}$, $z \neq 0$. Choose a point $u \in \partial\mathfrak{D}$ with $\|u\| = 1$, such that $z = tu$ for some $t \in (0, 1)$. Consider now the function $\varphi(\lambda) = \langle F(\lambda u), f\rangle$, holomorphic on the unit disk $\Delta = \{\lambda \in \mathbb{C} \, : \, |\lambda| < 1\}$, where $f \in \mathfrak{X}^*$ satisfies the conditions $\langle u, f\rangle = 1$ and $\|f\| = 1$ (such a functional does exist by a consequence of the Hahn-Banach Theorem). The function $\varphi$ satisfies, obviously, the conditions: $|\varphi(\lambda)| < 1$, for all $\lambda \in \Delta$, $\varphi(0) = 0$, and $\varphi'(0) = \langle Au, f\rangle$. Since $\varphi(t) = t \neq 0$, by a consequence of the one-dimensional Schwartz Lemma we obtain $\varphi(\lambda) = \lambda$, for all $\lambda \in \Delta$, hence $\varphi'(\lambda) = \langle Au, f\rangle = 1$. Since the space $\mathfrak{X}$ is strongly convex it follows that $Au = u$, and, therefore $Az = z$, i.e., $z \in \mathfrak{E}$. Thus we proved the inclusion $\mathfrak{F} \cap \mathfrak{D} \subseteq \mathfrak{E}$.

Conversely, let $z \in \mathfrak{E} \cap \mathfrak{D}$, $z \neq 0$. Then, for some function $\varphi$ as above, we have

$$\varphi(\lambda) = \langle A(\lambda u), f\rangle + \langle Q(\lambda u), f\rangle = \lambda\langle u, f\rangle + \langle Q(\lambda u), f\rangle,$$

where $Q(x) = F(x) - A(x)$ and $Q'(0) = 0$. Then $\varphi'(0) = 1$, hence $\varphi(\lambda) \equiv \lambda$. In particular, $\varphi(t) = t$, or

$$\langle t^{-1}F(tu), f\rangle = 1.$$

By the strong convexity of $\mathfrak{X}$ we get that $t^{-1}F(tu) = u$, or $Fz = z$. ▶

From the above proof it follows that if 0 is a regular point for an operator $F$ satisfying the conditions of the theorem, then 0 is the unique fixed point inside $\mathfrak{D}$. However, it is necessary to note that from Rudin's theorem we can not infer any relationship between the boundary points of the sets $\mathfrak{E}$ and $\mathfrak{F}$; there are examples when $\mathfrak{F} \cap \partial\mathfrak{D} \neq \mathfrak{E} \cap \partial\mathfrak{D}$. Nevertheless, an assertion analogous to Theorem 3.2 is also true in this case. Namely, the absence of fixed points on the boundary of the domain $\mathfrak{D}$ implies the existence of a unique fixed point inside $\mathfrak{D}$.

**COROLLARY 3.1.** *Let the domain $\mathfrak{D}$ and the operator $F$ be as in Theorem 3.3. If $\mathfrak{F} \cap \partial\mathfrak{D} = \emptyset$, then the point 0 is the unique fixed point for the operator $F$.*

◁ Consider the holomorphic operator $\Phi : \overline{\mathfrak{D}} \to \overline{\mathfrak{D}}$ defined by the equality $\Phi = \frac{1}{2}(x + F(x))$. Under our assumptions we have $\|F(x)\| \leqslant \|x\|$ and $F(x) \neq x$ for all $x \in \partial\mathfrak{D}$. Then, by the strong convexity of the space $\mathfrak{X}$, we find that $\|\Phi(x)\| < 1$

for any $x \in \partial\mathfrak{D}$. From Theorem 1.2 it follows that $\lambda = 1$ is not an eigenvalue of the operator $\Phi'(0)$, and this means that 0 is the unique fixed point for the operator $\Phi$. But the operators $F'(0)$ and $\Phi'(0)$, as well as the operators $F$ and $\Phi$, have the same fixed points, hence the assertion follows. ▶

**3.** In this subsection we consider strongly non-expansive operators, that is

$$\|F(x) - F(y)\| < \|x - y\|, \quad \text{for all } x \neq y \in \mathfrak{D}, \tag{3.2}$$

or non-expansive operators, that is

$$\|F(x) - F(y)\| \leqslant \|x - y\|, \quad \text{for all } x, y \in \mathfrak{D}, \tag{3.3}$$

acting on a complex Banach space $\mathfrak{X}$.

**THEOREM 3.4.** *Let $\mathfrak{D}$ be the unit ball of a complex Banach space $\mathfrak{X}$, centered at the origin, $F \in \mathcal{H}(\mathfrak{D}, \mathfrak{X}) \cap C(\overline{\mathfrak{D}}, \mathfrak{X})$ an operator satisfying condition (3.1), and $z \in \mathfrak{D}$ a fixed point for $F$. If the condition*

a) *the intersection between the spectrum $\sigma(A)$ of the operator $A = F'(z)$ and the unit disk of the complex plane $\mathbb{C}$ does not contain points form $\sigma_2(A)$,*

*together with at least one of the conditions (3.2) or*

b) *the space $\mathfrak{X}$ is strongly convex, the operator $F$ satisfies condition (3.3) and $F(0) \neq$ $\neq 0$,*

*are fulfilled, then $z$ is an s-fixed point for the operator $F$, and the sequence*

$$x_n = \theta x_{n-1} + (1 - \theta)F(x_{n-1}) \tag{3.4}$$

*converges to $z$ uniformly on $\mathfrak{D}$ for any $\theta \in [0,\, 1)$ and $x_0 \in \mathfrak{D}$.*

(See the theorems in §2 of Chapter V.)

◁ The operator $F(\theta) = \theta I + (1-\theta)F$ satisfies condition a), or both the conditions a) and b), simultaneously with the operator $F$. Indeed, from (3.2) or b) it follows that $\|A\| \leqslant 1$, hence $r(A) \leqslant 1$. By Lemma IV.1.1 there are no points from $\sigma_3(A)$ on the unit circle of the complex plane. Therefore, by a), on that circle we can find points from $\sigma_1(A)$ only. Let $B(\theta) = (F(\theta))'(0)$, and choose an arbitrary $t \in [0,\, 2\pi]$. Then $e^{it}I - B(\theta) = (1-\theta)\left(\dfrac{e^{it} - \theta}{1-\theta}I - A\right)$. Setting $\lambda = \dfrac{e^{it} - \theta}{1 - \theta}$ we obtain $|\lambda| \geqslant 1$. If $F$ satisfies condition a), then the operator $\lambda I - A$ is either a homeomorphism of $\mathfrak{X}$, or has a non-zero kernel. In the first case the operator $B(\theta) - e^{it}I$ is a homeomorphism of $\mathfrak{X}$, hence $e^{it}$ is a regular point of it, and in the second case $\mu = e^{it}$ is an eigenvalue of $B(\theta)$.

Thus it is enough to prove the uniform convergence to $z$ of the sequence of iterations $\{F^n(x_0)\}_{n\in\mathbb{N}}$, $x_0 \in \mathfrak{D}$, only.

Let $h \in \mathfrak{X}$ be such that $\|h\| = 1$ and $z = th$ for a certain $t \in [0, 1)$. Consider the vector-function $A(\lambda) = F'(\lambda h)$. By Lemma I.4.2, it is analytic in the open disk $\Delta$ of the complex plane $\mathbb{C}$. From conditions (3.2) or b) the inequality

$$\sup_{x\in\tilde{\mathfrak{D}}} \|F'(x)\| \leqslant 1$$

follows for any closed convex subset $\tilde{\mathfrak{D}} \subseteq \mathfrak{D}$. Therefore $r(A(\lambda)) \leqslant 1$, for all $\lambda \in \Delta$. According to the maximum principle for the spectral radius, if

$$r(A(\lambda_0)) < 1, \tag{3.5}$$

for some $\lambda_0 \in \Delta$, then $r(A(\lambda)) < 1$, for all $\lambda \in \Delta$. Thus $r(A(t)) = r(F'(z)) < 1$. The uniform convergence of the sequence of iterations $\{F^n(x_0)\}_{n\in\mathbb{N}}$, $x_0 \in \mathfrak{D}$, follows from this inequality and Theorem 1.1.

Thus, it remains to prove that condition a) together with one of conditions (3.2) or b) imply (3.5). We show that we can set $\lambda_0 = 0$, i.e., $r(A) < 1$.

1. Assume that conditions a) and (3.2) are fulfilled.

Then the operator $G$ defined by $G(x) = F(x) - F(0)$ satisfies the inequality

$$\|G(x)\| < \|x\|, \quad 0 \neq x \in \mathfrak{D}. \tag{3.6}$$

Observe that $G'(0) = A$ and let us show that the operator $A$ has no eigenvalues on the unit circle of the complex plane. Suppose, on the contrary, that the equality $Au = e^{it}u$ is fulfilled for a number $e^{it}$, where $t \in [0, 2\pi]$, and a certain $u \in \mathfrak{D}$ with $u \neq 0$. By the Hahn-Banach-Suchomlinoff Theorem we can choose a functional $f \in \mathfrak{X}^*$ such that $\|f\| = 1$ and $\langle u, f\rangle = \|u\| = r < 1$, and let us consider the function

$$\begin{gathered} g(\lambda) = \langle r^{-1}G(\lambda u), f\rangle = \\ = \langle r^{-1}A(\lambda u), f\rangle + \langle r^{-1}H(\lambda u), f\rangle = \lambda e^{it} + r^{-1}\langle H(\lambda u), f\rangle. \end{gathered} \tag{3.7}$$

This function is analytic on the closed disk $\overline{\Delta}$, and, by (3.6), the inequality $|g(\lambda)| < 1$ holds for all $\lambda \in \overline{\Delta}$. From (3.7) it follows that $g'(0) = e^{it}$. According to Schwartz Lemma we obtain $g(\lambda) = \lambda e^{it}$. Then $|\langle r^{-1}G(u), f\rangle| = |g(1)| = 1$, an equality which contradicts (3.6). Hence the unit circle consists of regular points for the operator $A$, so that, by Corollaries IV.1.1 and IV.2.2, we obtain $r(A) < 1$.

2. Assume that conditions a) and b) are fulfilled. Then, for all $x \in \mathfrak{D}$ and $t \in [0, 2\pi]$, the inequality $\|e^{it}G(x)\| \leqslant \|x\|$ is true. By Theorem 3.3 this inequality and the strong convexity of the space $\mathfrak{X}$ imply that the set of all fixed points for the operator $e^{-it}G$ which lie inside $\mathfrak{D}$ coincides with the set of all fixed points for the linear operator $e^{-it}A = e^{-it}G'(0)$. Therefore, if $e^{it} \in \sigma_1(A)$, i.e., there exists $v \in \mathfrak{X}$ with $\|v\| = 1$ such that $v = e^{-it}Av$, then $\lambda v = e^{it}G(\lambda v)$, for all $\lambda \in \Delta$. It follows that $F(\lambda v) = G(\lambda v) + F(0) = \lambda w + s$, where $w = e^{-it}v$ and $s = F(0)$. As in the first part of the proof, we choose $f \in \mathfrak{X}^*$ with $\|f\| = 1$ such that $\langle w, f\rangle = \|w\| = 1$, and consider the function $\psi(\lambda) = \langle F(\lambda v), f\rangle$ which is analytic in $\Delta$. It is easy to see that $|\psi(\lambda)| \leqslant 1$, for all $\lambda \in \Delta$. On the other hand

$$\psi(\lambda) = \lambda\langle w, f\rangle + \langle s, f\rangle = \lambda + \langle s, f\rangle.$$

Hence $|\lambda + \langle s, f\rangle| \leqslant 1$, for all $\lambda \in \overline{\Delta}$, and, therefore, $\langle s, f\rangle = 0$. But then

$$\langle w + s, f\rangle = \langle w, f\rangle = 1. \tag{3.8}$$

Since $\|\lambda w + s\| = \|F(\lambda v)\| \leqslant 1$, for all $\lambda \in \Delta$, taking the limit as $\lambda \to 1$ we obtain $\|w + s\| \leqslant 1$. By Theorem 0.4.7, this inequality, (3.8) and the strong convexity of the space $\mathfrak{X}$ imply that $s$ $(= F(0)) = 0$, which contradicts b). As above, we conclude that $r(A) < 1$. ▶

**REMARK 3.1.** If the operator $F$ is completely continuous, then, by Theorem I.8.1, the operator $A = F'(0)$ is completely continuous, too; so, by Theorem IV.4.3, condition a) is fulfilled. The converse is not true. Some examples of non-completely continuous operators for which the Fréchet derivative is completely continuous are presented, for instance, in L. V. Kantorovic and G. P. Akilov [1]. Note also — as Example V.2.4 shows — that condition a) is essential. In the case of a reflexive space $\mathfrak{X}$ it is sufficient to assume that the operator $e^{i\eta}I - A$ is normally solvable for all $\eta \in [0, 2\pi]$, in order to accomplish condition a). This follows from the Statistical Ergodic Theorem of K. Yoshida (see §4 of Chapter IV).

**REMARK 3.2.** It is not difficult to observe that the assertions in Theorem 3.4 remain true if one replaces the operator $A$ with the operator $A^n$, where $A^n$ is a certain power of the operator $A$, $n = 1, 2, \ldots$. In many cases this may be useful, as the next example shows.

**EXAMPLE 3.1.** Let $\mathfrak{X} = \boldsymbol{m}$ be the space of all bounded sequences $x = (x_1, x_2, \ldots, x_k, \ldots)$ of complex numbers, with the norm $\|x\| = \sup_k |x_k|$, and let $\mathfrak{D}$ be the open unit ball of $\mathfrak{X}$. Consider the operator $F : \mathfrak{D} \to \mathfrak{D}$ given by $Fx = (z_1, z_2, \dots, z_k, \dots)$ where

$$z_1 = \frac{1}{2}x_1^2 + \frac{3}{8},$$
$$z_k = 0, \quad \text{for } k = 2m+1,\ m = 1, 2, \dots ,$$
$$z_k = (k-2)(k-1)^{-1}x_{k-1} \quad \text{for } k = 2m,\ m = 1, 2, \dots .$$

It is easy to see that $F$ satisfies condition (3.3) on $\overline{\mathfrak{D}}$ and condition (3.2) inside $\mathfrak{D}$. Moreover, obviously, the point $x^* = (\frac{1}{2}, 0, 0, \dots)$ is a fixed point for the operator $F$. Further, the operator $F$ is Fréchet differentiable and the operator $A = F'(0)$ is defined by $Ax = (p_1, p_2, \dots, p_k, \dots)$, where

$$p_k = 0 \quad \text{for } k = 2m+1,\ m = 0, 1, 2, \dots ,$$
$$p_k = (k-2)(k-1)^{-1}x_{k-1} \quad \text{for } k = 2m,\ m = 1, 2, \dots .$$

The operator $A$ is not compact but it is nilpotent, since $A^2 = 0$; hence condition a) is fulfilled for the operator $A^2$. Consequently, the sequence (3.4) converges to $x^*$ for any $x_0 \in \mathfrak{D}$.

## §4. A priori estimates and the extension of an s-solution to the boundary of the domain

The criterions for the existence of a fixed point considered in the previous section were based, essentially, on the following approach. Given the operator $F(\cdot) : \mathfrak{D} \to \mathfrak{D}$, we first construct an operator $\Phi(\cdot, \cdot) : \mathfrak{D} \times \Delta \to \mathfrak{D}$, where $\Delta = \{\lambda \in \mathbb{C} \ : \ |\lambda| < 1\}$, satisfying the conditions in Theorem 2.1, and then we take the limit $x(\lambda) \to x^*$ for $\lambda \to 1$, where $x(\lambda)$ is an s-solution of the equation

$$x(\lambda) = \Phi(x(\lambda), \lambda). \tag{4.1}$$

This second step is possible because of additional properties of the operator $F$ such as, for instance, the properness of $I - F$ or the compactness of $F$, and so on.

In the theory of linear equations the method of extending a solution with respect to a parameter is well-known. Roughly speaking, the method consists in establishing a priori estimates of the possible solutions for equation (4.1) — with $\Phi$ a linear operator —, for all $\lambda \in [0, 1)$ (let say, upper bounds for the norm $\|x(\lambda)\|$) and then proving that such a solution can be extended on the interval [0, 1] (see, for instance, V. A. Trenogin [1]).

We will use an analogous approach for equation (4.1) when $\Phi$ is a holomorphic operator. To this end, it is enough for our purposes, to establish a priori estimates of the solutions for a linear equation associated to the given operator $\Phi$.

**THEOREM 4.1.** *Under the conditions in Theorem 2.1 let $\Omega$ be the open disk $\Delta = \{\lambda \in \mathbb{C} : |\lambda| < 1\}$ and assume that the operator $\Phi(\cdot, \lambda) : \mathfrak{D} \to \mathfrak{D}$ is continuous on $\overline{\mathfrak{D}}$ for any fixed $\lambda \in \Delta$. Moreover, suppose that the partial derivatives of the operator $\Phi$ satisfy the conditions:*

1) $\|\Phi'_\lambda(x, \lambda)\| \leqslant L < \infty$, $(x, \lambda) \in \mathfrak{D} \times \Delta$;

2) *there exists $\alpha$ with $0 \leqslant \alpha < 1$ such that for all possible solutions of the linear equation*

$$(I - A(\lambda))z = y, \quad z, y \in \mathfrak{X},$$

*where $A(\lambda) = \Phi'_x(x(\lambda), \lambda)$, and $x(\lambda)$ is a possible solution for equation (4.1) on $\Delta$, the a priori estimate*

$$\|z\| \leqslant M(1 - |\lambda|)^{-\alpha}\|y\| \quad (|\lambda| < 1)$$

*is true.*

*Then the vector-function $x(\lambda)$ — a solution of equation (4.1) — is defined and continuous on the closed disk $\overline{\Delta} = \{\lambda \in \mathbb{C} : |\lambda| \leqslant 1\}$ and satisfies on this disk the Hölder-type condition*

$$\|x(\lambda) - x(\lambda')\| \leqslant K|\lambda - \lambda'|^{1-\alpha}, \tag{4.2}$$

*where*

$$K = 3\left(\frac{2}{3}\right)^{\alpha}\left(\frac{\pi}{2}\right)^{1-\alpha}\max\left\{2R, ML\left(\frac{2}{1-\alpha} + 1\right)\right\},$$

*and $R$ is the radius of a ball centered at the origin and containing the domain $\mathfrak{D}$.*

◁ According to Theorem 2.1, the vector-function $x(\lambda)$ is defined and holomorphic for all $\lambda \in \Delta$ and, moreover, the spectrum of the operator $A(\lambda) = \Phi'_x(x(\lambda), \lambda)$ for a fixed $\lambda \in \Delta$ lies completely inside the unit disk. Therefore the operator $I - A(\lambda)$ is continuously invertible for each $\lambda \in \Delta$. Then, from (4.1) it follows that

$$x'(\lambda) = [I - A(\lambda)]^{-1}\Phi'_\lambda(x(\lambda), \lambda),$$

and, from conditions 1), 2), we obtain

$$\|x'(\lambda)\| \leqslant ML(1 - |\lambda|)^{-\alpha}.$$

Therefore

$$\left\| (\mathrm{P}) \int_0^1 x'(r\mathrm{e}^{\mathrm{i}\varphi})\mathrm{d}r \right\| \leqslant ML \left| \int_0^1 \frac{\mathrm{d}r}{(1-r)^\alpha} \right| = \frac{ML}{1-\alpha},$$

and, consequently, the limit

$$\lim_{r\to 1} x\left(r\mathrm{e}^{\mathrm{i}\theta}\right) = (\mathrm{P}) \int_0^1 x'\left(r\mathrm{e}^{\mathrm{i}\varphi}\right) \mathrm{d}r$$

exists and is finite for all $\varphi \in [0, 2\pi]$. Thus the vector-function $x(\lambda)$ is defined for all $\lambda \in \Gamma = \partial\Delta$.

Let us show first that (4.2) is fulfilled for all $\lambda, \lambda' \in \Gamma$, with a suitable constant $K$. This will prove also the continuity of $x(\lambda)$ on $\Gamma$. Without losing the generality, we may assume that $|\arg\lambda - \arg\lambda'| \leqslant \pi$. Consider the vector-function $\tilde{x}(\lambda) = (2R)^{-1}x(\lambda)$. Using the relation

$$|\varphi - \varphi'| \leqslant \pi \left| \sin\frac{\varphi-\varphi'}{2} \right| = \frac{\pi}{2} \left| \mathrm{e}^{\mathrm{i}\varphi} - \mathrm{e}^{\mathrm{i}\varphi'} \right|,$$

for all $|\varphi - \varphi'| \leqslant \pi$, we obtain that it is enough to prove the inequality

$$\|\tilde{x}(\lambda) - \tilde{x}(\lambda')\| \leqslant K_1 |\arg\lambda - \arg\lambda'|^{1-\alpha} \tag{4.3}$$

with a suitable constant $K_1$. To this end we may assume that $|\arg\lambda - \arg\lambda'| < 1$, since, on the contrary, (4.3) with $K_1 = 1$ is obvious. Let us represent the left hand side of (4.3) as

$$\tilde{x}(\lambda) - \tilde{x}(\lambda') = (\mathrm{P}) \int_l \tilde{x}'(\zeta)\mathrm{d}\zeta,$$

where $l$ is the piece-wise smooth curve consisting of the line segments $[\lambda, t\lambda]$ and $[t\lambda', \lambda']$, and of the arc which connects $t\lambda$ and $t\lambda'$ along the circle $|\zeta| = t = 1 - |\arg\lambda - \arg\lambda'| < 1$. Then

$$\|\tilde{x}(\lambda) - \tilde{x}(\lambda')\| \leqslant \int_t^1 (\|\tilde{x}'(r\lambda)\| + \|\tilde{x}'(r\lambda')\|)\mathrm{d}r + \left| \int_{\arg\lambda}^{\arg\lambda'} t \left\|\tilde{x}'\left(t\mathrm{e}^{\mathrm{i}\varphi}\right)\right\| \mathrm{d}\varphi \right| \leqslant$$

$$\leqslant 2 \int_t^1 \frac{ML}{2R}(1-r)^{-\alpha}\mathrm{d}r + \left| \int_{\arg\lambda}^{\arg\lambda'} \frac{MLt}{2R}(1-t)^{-\alpha}\mathrm{d}\varphi \right| =$$

$$= ML(2R)^{-1}(2(1-\alpha)^{-1} + 1)|\arg\lambda - \arg\lambda'|^{1-\alpha}.$$

Thus (4.3) is fulfilled for all $\lambda, \lambda' \in \Gamma$, with

$$K_1 = \max\left\{1, \frac{ML}{2R}\left(\frac{2}{1-\alpha}+1\right)\right\}.$$

This means that (4.2) is fulfilled, for the same $\lambda, \lambda'$, with

$$K = K_2 = \left(\frac{\pi}{2}\right)^{1-\alpha} \max\left\{2R, ML\left(\frac{2}{1-\alpha}+1\right)\right\}.$$

We fix now the points $\lambda, \lambda' \in \Gamma$ and consider the vector-function $x_1(\zeta) = x(\zeta\lambda) - x(\zeta\lambda')$, analytic in the disk $|\zeta| < 1$ and continuous on the closure of this disk. Since $x_1(0) = 0$, by Schwartz Lemma we obtain

$$\|x_1(\zeta)\| \leqslant \sup_{|\xi|=1} \|x_1(\xi)\| \; |\zeta|.$$

It follows that for any $r \in [0,\, 1]$ we have

$$\begin{aligned} \|x(r\lambda) - x(r\lambda')\| &\leqslant r \sup_{|\xi|=1} \|x(\xi\lambda) - x(\xi\lambda')\| \leqslant \\ \leqslant r K_2 \sup_{|\xi|=1} |\xi\lambda - \xi\lambda'|^{1-\alpha} &= K_2 r^{\alpha} |r(\lambda - \lambda')|^{1-\alpha} \leqslant \\ &\leqslant K_2 |r\lambda - r\lambda'|^{1-\alpha}. \end{aligned} \tag{4.4}$$

Finally, let us prove inequality (4.2) — with an appropriate constant $K$ —, for all $\lambda, \lambda' \in \overline{\Delta}$ lying on the same ray, that is, $\arg\lambda = \arg\lambda'$. In this case

$$\begin{aligned} \|x(\lambda) - x(\lambda')\| &\leqslant \left| \int_{|\lambda|}^{|\lambda'|} \left\| x'\left(t e^{\mathrm{i}\arg\lambda}\right)\right\| \mathrm{d}t \right| \leqslant \\ \leqslant \left| \int_{|\lambda|}^{|\lambda'|} \frac{ML \mathrm{d}t}{(1-t)^{\alpha}} \right| &= \frac{1}{1-\alpha} ML[(1-|\lambda|)^{1-\alpha} + (1-|\lambda'|)^{1-\alpha}] \leqslant \\ &\leqslant \frac{2^{\alpha}}{1-\alpha} ML|\lambda - \lambda'|^{1-\alpha}. \end{aligned} \tag{4.5}$$

Now let $\lambda$ and $\lambda'$ be arbitrary elements of $\overline{\Delta}$. Choose a point $\lambda''$ such that $|\lambda''| = |\lambda|$ and $\arg\lambda'' = \arg\lambda'$. It is easy to show that in this case we have

$$|\lambda - \lambda''| + |\lambda'' - \lambda| \leqslant 3|\lambda - \lambda'|.$$

From inequalities (4.4) and (4.5) it follows that

$$\begin{aligned}\|x(\lambda)-x(\lambda')\| &\leqslant \|x(\lambda)-x(\lambda'')\|+\|x(\lambda'')-x(\lambda')\| \leqslant \\ &\leqslant K_2|\lambda-\lambda''|^{1-\alpha}+\frac{2^\alpha}{1-\alpha}ML|\lambda''-\lambda'|^{1-\alpha} \leqslant \\ &\leqslant 3^{1-\alpha}2^\alpha \max\left\{K_2, \frac{2^\alpha}{1-\alpha}ML\right\}|\lambda-\lambda'|^{1-\alpha} \leqslant \\ &\leqslant 3\left(\frac{2}{3}\right)^\alpha K_2|\lambda-\lambda'|^{1-\alpha}.\end{aligned}$$

Setting $K = 3(2/3)^\alpha K_2$ we get (4.2). ▶

**COROLLARY 4.1.** *Let $F \in \mathcal{H}(\mathfrak{D},\mathfrak{X}) \cap C(\overline{\mathfrak{D}},\mathfrak{X})$, where $\mathfrak{D}$ is a $\mathbb{C}$-star-shaped domain in $\mathfrak{X}$, and assume that for all possible solutions of the equation*

$$B(\lambda)z = y, \quad \lambda \in [0,\,1],\ y \in \mathfrak{X}, \tag{4.6}$$

*where $B(\lambda) = I - \lambda F'(x)$ and $x$ is a fixed element of $\mathfrak{D}$, the a priori estimate*

$$\|z\| \leqslant \gamma\|y\| \tag{4.7}$$

*is true. Then the operator $F$ has a fixed point $x^*$ which can be approximated by the sequence $x_t$ of the s-fixed points for the operators $tF$ with $t \in [0,\,1)$, and the rate of convergence is determined by the estimate*

$$\|x^* - x_t\| \leqslant \kappa(1-t),$$

*where $\kappa = (3/2)\pi R \cdot \max\{2, 3\gamma\}$ and $R$ is the radius of a ball centered at the origin and containing $\mathfrak{D}$. Moreover, if the point $x^*$ belongs to $\mathfrak{D}$, then it is regular.*

◁ Indeed, the existence of the point $x^* \in \overline{\mathfrak{D}}$ and the estimate of the convergence rate follow straightforwardly from Theorem 4.1 if we set $\Phi(x,\lambda) = \lambda F x$, $L = R$, $M = \gamma$ and $\alpha = 0$. If $x^*$ is an interior point of $\mathfrak{D}$ then we can define the operator $B = B(1) = I - F'(x^*)$. By estimate (4.7) and by the extension with respect to a parameter principle for the solutions of a linear equation (see V. A. Trenogin [1]), equation (4.6) has a unique solution for $x = x^*$, $\lambda \in [0,\,1]$ and $y \in \mathfrak{X}$. Consequently, the operator $B$ is continuously invertible and $\lambda = 1$ is not an eigenvalue of the operator $F'(x^*)$. ▶

As an example, let us consider the integral equation

$$x(t) = \alpha \int_0^1 \sum_{n=2}^{\infty} K_n(t,s)x^n(s)ds + (1-\alpha)y(t), \tag{4.8}$$

where $K_n(t,s)$ and $y(t)$ are continuous complex-valued functions on [0, 1] satisfying the conditions $|y(t)| \leqslant t$, $0 \leqslant t \leqslant 1$, and

$$\sum_{n=2}^{\infty} |K_n(t,s)| \leqslant 1.$$

We write equation (4.8) in the operatorial form

$$x = \alpha F(x) + (1-\alpha)y,$$

where $F$ is the operator defined by the integral in the right hand side of (4.8), satisfying, obviously, the conditions

$$F \in \mathcal{H}(\mathfrak{D},\mathfrak{X}) \cap C(\overline{\mathfrak{D}},\mathfrak{X}),$$
$$F(\overline{\mathfrak{D}}) \subseteq \overline{\mathfrak{D}}, \quad F(0)=0 \text{ and } F'(0)=0,$$

where $\mathfrak{X} = \tilde{C}[0,\,1]$ and $\mathfrak{D} = \left\{ x \in \mathfrak{X} \;:\; \|x\| = \max_{0 \leqslant t \leqslant 1} |x(t)| < 1 \right\}$.

Let

$$\gamma = \left( \sup_{\|x\| \leqslant 1} \sum_{n=2}^{\infty} n\, |K_n(t,s) x^n(s)| \right)^{-1}.$$

It is obvious that for all $\alpha \in [0, \gamma)$ the operator $\Phi x = \alpha F(x) + (1-\alpha)y$ acts from $\overline{\mathfrak{D}}$ into $\overline{\mathfrak{D}}$ and satisfies $\|\Phi'(x)\| \leqslant \alpha\gamma^{-1} < 1$.

According to a consequence of the Banach principle it follows that equation (4.8) has a unique solution $x(t)$ with $\|x(t)\| < 1$.

Let us also show that — independently of the value of $\gamma$ — if $\alpha \in \left(\frac{1}{2},\, 1\right]$, then equation (4.8) has a unique solution. Indeed, using Schwartz Lemma, we obtain

$$\|\Phi(x)\| \leqslant \alpha\|F(x)\| + (1-\alpha)\|y\| \leqslant \alpha\|x\|^2 + (1-\alpha)\|y\|,$$

for any $x$ with $\|x\| \leqslant r \leqslant 1$. Note that for any $r \in [(1-\alpha)\alpha^{-1},\, 1]$ with $1/2 < \alpha \leqslant 1$, there exists $q \in [0,\,1)$ such that $\alpha r^2 + (1-\alpha) < qr$. Hence $\|\Phi(x)\| < qr$, for all $x$ with $\|x\| \leqslant r$. So our assertion follows straightforwardly from Theorem 2.2. In addition, the estimate $\|x^*\| < (1-\alpha)\alpha$ is true. Thus we proved that equation (4.8) has solutions for $\alpha \in [0,\,\gamma) \cup \left(\frac{1}{2},\, 1\right]$. Finally we show that when $\gamma = 1/2$ there exists a solution for all $\alpha \in [0,\,1]$. To this end we consider the operator $G(\cdot,\cdot)$ defined by $G(\lambda, x) = \alpha F(x) + \lambda(1-\alpha)y$, where $\lambda \in \Delta = \{\lambda \in \mathbb{C} \;:\; |\lambda| < 1\}$. Obviously, $G(\Delta \times \mathfrak{D}) \subseteq \mathfrak{D}$ and the operator $G(0,\cdot) = \alpha F$ has the s-fixed point 0 for any $\alpha \in [0, 1]$. Let us show that if $\gamma = 1/2$, then the operator $G(\cdot, \cdot)$ and the s-solution $x(\lambda)$ of equation (4.1) satisfy the conditions in Theorem 4.1 for any $\alpha \in [0, 1]$. Condition 1) is obviously fulfilled.

We prove now condition 2). Together with the operator $G$ consider the real--valued function $f(s,t) = \alpha s^2 + (1-\alpha)t$, $s,t \in [0, 1]$. One of the solutions for the quadratic equation $s = f(s,t)$ is the function

$$s(t) = \frac{1 - \sqrt{1 - 4\alpha(1-\alpha)t}}{2\alpha},$$

relatively to which we assert that

$$\|x(\lambda)\| \leqslant s(t), \quad |\lambda| \leqslant t < 1. \tag{4.9}$$

Indeed, set $x_0(\lambda) = \lambda(1-\alpha)x_0$ and

$$x_{n+1}(\lambda) = G(\lambda, x_n(\lambda)), \quad n = 0, 1, 2, \dots ,$$

where $x_0$ is an arbitrary element of $\mathfrak{D}$. Obviously $\|x_0(\lambda)\| \leqslant s(t)$, for $|\lambda| \leqslant t$, and assuming that $\|x_n(\lambda)\| \leqslant s(t)$ for $|\lambda| \leqslant t$, then by induction we obtain

$$\|x_{n+1}(\lambda)\| \leqslant \alpha\|x_n(\lambda)\|^2 + |\lambda|(1-\alpha) \leqslant \alpha s^2(t) + t(1-\alpha) = s(t).$$

Taking the limit for $n \to \infty$ we get (4.9). Hence it follows that for $|\lambda| = t$ we have

$$\begin{aligned}\|I - A(\lambda)\| &= \|I - G'_x(\lambda, x(\lambda))\| \geqslant |1 - \|G'_x(\lambda, x(\lambda))\|\,| = \\ &= |1 - \alpha\|F'(x(\lambda))\|\,| \geqslant |1 - 2\alpha\|x(\lambda)\|\,| \geqslant \\ &\geqslant \left|1 - 1 + \sqrt{1 - 4\alpha(1-\alpha)t}\right| \geqslant \sqrt{1 - |\lambda|},\end{aligned}$$

since $4\alpha(1-\alpha) \leqslant 1$, for $\alpha \in [0, 1)$.

By Theorem 4.1, the limit $\lim_{\lambda \to 1} x(\lambda) = x^*$ exists and, obviously, $x^*$ is a solution of equation (4.8). For $\alpha = 1$, equation (4.8) has the solution $x^* = 0$.

## §5. Local inversion of holomorphic operators and a posteriori error estimates

Let us consider the equation

$$\Phi(x) = 0 \tag{5.1}$$

associated with an operator $\Phi : \mathfrak{X} \to \mathfrak{X}$, and assume that $x_0 \in \mathfrak{X}$ is an approximate solution of this equation. There are a lot of methods for determining the value of $x_0$, but many of them involve computational difficulties. Therefore in practice it is important to know — as accurate as possible — a posteriori estimates for the errors appearing in the approximate computation of $x_0$. (See M. A. Krasnoselskii [2], [3], and L. V. Kantorovic and G. P. Akilov [1].) The notion of error estimate, needs a more precise definition. We will state a definition introduced by M. A. Krasnoselskii.

**DEFINITION 5.1.** An *error estimate* for an approximate solution $x_0$ of equation (5.1) is an arbitrary domain $\mathfrak{V} \subset \mathfrak{X}$ containing $x_0$, provided that this equation has at least one solution $x^*$ in $\mathfrak{V}$. Let $\mathfrak{F}$ be the set of all solutions for equation (5.1). The *error* of an approximate solution $x_0$ is defined by

$$\delta(x_0) = \inf_{x \in \mathfrak{F}} \|x_0 - x\|.$$

Let us assume that the operator $\Phi$ is Fréchet differentiable in a neighborhood of the point $x_0$. Without restricting the generality, we may set $x_0 = 0$; on the contrary, equation (5.1) can be always replaced by the equivalent equation $\Phi(x_0 + h) = 0$, and one has to look for the exact solutions of the last equation in a neighborhood of the approximate solution $h_0 = 0$.

The value $\Phi(0) \in \mathfrak{X}$ is called the *unbinding* of equation (5.1).

In many computing proccesses the point $x_0 = 0$ is considered to be a sufficiently "good" approximate solution of equation (5.1) if the norm $\|\Phi(0)\| = \varepsilon$ of the unbinding is small enough. Generally speaking, this works well in two cases: either in connection with a certain problem, when we are interested only in an approximate equation $\Phi(x) \approx 0$ rather than equation (5.1), or, when the error $\delta(0)$ of the approximate solution $x_0$ approaches 0 as $\varepsilon$ approaches 0. Let us note that for a continuous operator $\Phi$ the converse is always true, namely, $\varepsilon \to 0$ for $\delta(0) \to 0$. Therefore we are interested in estimating the value of the error $\delta(0)$ in terms of the norm $\varepsilon = \|\Phi(0)\|$. To this end we represent the operator $\Phi$ as

$$\Phi(x) = \Phi(0) + \Phi'(0)x + \omega(x),$$

where $\omega(x) = o(\|x\|)$.

In this section we are dealing with the so-called regular case, when the operator $A = \Phi(0)$ is continuously invertible. Then equation (5.1) is equivalent to

$$x = Q(x) + y, \tag{5.2}$$

where $Q(x) = -A^{-1}\omega(x)$ and $y = -A^{-1}\Phi(0)$.

Since the operator $F = I - Q$ satisfies the condition $F'(0) = I$, then, by the Local Inversion Theorem, there exists an operator $G$ defined for all $y \in \mathfrak{X}$ with a sufficiently small norm, which is a solution for equation (5.2). Our aim is to obtain estimates — as precise as possible — for the domain of existence and continuity of the operator $G$. These estimates will enable us to find solutions for the above considered problem.

Throughout this section we will assume again that $\mathfrak{X}$ is a complex Banach space. To begin with, let us suppose that the operator $Q$ is defined and differentiable on the ball $\mathfrak{D}$ centered at $x_0$ and with radius $R$ (clearly the operator $F = I - Q$ has the same properties). The next theorem is useful in obtaining the above mentioned estimates.

**THEOREM 5.1.** *Let $\varphi(t)$ be a monotone increasing real-valued function defined on the interval $[0,\ R_0]$ with $R_0 \leqslant R$, which satisfies the condition*

$$\|Q(x)\| \leqslant \varphi(\|x\|), \qquad x \in \mathfrak{X}.$$

*Suppose that the function $\psi(t) = t - \varphi(t)$ is a homeomorphism from the interval $[0,\ R_0]$ onto an interval $[0,\ r]$ with $r > 0$. Then*

*1) the operator $G = F^{-1}$ is defined and differentiable on the ball $\mathfrak{V}$ centered at the origin and with radius $r$;*

*2) the estimate $\|G(y)\| \leqslant \psi^{-1}(\|y\|)$ is valid for all $y \in \mathfrak{V}$;*

*3) the operator $G$ can be obtained as the limit of the sequence of operators*

$$G_0(y) \equiv 0, \quad G_{n+1}(y) = y + QG(y), \quad n = 1, 2, \dots .$$

◁ Assume, by induction, that the operator $G_n$ is defined and differentiable on the ball $\mathfrak{V}$ and satisfies the relation

$$\|G_n(y)\| \leqslant \psi^{-1}(\|y\|) < R_0, \quad \|y\| < r. \tag{5.3}$$

By the recurrence rule, the operator $G_{n+1}$ is also defined and differentiable on $\mathfrak{V}$ and satisfies an analogous estimate. Indeed, since the function $\varphi$ is increasing, then

$$\|G_{n+1}(y)\| \leqslant \|y\| + \|QG_n(y)\| \leqslant \|y\| + \varphi(\|G_n(y)\|) \leqslant$$
$$\leqslant \|y\| + \varphi(\psi^{-1}(\|y\|)) = \psi^{-1}(\|y\|) < R_0.$$

Thus the sequence $\{G_n\}_{n \geqslant 0}$ of operators defined on the ball $\mathfrak{V}$ is uniformly bounded by the constant $R_0$. Moreover, by the Local Inversion Theorem, this sequence converges in the strong topology of the space $\mathfrak{X}$ on some neighborhood $\mathfrak{V}_1 \subseteq \mathfrak{V}$ of the point $y = 0$. By Vitali Theorem the sequence converges to a differentiable operator $G : \mathfrak{V} \to \mathfrak{D}$ on the whole domain $\mathfrak{V}$. Thus assertion 3) is proved and by (V.5.3) assertion 2) follows immediately. Further, by continuity, the operator $G$ satisfies the identity $G(y) = y + QG(y)$ for all $y \in \mathfrak{V}$. Hence $FG(y) = y$ and so assertion 1) is proved, too. ▶

If we set now $M = \sup_{x \in \mathfrak{D}} \|Q(x)\| < \infty$, then, by Schwartz Lemma, we get $\|Q(x)\| \leqslant MR^{-2}\|x\|^2$. Therefore we may choose $\varphi$ to be the function $\varphi(t) = MR^{-2}t^2$. It is easy to see that the corresponding function $\psi(t) = t - \varphi(t)$ provides a homeomorphism from the interval $[0, R^{-2}(2M)^{-1}]$ onto the interval $[0, R^{-2}(4M)^{-1}]$. Then, by Theorem 5.1, the next result follows.

**COROLLARY 5.1.** *Let $Q$ be an operator defined and differentiable on the ball $\mathfrak{D} = \{x : \|x\| < R\}$ and satisfying the conditions $Q(0) = 0$, $Q'(0) = 0$. Let $F = I - Q$. Then the operator $G = F^{-1}$ is defined on the ball $\mathfrak{V}$ with center at the origin and radius $r$, where*

$$r = \min\left\{\frac{R}{2}, \frac{R^2}{4M}\right\} \quad \text{and} \quad M = \sup_{\|x\|<R} \|Q(x)\|.$$

*Moreover,*

$$\|G(y)\| \leqslant \min\left\{R, \frac{R^2}{2M}\right\} \cdot \|y\|, \qquad \|y\| < r,$$

*and the sequence $\{G_n\}_{n\in\mathbb{N}}$, with $G_0(y) \equiv 0$ and $G_{n+1}(y) = y + QG_n(y)$, $n = 1, 2, \dots$, converges to the operator $G$ for all $y \in \mathfrak{V}$.*

In the finite-dimensional case this result was obtained by A. P. Yuzhakov using the theory of logarithmic residues (see [1], [2]).

◁ For a complete proof it is enough to observe that whenever $R \leqslant R^2(2M)^{-1}$, then

$$\psi(R) \geqslant R - M \geqslant R - \frac{R}{2} = \frac{R}{2}.$$ ▶

Let us return now to the initial equation (5.1) and assume that the operator $\Phi$ is differentiable on the ball of radius $R$ with center at the origin, and let $m = \|A^{-1}\|$. From the already obtained results it follows that if the unbinding $\Phi(0)$ satisfies the estimate

$$\varepsilon = \|\Phi(0)\| \leqslant \frac{R}{2m} \min\{1, R(2M)^{-1}\},$$

where

$$M = m \cdot \sup_{\|x\|<R} \|\Phi(x) - \Phi'(0)x - \Phi(0)\|,$$

then the error of the approximate solution $x_0$ satisfies the relation

$$\delta(0) \leqslant \varepsilon \cdot m \cdot R \cdot \min\{1, R(2M)^{-1}\}.$$

This obviously implies that $\delta(0) \to 0$ as $\varepsilon \to 0$.

**REMARK 5.1.** It is worth to emphasize the qualitative aspect of the estimate found in Corollary 5.1. In the usual estimates obtained for arbitrary Banach spaces (see, for instance, M. A. Krasnoselskii [2], [3]) the constants $r$ and $M$ are related — as a rule — by a condition of the form $\alpha r + \beta M \leqslant R$, $\alpha, \beta \geqslant 0$. Whence it follows that for the existence of a unique solution of the equation (5.2) both $r$ and $M$ are necessarily bounded (at least by the number $R$). In our case it is clear that if $M$ is sufficiently small, then $r$ may be chosen as close to $R/2$ as we want, and, conversely, if $r$ is sufficiently small, then the solution of equation (5.2) exists for arbitrary large values of $M$.

We illustrate these remarks by an example.

Consider the Hammerstein equation

$$x(t) = \int_a^b K(s,t) f(s, x(s)) \mathrm{d}s + y(t), \tag{5.4}$$

where:

a) $K(s,t)$ is a continuous function on the rectangle $[a, b] \times [a, b]$, and

b) the function $f(s,t)$ is analytic with respect to $u$ in the disk $|u| < R$ of the complexe plane for a fixed $s \in [a, b]$, is continuous in $s \in [a, b]$ for a fixed $u$, and satisfies the conditions

$$f(s,0) = 0, \quad f'_u(s,0) = 0.$$

Some conditions for the existence of a unique solution of equation (5.4) were obtained in §4 of Chapter V, using the Banach principle. They were

$$\begin{cases} \kappa M(b-a) + \|y\| \leqslant R_0 \\ L\kappa(b-a) < 1, \end{cases}$$

where

$$R_0 < R,$$

$$\kappa = \max_{a \leqslant s,\ t \leqslant b} |K(s,t)|,$$

$$M = \max_{\substack{a \leqslant s \leqslant b \\ |u| \leqslant R}} |f(s,u)|, \quad \text{and} \quad L = \sup_{\substack{a \leqslant s \leqslant b \\ |u| \leqslant R_0}} |f'_u(s,u)|.$$

These inequalities lead to the next relations

$$M < \frac{R_0}{\kappa(b-a)}, \quad L < \frac{1}{\kappa(b-a)}.$$

Since equation (5.4) is a particular case of equation (5.2), from Corollary 5.2 we obtain a different condition for the existence of a unique complex-valued solution of equation (5.4), i.e.,

$$\|y\| M\kappa(b-a) < R \min\left\{\frac{1}{2}, \frac{R}{4}\right\}, \tag{5.5}$$

where, as a matter of fact, the constant $L$ does not appear, and the constant $M$ may be as large as we want if the norm $\|y\|$ of $y$ is sufficiently small. Let us note also that condition (5.5) is simpler than condition (V.5.20) obtained in §5 of Chapter V using the Cauchy-Goursat method of majorant power series.

To conclude this section, we state a consequence of the already obtained estimates, based on the so-called covering theorems. These theorems play an important role in the finite-dimensional complex analysis (see Goluzin [1], Shabat [3], Lebedev [1]).

**COROLLARY 5.2.** *Let $F$ be a bounded analytic operator defined on some domain $\mathfrak{D}$ and such that $F'(x_0)$ is continuously invertible at least for one point $x_0 \in \mathfrak{D}$. Further, let $\{F_n\}_{n\in\mathbb{N}}$ with $F_n \in \mathcal{H}(\mathfrak{D}, \mathfrak{X})$, be a sequence of analytic operators which converges uniformly to the operator $F$ on some neighborhood of the point $x_0$. Then there exist a domain $\mathfrak{V}$ and an integer $n_0$ such that*

$$\bigcap_{n>n_0} F_n(\mathfrak{D}) \supset \mathfrak{V}.$$

## §6. Single-valued small solutions in some degenerate cases

Let $\mathfrak{X}$, $\Lambda$ and $\mathfrak{Y}$ be complex Banach spaces. Consider an operator $F$, with values in $\mathfrak{Y}$, defined and Fréchet differentiable on a neighborhood of the point $(0,0) \in \mathfrak{X} \times \Lambda$.

Assume that

$$F(0,0) = 0. \tag{6.1}$$

If on some neighborhood $\Omega \subset \Lambda$ of $0 \in \Lambda$ there exists an operator $x = x(\lambda)$ defined by the equation

$$F(x(\lambda), \lambda) = 0, \tag{6.2}$$

and such that $x(\lambda) \to 0$ as $\lambda \to 0$, then $x$ is called a *small solution* of equation (6.2).

Whenever the linear operator $F'_x(0,0)$ is continuously invertible (the non-degenerate case), in view of the Implicite Function Theorem, a local implicit operator $x = x(\lambda)$ exists and, obviously, it is a small solution due to its continuity.

In the degenerate case, when 0 is a point in the spectrum of the operator $F'_x(0,0)$, some other possibilities occur also: either there are no solutions at all, or there is a solution but it is not single-valued (in this case we have to deal with branches of the solution), or, finally, there exists a single-valued solution, but it is not small (as, for instance, when $x(0) \neq 0$). These situations are generically called critical. They will be considered in detail in Chapter VIII.

Here we will be concerned with criterions which give some sufficient conditions to avoid critical situations when the operator $F'_x(0,0)$ is degenerated, i.e., criterions for the existence of a single-valued small solution in the degenerate case.

Unfortunately the criterions which will be presented below have merely a particular feature: they provide neither sufficient nor necessary conditions to handle a general non-critical situation in a degenerate case. However they are extremely important in specific problems, since any small solution is stable.

**1.** SMALL SOLUTIONS IN A GENERALIZED SENSE. THE FINITE-DIMENSIONAL CASE. Let us note that, as long as we are concerned with stability, in many problems it is enough to check it only along all linear one-dimensional directions, i.e., to show that small perturbations of the parameter $\lambda$ along any one-dimensional linear subspace yield small changes in the solution $x(\lambda)$. In this respect, the next notion of "generalized smallness" is useful.

**DEFINITION 6.1.** A solution $x(\lambda)$ of equation (6.2), where the operator $F$ satisfies condition (6.1), will be called *small in the generalized sense* if for any one-dimensional subspace $\mathfrak{E} \subseteq \Lambda$ there exists a neighborhood $\mathfrak{U} \subset \mathfrak{E}$ of $0 \in \mathfrak{E}$ on which the vector-function $x(\lambda)$ is defined and satisfies the condition $x(\lambda) \to 0$ as $\lambda \to 0$, $\lambda \in \mathfrak{U}$.

Obviously if an operator $x(\cdot)$ is Gâteaux differentiable on a neighborhood of the point 0 and satisfies equation (6.2) and condition $x(0) = 0$, then it is a small solution in the generalized sense. (Therefore a small solution in the generalized sense may not be always continuous on a whole neighborhood of 0.) Of course when the space $\Lambda$ is one-dimensional, a small solution in the generalized sense is also small in the usual sense and therefore the considered set-up is interesting only when the dimension of $\Lambda$ is greater than 1.

The above remarks lead to the conclusion that a possible approach towards some criterions for the existence of small solutions in the generalized sense, in a degenerate case, is to look for such solutions not in the class of Fréchet differentiable operators but in the class of Gâteaux differentiable operators (or in the class of $\rho$-holomorphic operators if $\mathfrak{X}$, $\Lambda$, $\mathfrak{Y}$ are complex spaces). In this respect, Theorem 2.1 turns out to be useful. Until now we used neither that the operator $\Phi(\cdot,\cdot)$ is $\rho$-holomorphic in the variable $\lambda$, nor the $\rho$-holomorphicity of its fixed point $x = x(\lambda)$ (see §2 of this chapter).

In this subsection we will consider only the finite-dimensional case to illustrate how Theorem 2.1 can be used.

Let $f$ be an arbitrary function, holomorphic in a neigborhood of $0 \in \mathbb{C}^n$, with $f(0) = 0$. For any subspace $\mathfrak{L} \subseteq \mathbb{C}^n$, we will denote by $\mu_0(f \mid \mathfrak{L})$ the multiplicity of 0 for the function $f \mid \mathfrak{L}$ — the restriction of $f$ to $\mathfrak{L}$ (see L. A. Aizenberg and A. P. Yuzhakov [1]).

Let $\Lambda = \mathbb{C}^n$, $\mathfrak{X} = \mathfrak{Y} = \mathbb{C}$ and assume that $F$ is a function which is holomorphic in a neighborhood $\mathfrak{U}_1 \times \mathfrak{U}_2$ of the point $(0,0) \in \mathbb{C}^{n+1}$ and which admits the following representation

$$F(x, \lambda) = P(\lambda)x + Q(x, \lambda), \tag{6.3}$$

where $P$ and $Q$ are holomorphic in $\mathfrak{U}_2 \subseteq \mathbb{C}^n$ and $\mathfrak{U}_1 \times \mathfrak{U}_2 \subseteq \mathbb{C}^{n+1}$, respectively, and $P(0) = Q'(0,0) = 0$.

Obviously, $F'_x(0,0) = 0$ and so we find ourselves in a degenerate case.

**THEOREM 6.1.** *Assume that the functions $P$ and $Q$ in (6.3) satisfy the condition*

$$\mu = \sup_{\mathfrak{L} \subseteq \mathbb{C}^n} \mu_0(P \mid \mathfrak{L}) < \mu_0(Q \mid \mathbb{C}^n) < \infty.$$

*Then, for any one-dimensional subspace $\mathfrak{C} \subset \mathbb{C}^n$ there exist a neighborhood $\mathfrak{V} \subseteq \mathfrak{C}$ of $0 \in \mathfrak{C}$ and a unique function $x(\lambda)$, holomorphic in $\mathfrak{V}$, implicitely defined by equation (6.2) and such that $x(0) = 0$. Thus $x(\lambda)$ is a small solution of equation (6.2) in the generalized sense.*

*Moreover, if the function*

$$f(\lambda) = Q(0, \lambda)[P(\lambda)]^{-2} \tag{6.4}$$

*is holomorphic on $\mathfrak{U}_2$, then there exists a neighborhood $\Sigma \subseteq \mathbb{C}^n$ of the point $0 \in \mathbb{C}^n$ in which the function $x(\lambda)$ is holomorphic, and, consequently, $x(\lambda)$ is a small solution in the usual sense.*

◁ Consider the function $\Phi : \mathbb{C}^{n+1} \to \mathbb{C}$ defined by the equality

$$\Phi(x,\lambda) = \begin{cases} x - \dfrac{F(x,\lambda)}{P(\lambda)}, & \lambda \neq 0 \\ 0, & \lambda = 0. \end{cases}$$

Fix an arbitrary element $h \in \mathbb{C}^n$ with $\|h\| = 1$. Taking into account the assumptions of the theorem we obtain the representation

$$\Phi(x,\zeta h) = \zeta^k g(x,\zeta h), \tag{6.5}$$

for all $\zeta \in \mathbb{C}$ with $|\zeta| < \rho(h)$, where $\rho(h)$ is sufficiently small, $k = \mu_0(Q \mid \mathbb{C}^m) - \mu \geqslant 1$, and the function $g(x,\zeta h)$ is holomorphic in $\zeta$ for any fixed $x \in \mathfrak{U}_1$, i.e., the function $\Phi$ is $\rho$-holomorphic in some $\rho$-open subset $\Omega \ni 0$ (we can choose $\Omega = (\mathfrak{U}_2 \setminus \{\lambda \in \mathbb{C}^n : : P(\lambda) = 0\}) \cup \{0 \in \mathbb{C}^n\})$. Thus equation (6.2) is equivalent to

$$x = \Phi(x,\lambda)$$

considered on the set $\Omega$.

Further, from (6.5) it is obvious that for any $h$ with $\|h\| = 1$ we may choose $\rho(h)$ such that $\Phi(x,\zeta h) \in \mathfrak{U}_2$ for all $\zeta$ with $|\zeta| < \rho(h)$ and any $x \in \mathfrak{U}_1$. Therefore there exists a $\rho$-open subset $\Omega_1 \ni 0$ in $\Omega$ having the property that $\Phi(x,\lambda) \in \mathfrak{U}_1$ for $(x,\lambda) \in \mathfrak{U}_1 \times \Omega_2$.

Since the point $x = 0$ is an s-fixed point for the operator $\Phi(\cdot,0) : \mathbb{C} \to \mathbb{C}$ (recall that $\Phi(\cdot,0) \equiv 0$), by Theorem 2.1 it follows that the last equation has a unique s-solution $x(\lambda)$ which is $\rho$-holomorphic in $\Omega_1$.

So the first assertion of the theorem is proved. For the proof of the second one let us remark that, by condition (6.4), the function $v_1(\lambda) = \Phi(0,\lambda) = P(\lambda)f(\lambda)$ is holomorphic in some neighborhood of the point $0 \in \mathbb{C}^n$. Therefore the functions $v_{k+1}(\lambda) = \Phi(v_k(\lambda),\lambda)$ are also holomorphic in a neighborhood $\Sigma$ of the point $0 \in \mathbb{C}^n$. Thus the orbit of the element $v_0 \equiv 0$ ($v_0 \in \mathcal{H}(\Sigma,\mathbb{C})$) is completely contained in $\mathcal{H}(\Sigma,\mathbb{C})$. Hence $x = x(\cdot) \in \mathcal{H}(\Sigma,\mathbb{C})$ (see Remark 2.1). ▶

**EXAMPLE 6.1.** Let $\Lambda = \mathbb{C}^2$, $\mathfrak{X} = \mathfrak{Y} = \mathbb{C}$, and $F(x,\lambda) = \lambda_1^3 x^2 + (\lambda_1^2 - \lambda_2)x + \lambda_1^3 - -\lambda_1\lambda_2$, where $(\lambda_1,\lambda_2) = \lambda \in \mathbb{C}^2$. Obviously the function $F$ admits representation (6.3) with $P(\lambda) = \lambda_1^2 - \lambda_2$, $F'_x(0,0) = 0$, and satisfies the first requirement in Theorem 6.1. Therefore equation (6.2) has a unique small in the generalized sense solution $x(\lambda)$ which is $\rho$-holomorphic in some set $\Omega \subseteq \mathbb{C}^2$, $\Omega \ni 0$. We deduce easily this conclusion by a straightforward computation. Indeed, solving the quadratic equation $F(x,\lambda) = 0$

and setting $\lambda = \zeta h$, where $h \in \mathbb{C}^2$ with $\|h\| = 1$, we obtain two solutions, $x_1(\lambda)$ and $x_2(\lambda)$. One of these solutions is given by

$$x_1(\lambda) = x_1(\zeta h) = -\zeta h_1 - \frac{1}{2 \cdot 2!} \frac{\zeta^4 h_1^5}{\zeta h_1^2 - h_2} - \frac{1}{2^2 \cdot 3!} \frac{\zeta^7 h_1^9}{(\zeta h_1^2 - h_2)^2} - \cdots$$

and, therefore, it is small in the generalized sense, while the second one satisfies the condition $\lim_{\zeta \to 0} x_2(\zeta h) = \infty$ and, consequently, it is not small. Let us note that $x_1(\lambda)$ is a continuous function on some $\rho$-open set $\Omega \in 0$, whose intersection with $\mathbb{R}^2$ is the set $\tilde{\Omega}$ represented in Figure 2 and having the form of a helix.

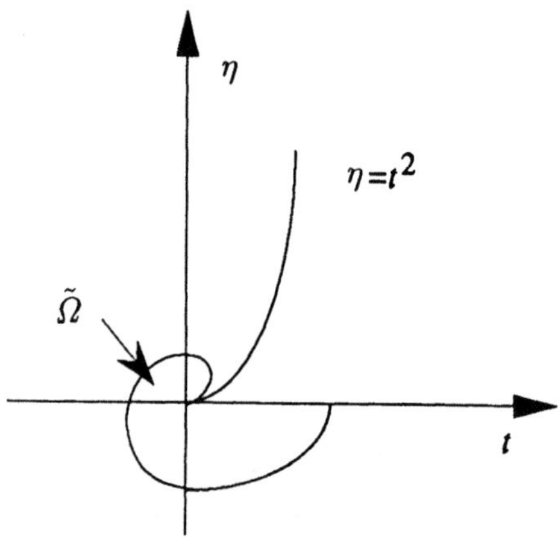


Figure 2

**EXAMPLE 6.2.** For the same $\Lambda$, $\mathfrak{X}$ and $\mathfrak{Y}$ let now $F$ be the function given by

$$F(x, \lambda) = \lambda_1^3 x^2 + (\lambda_1^2 - \lambda_2)x + (\lambda_1^2 - \lambda_2)^2 g(\lambda),$$

where $g(\lambda)$ is an arbitrary holomorphic function. Obviously $F(0,0) = 0$ and $F'_x(0,0) = 0$, and, consequently, in this case, we may not use the classical Implicit Function Theorem. However, the function $F$ satisfies both the first and the second requirements in Theorem 6.1. Hence the equation $F(x, \lambda) = 0$ has a unique holomorphic solution $x(\lambda)$ such that $x(0) = 0$. Indeed, solving straightforwardly this equation we obtain

$$x(\lambda) = -(\lambda_1^2 - \lambda_2) \sum_{n=0}^{\infty} 2^n \lambda_1^{3n} [g(\lambda)]^{n+1},$$

where the series in the right hand side converges uniformly on some neighborhood of the point $\lambda = 0$.

**2.** SMALL SOLUTIONS FOR AN OPERATOR LINEAR IN THE PARAMETER. THE INFINITE DIMENSIONAL CASE. We return once more to a complex Banach space $\mathfrak{X}$. Now we will suppose that equation (6.2) has the next special form

$$Bx = \lambda G(x) + y, \tag{6.6}$$

where: a) $B : \mathfrak{X} \to \mathfrak{X}$ is a linear normally solvable operator such that

(*) the kernel $N(B) = \ker B$ of the operator $B$ is isomorphic with the linear space $\mathfrak{H} = \mathfrak{X} \dot{-} R(B)$ where $R(B)$ is the range of the operator $B$,

b) $y$ is an arbitrary fixed element in $R(B)$,

c) $G$ is an operator anaytic in the ball $\mathfrak{D} = \{x \in \mathfrak{X} \ : \ \|x - z\| < R\}$, where $z$ is a solution of the equation $Bz = y$, and

d) $\lambda \in \mathbb{C}$.

In this setting we have to find conditions for the existence of an analytic solution $x(\lambda)$ of the equation (6.6) satisfying the relation $x(0) = z$, and to indicate the domain of analyticity for this vector-function.

Equations as (6.6) above arise in connection with a series of problems in nonlinear mechanics (see for example P. P. Rybin [1], [2], M. M. Vainberg [1] and M. M. Vainberg and V. A. Trenogin [1]). Moreover, equation (6.6) with an integral operator $G$, is related to the well-known Duffing equation, which has applications in electrotechnics.

When $N(B) = \{0\}$, the operator $B$ is continuously invertible and, consequently, we find ourselves under the conditions in the Implicit Function Theorem. When $N(B) \neq \{0\}$, it is convenient to assume that $B$ is finitely degenerated, i.e., $\operatorname{ind} B = \dim N(B) - \dim N(B^*) = 0$ ($B^*$ is the adjoint of $B$). In this case, conditions for the existence of a unique analytic solution $x(\lambda)$ have been obtained for the first time apparently by K. T. Ahmedov [1]. To be more specific, according to the Nekrasov-Nazarov method, one looks for a solution represented by a power series whose convergence is established using a majorant numerical series. This approach enables one to indicate an estimate for the analyticity domain of the solution $x(\lambda)$ and this turns out to be important in many applications.

Our aim in what follows is to provide conditions for the existence of a unique solution for equation (6.6) in the case of an arbitrary space $N(B)$ which is isomorphic with the space $\mathfrak{H}$ (this property is obvious if the operator $B$ is finitely degenerate). These conditions are obtained by reducing our problem to finding an s-solution for an equation of type (2.1). This approach enables us to point out a different way for approximating $x(\lambda)$, and also to establish a more convenient and more precise estimate for the domain of analyticity.

First of all we observe that by the Banach Theorem the operator $B \mid \mathfrak{X} \dot{-} N(B)$ — the restriction of $B$ on the subspace $\mathfrak{X} \dot{-} N(B)$ — has a continuous inverse operator $\Gamma : R(B) \to \mathfrak{X} \dot{-} N(B)$.

Setting $x = z + h$ we write $G(x)$ as

$$G(z+h) = G(z) + Ah + Q(z,h), \tag{6.7}$$

where $A$ is the Fréchet derivative of the operator $G$ at the point $z$ and $\|Q(z,h)\| = o(\|h\|)$.

Let us introduce the notations

$$M(G) = \sup_{x \in \mathfrak{D}} \|G(x)\|;$$
$$M(Q) = \sup_{\|h\|<R} \|Q(z,h)\|;$$
$$P : \mathfrak{X} \to \mathfrak{H} \text{ — the projection from } \mathfrak{X} \text{ onto } \mathfrak{H};$$
$$\hat{A} = A \mid N(B) \text{ — the restriction of the operator } A \text{ on } N(B).$$

**THEOREM 6.1.** *If $\hat{A}$ is an isomorphism from $N(B)$ onto $\mathfrak{H}$, and $G(z) \in R(B)$, then equation (6.6) has a unique analytic solution $x(\lambda)$ for all $\lambda$ such that*

$$|\lambda| < \delta = \frac{R}{2\|\Gamma\| M(G)} \min\left\{1, \frac{dR}{2M(Q)}, \frac{d}{2\|A\|}, \frac{Rd^2}{8M(Q)\|A\|}\right\}, \tag{6.8}$$

*where $d = \|\hat{A}^{-1}P\|^{-1}$. Moreover, $x(\lambda)$ can be approximated in the norm topology of $\mathfrak{X}$ by the sequence $\{x_n(\lambda)\}_{n\in\mathbb{N}}$ which is defined by the relations*

$$x_n(\lambda) = z + \lambda \Gamma G(x_{n-1}(\lambda)) + v_{n-1}(\lambda)), \tag{6.9}$$

$$v_n(\lambda) = -[\lambda \hat{A} P A \Gamma G(x_{n-1}(\lambda)) + \hat{A} P Q(z, x_n(\lambda) - z)], \tag{6.10}$$

*$n = 1, 2, \dots$, where $x_0(\lambda)$ is an arbitrary element of $\mathfrak{X}$ such that*

$$\|x_0(\lambda) - z\| < \hat{r} = \min\left\{R, \frac{d}{2M(Q)}\right\},$$

*and $v_0(\lambda)$ is an arbitrary element of $N(B)$ such that*

$$\|v_0(\lambda)\| < \frac{1}{2}\hat{r}.$$

◁ Setting $x = u + v$, where $u \in \mathfrak{X} \dot{-} N(B)$, $v \in N(B)$, and projecting equation (6.6) on the subspaces $\mathfrak{H}$ and $R(B)$, respectively, we obtain a system of equations

$$PG(u+v) = 0, \tag{6.11}$$

$$Bu = \lambda G(u+v) + y, \tag{6.12}$$

which is equivalent to equation (6.6). Applying the operator $\Gamma$ to the both sides of (6.12) and taking into account the equalities $u = x - v$ and $\Gamma y = z$ we obtain

$$PG(x) = 0, \tag{6.13}$$

$$x = z + \lambda \Gamma G(x) + v = F_1(\lambda, x, v). \tag{6.14}$$

We perform now a few formal transformations, the substantiation of which will become apparent below. Let us replace $h$ in (6.7) by $\lambda \Gamma G(x) + v$. Based on (6.14) we get

$$G(x) = G(z) + \lambda A \Gamma G(x) + Av + Q(z, \lambda \Gamma G(x) + v).$$

Applying the operator $P$ to the both sides of the last equality, and taking into account the equalities (6.13), $PG(z) = 0$ and $PAv = Av$ (recall that $Av \in \mathfrak{H}$) we obtain

$$0 = Av + \lambda PA\Gamma G(x) + PQ(z, \lambda \Gamma G(x) + v),$$

or

$$Av = -[\lambda PA\Gamma G(x) + PQ(z, \lambda \Gamma G(x) + v)],$$

whence

$$v = -[\lambda \hat{A}^{-1} PA\Gamma G(x) + \hat{A}^{-1} PQ(z, \lambda \Gamma G(x) + v)] = F_2(\lambda, x, v). \tag{6.15}$$

Consider now the Banach space $\mathfrak{E} = \mathfrak{X} \dot{+} N(B)$ with the norm given by $\|w\| = \max\{\|x\|, \|v\|\}$, for any $w = (x, v) \in \mathfrak{E}$. Denoting $F = (F_1, F_2)$ we may rewrite the system (6.14), (6.15) as

$$w = F(\lambda, w). \tag{6.16}$$

We show that the mapping $F$ is defined and analytic in some domain in $\mathfrak{E}$, and for all $\lambda \in \Omega = \{\lambda \in \mathbb{C} \ : \ |\lambda| < \delta\}$ (see (6.8)) it satisfies the conditions in Theorem 2.1.

The mappings $F_1$ and $F_2$ are defined and analytic if

$$\|h\| \ (= \|\lambda G(x) + v\|) < r \leqslant R. \tag{6.17}$$

Consider the domain $\mathfrak{D}_r = \{w : \|x - z\| < r, \ \|v\| < r\varepsilon\}$ included in $\mathfrak{E}$. Inequality (6.17), for all $w = (x, v)$, is obviously fulfilled under the following conditions: $0 < \varepsilon < 1$ and

$$|\lambda| < \frac{r(1-\varepsilon)}{\|\Gamma\| M(G)}. \tag{6.18}$$

Let us establish the existence of such an $r = \hat{r} > 0$ and of a function $\delta(r) : [0, R] \to$ $\to [0, \delta]$ so that the mapping $F$ leaves the domain $\mathfrak{D}_{\hat{r}}$ invariant for $\lambda$ with $|\lambda| < \delta(\hat{r})$. From (6.14) and (6.18) it follows that $\|F_1(\lambda, w)\| < r$ for any $r$ with $0 < r < R$ and $|\lambda| < \rho(r)$. Consider the inequality

$$\|F_2(\lambda, w)\| < r\varepsilon. \tag{6.19}$$

It is not difficult to convince ourselves that inequality (6.19) follows from

$$|\lambda| < \frac{dr\varepsilon - r^2 M(Q) R^{-2}}{\|A\Gamma\| M(G)} = \varphi(r),$$

for any $r$ such that $\varphi(r) > 0$. A direct computation shows that

$$\max_{0<r\leqslant R} \varphi(r) = \varphi(\hat{r}) = \min\left\{\frac{Rd\varepsilon}{2\|A\Gamma\|M(G)}, \frac{(Rd\varepsilon)^2}{4\|A\Gamma\|M(G)M(Q)}\right\},$$

where

$$\hat{r} = \min\left\{R, \frac{d}{2M(Q)}\right\}.$$

Thus the mapping $F$ is analytic and $F(\lambda, w) \in \mathfrak{D}_{\hat{r}}$ for all $w \in \mathfrak{D}_{\hat{r}}$ and all $|\lambda| < \delta(\hat{r}) = \min\{\rho(\hat{r}), \varphi(\hat{r})\}$.

Clearly $\delta(\hat{r})$ depends also on $\varepsilon$, and, as one can easily see, it attains its maximum value for $\varepsilon = 2^{-1}$, whence (6.8) follows. From (6.14) and (6.15) it is not difficult to remark that the operator $F(0, \cdot)$ is a $q$-contraction on some neighborhood of the point $w_0 = (z, 0)$ for a certain $q$ with $0 < q < 1$. By Theorem 2.1 we obtain an s-solution $w(\lambda) = (x(\lambda), v(\lambda))$ of equation (6.16) and the iterative sequence $w_n(\lambda) = F(\lambda, w_{n-1}(\lambda)) = (x_n(\lambda), v_n(\lambda))$ which appears in formulas (6.9), (6.10). ▶

To conclude this section let us show what the conditions in the theorem above mean in the case of a finitely degenerate operator $B$.

Assume that $N(B)$ has dimension $m$ and $\operatorname{ind} B = 0$. Futher, let $\{v_1, \ldots, v_m\}$ be a basis of $N(B)$ and let $\mathfrak{H}_k = \operatorname{clin}\{Av_1, \ldots, Av_{k-1}, Av_{k+1}, \ldots, Av_m\}$ be the linear hull of the elements $Av_1, \ldots, Av_{k-1}, Av_{k+1}, \ldots, Av_m$. Consider the spaces $\mathfrak{L}_k =$ $= R(B) \dot{+} \mathfrak{H}_k$ which are proper subspaces of the space $\mathfrak{X}$.

**PROPOSITION 6.1.** *The following conditions are equivalent:*

1) *the operator* $\hat{A} = A \mid N(B)$ *is an isomorphism from* $N(B)$ *onto* $\mathfrak{H}$;

2) $d_k = \rho(Av_k, \mathfrak{L}_k) = \inf_{x \in \mathfrak{L}_k} \|Av_k - x\| > 0$, $k = 1, \dots, m$;

3) *the equalities*

$$\langle \psi_i, Av_j \rangle = \delta_{ij}, \quad i, j = 1, \dots, m,$$

*are fulfilled for a certain basis* $\{\psi_i\}_{i=1}^m$ *of the space* $N(B^*)$ ($B^*$ *is the adjoint of* $B$).

$\triangleleft$ From 1) it follows that the system of elements $\{Av_i\}_{i=1}^m$ forms a basis of $\mathfrak{H}$. Assume that condition 2) fails for some $k = 1, \dots, m$. Then there exists a non-zero element $y \in R(B)$ such that

$$Av_k = y + \sum_{\substack{i \neq k \\ 1 \leqslant i \leqslant m}} \alpha_i Av_i, \quad \alpha_i \in \mathbb{C},$$

or

$$y = \sum_{i=1}^m \beta_i Av_i,$$

where $\beta_i = -\alpha_i$ if $i \neq k$, $1 \leqslant i \leqslant m$, and $\beta_k = 1$, whence $y \in \mathfrak{H}$. This is impossible since $y \neq 0$ and $\mathfrak{H} \cap R(B) = \{0\}$. The implication 1) $\Rightarrow$ 2) is proved.

Assume now that condition 2) is fulfilled. By a consequence of the Hahn-Banach Theorem (see, for instance, Chapter 0) there exist the elements $\psi_k \in \mathfrak{X}^*$, $k = 1, \dots, m$, such that

$$\psi_k \perp \mathfrak{L}_k, \tag{6.20}$$

$$\langle \psi_k, Av_k \rangle = 1, \tag{6.21}$$

$$\|\psi_k\| = d_k^{-1}, \tag{6.22}$$

for all $k = 1, \dots, m$. From (6.20) it follows that $\psi_k \in N(B^*)$. Indeed, since $\mathfrak{L}_k \supset R(B)$ then for any $x \in \mathfrak{X}$ the equality

$$\langle \psi_k, Bx \rangle = \langle B^* \psi_k, x \rangle = 0$$

holds, whence – in view of the fact that $x$ is arbitrary — we have $B^* \psi_k = 0$. Moreover, taking account of equality (6.21) we obtain that

$$\langle \psi_i, Av_j \rangle = \delta_{ij}, \quad i, j = 1, \dots, m.$$

In order to prove 2) $\Rightarrow$ 3) it remains to establish that the system of elements $\{\psi_k\}_{i=1}^m$ is linearly independent. Assume that

$$\sum_{k=1}^{m} \gamma_k \psi_k = 0,$$

for some complex numbers $\gamma_k \in \mathbb{C}$. Then for each $1 \leqslant i \leqslant m$ we obtain

$$0 = \left\langle \sum_{k=1}^{m} \gamma_k \psi_k, Av_i \right\rangle = \sum_{k=1}^{m} \gamma_k \langle \psi_k, Av_i \rangle = \gamma_k.$$

Let us prove now implication 3) $\Rightarrow$ 1). To this end it is sufficient to show that, if condition 3) holds, then the system of elements $\{Av_i\}_{i=1}^m$ is a basis of $\mathfrak{H}$. Indeed, from 3) it follows that the system $\{Av_i\}_{i=1}^m$ is linearly independent. It remains to show that the elements $Av_i$, $i = 1, \ldots, m$, lie in $\mathfrak{H}$. Assuming the contrary, we obtain that for some $1 \leqslant j \leqslant m$ the element $Av_j$ lies in $R(B)$. From the Fredholm alternative it follows that $Av_j$ is orthogonal to the space $N(B^*)$, which is impossible since

$$\langle \psi_j, Av_j \rangle = 1, \quad \psi_j \in N(B^*),$$

for all $j = 1, \ldots, m$. ▶

# Chapter VII

# Banach manifolds

## §1. Basic definitions

Let $\mathfrak{H}$ be a Hausdorff tolopogical space and let $\mathfrak{Y}$ be a Banach space.

**DEFINITION 1.1.** The space $\mathfrak{H}$ is called a *manifold modelled on* $\mathfrak{Y}$, if for any point $x \in \mathfrak{H}$ there exist an open neighborhood $\mathfrak{U}_x \subseteq \mathfrak{H}$ of $x$ and a homeomorphism $\varphi_x : \mathfrak{U}_x \to \mathfrak{V}_x \subseteq \mathfrak{Y}$, where $\mathfrak{V}_x$ is an open set in $\mathfrak{Y}$.

If the space $\mathfrak{Y}$ is the $n$-dimensional Euclidean space $\mathbb{K}^n$, then $\mathfrak{H}$ is called simply an *$n$-dimensional manifold.*

A pair $(\mathfrak{U}_x, \varphi_x)$ as above is called a *chart* (sometimes, a *local chart*) of the manifold $\mathfrak{H}$. A family of charts $\{(\mathfrak{U}_x, \varphi_x)\}_x$ such that $\bigcup\limits_x \mathfrak{U}_x \supseteq \mathfrak{H}$ is called an *atlas* of the manifold $\mathfrak{H}$.

**DEFINITION 1.2.** We say that two charts $(\mathfrak{U}_x, \varphi_x)$ and $(\mathfrak{U}_y, \varphi_y)$ satisfy the condition of $C^m$- compatibility, $m = 0, 1, 2, \ldots, \infty$ (respectively, of $O$-compatibility), if whenever $\mathfrak{U}_x \cap \mathfrak{U}_y \neq \emptyset$, the mappings $\varphi_x \circ \varphi_y^{-1} : \varphi_y(\mathfrak{U}_x \cap \mathfrak{U}_y) \to \varphi_x(\mathfrak{U}_x \cap \mathfrak{U}_y)$ and $\varphi_y \circ \varphi_x^{-1} : \varphi_x(\mathfrak{U}_x \cap \mathfrak{U}_y) \to \varphi_y(\mathfrak{U}_x \cap \mathfrak{U}_y)$ are of class $C^m$, $m = 0, 1, 2, \ldots, \infty$ (respectively, of class $O$ — the class of analytic mappings).

If all the charts belonging to a given atlas are $C^m$-compatible, $m = 0, 1, 2, \ldots,$ $\infty$ (respectively, $O$-compatible), where $m$ does not depend on the charts, then we say that the manifold is of class $C^m$ (respectively, of class $O$).

A manifold of class $C^0$ is called a *topological manifold.* A manifold of class $O$ is called *analytic* (if $\mathfrak{Y}$ is a complex space, the manifold is called *complex-analytic*).

The manifolds of class $C^m$, $m \geqslant 1$, are also called *smooth manifolds.*

**REMARK 1.1.** In connection with the above introduced definitions the following questions arise.

1) Can we speak about an unique manifold if the space $\mathfrak{H}$ is endowed with two different atlases?

An answer to this question depends, of course, on the convention we agree upon. The advisability of a convention is determined by the applications which may occur later on.

Usually two atlases on the same space $\mathfrak{H}$ are called *equivalent* if any chart of one of the atlases is compatible (relatively to a fixed class $C^m$, respectively $O$) with any chart of the other atlas. In this case we say that both atlases define one and the same structure of a Banach manifold on the space $\mathfrak{H}$.

2) Are there manifolds which coincide as topological spaces but have different models? In other words, if $\mathfrak{H}$ is a manifold with the model space $\mathfrak{Y}$, can $\mathfrak{H}$ be a manifold with a model space $\mathfrak{Y}' \neq \mathfrak{Y}$?

An answer to this question will be given in the sequel. However, even at this moment, it is obvious that, if this is the case, the spaces $\mathfrak{Y}$ and $\mathfrak{Y}'$ must contain some homeomorphic open sets.

Any open subset of some Banach space $\mathfrak{X}$ (in particular the space $\mathfrak{X}$) offers a simple example of a manifold. Indeed, for such a subset $\mathfrak{U} \subseteq \mathfrak{X}$, the model space is the space $\mathfrak{X}$ itself, and the atlas consists of one chart defined by the whole set $\mathfrak{U}$ and the identity homeomorphism. These manifolds are smooth (of class $C^\infty$) and even analytic. If $\mathfrak{X}$ is the $n$-dimensional Euclidean space, then these manifolds are $n$-dimensional.

Any subset of a space $\mathfrak{X}$ consisting of isolated points is a zero-dimensional manifold.

A more interesting example is offered by the circle $\{(x,y) \in \mathbb{R}^2 \,:\, x^2+y^2=1\}$. Indeed, let us choose the open cover of the circle given by the four semicircles $\{\mathfrak{U}_i\}_{i=1}^4$ obtained intersecting the circle with the half-spaces $y>0$, $x>0$, $y<0$, $x<0$, respectively. Take a point $(x,y)$ with $y>0$ on the semicircle $\mathfrak{U}_1$. The mapping $\varphi_1$ which associates to $(x,y) \in \mathfrak{U}_1$ the point $x \in ]-1,1[ \subseteq \mathbb{R}^1$ is a homeomorphism. The inverse mapping $\varphi_1^{-1}$ is given by $\varphi_1^{-1}(x) = \left(x, \sqrt{1-x^2}\right)$. The pair $(\mathfrak{U}_1, \varphi_1)$ defines a chart. The charts $(\mathfrak{U}_2, \varphi_2)$, $(\mathfrak{U}_3, \varphi_3)$, $(\mathfrak{U}_3, \varphi_3)$ are obtained analogously. Thus the circle is a manifold with the model space $\mathbb{R}^1$, or, a manifold of dimension 1. Further, let us remark that the indicated atlas gives a smooth structure on the circle as a manifold. Indeed, consider, for instance, the neighborhood $\mathfrak{U}_{12} = \mathfrak{U}_1 \cap \mathfrak{U}_2$ consisting of that part of the circle which lies in the positive quadrant $y > 0$, $x > 0$. There are two homeomorphisms defined on $\mathfrak{U}_{12}$, namely, $\varphi_1 : (x, y) \to x$ and $\varphi_2 : (x, y) \to y$. The overlap mapping $\varphi_1 \circ \varphi_2^{-1} : y \mapsto x$ is given by the equality $x = +\sqrt{1 - y^2}$ and is a homeomorphism from the interval $]0, 1[$ onto itself.

Analogously we can show that the sphere is a smooth manifold of dimension 2. In the above mentioned examples it is worth to remark that the circle and the sphere are subsets of the spaces $\mathbb{R}^2$ and $\mathbb{R}^3$, respectively, which, on their turn, are also manifolds. We will be concerned with such aspects in §3.

## §2. Smooth mappings

The main reason for introducing the notion of manifolds consists in the possibility they offer to differentiate and integrate functions or mappings defined on "nonlinear" spaces. This is done by identifying locally these spaces with theirs models. Thus, for example, a function defined on a circle can be considered locally as a function given on an interval of the real line. Therefore it is convenient to think of a circle not as a set of points in the two-dimensional space $\mathbb{R}^2$, but as a set of points which is identified locally with subsets of the one-dimensional space $\mathbb{R}^1$.

Let us give the precise definitions.

**DEFINITION 2.1.** Let $\mathfrak{H}$ be a manifold of class $C^m$, $m \geqslant 1$, modelled on the space $\mathfrak{Y}$, and let $f$ be a mapping from $\mathfrak{H}$ into a Banach space $\mathfrak{Z}$. The mapping $f$ is called *m-times continuously Fréchet differentiable* at a point $a \in \mathfrak{H}$ if there exists a chart $(\mathfrak{U}_a, \varphi_a)$ with $a \in \mathfrak{U}_a \subseteq \mathfrak{H}$ and $\varphi_a : \mathfrak{U}_a \to \mathfrak{V}_a \subseteq \mathfrak{Y}$, such that the mapping $f \circ \varphi_a^{-1} : : \mathfrak{V}_a \to \mathfrak{Z}$ is $m$-times continuously Fréchet differentiable, in the usual sense, at the point $b = \varphi_a(a) \in \mathfrak{Y}$.

For the correctness of this definition it has to be shown that the differentiability of $f$ is independent of the choice of a compatible chart. Indeed, let $(\mathfrak{U}'_a, \varphi'_a)$ with $a \in \mathfrak{U}'_a$ be another chart, compatible with $(\mathfrak{U}_a, \varphi_a)$ for which the mapping $f \circ \varphi_a^{-1}$ is $m$-times differentiable at the point $b = \varphi_a(a) \in \mathfrak{Y}$. Then the mapping $f \circ {\varphi'}_a^{-1} = = f \circ \varphi_a^{-1} \circ \varphi_a \circ {\varphi'}_a^{-1}$ is $m$-times differentiable at the point $b' = \varphi'_a(a) \in \mathfrak{Y}$ as the composition of two $m$-times differentiable mappings, since the mapping $\varphi_a \circ {\varphi'}_a^{-1}$ is $m$-times differentiable by the compatibility of the charts.

In particular, if $\mathfrak{Y} = \mathbb{K}^n$, then the differentiability of $f$ means the existence of the partial derivatives of the mapping $f \circ \varphi_a^{-1}$, with respect to the variables $(x_1, x_2, \ldots, x_n) \in \mathbb{K}^n$, at the point $x_a = (\varphi_{1a}(a), \ldots, \varphi_{na}(a))$. Therefore the homeomorphism $\varphi_a$ is often called a *local system of coordinates* for the manifold $\mathfrak{H}$ at the point $a$, and the corresponding neighborhood $\mathfrak{U}_a \subseteq \mathfrak{H}$ is called a *local coordinate neighborhood.*

In other words, a chart in an atlas of the manifold $\mathfrak{H}$ consists of a local coordinate neighborhood and a local system of coordinates.

**DEFINITION 2.2.** Let $\mathfrak{H}$ and $\mathfrak{M}$ be two manifolds of class $C^m$, $m \geqslant 1$, modelled on spaces $\mathfrak{Y}$ and $\mathfrak{Z}$, respectively, and let $g$ be a mapping from $\mathfrak{H}$ into $\mathfrak{M}$. The mapping $g$ is called *m-times Fréchet differentiable* at a point $a \in \mathfrak{H}$ if there exists a local system of coordinates $\psi_b$ at $b = g(a) \in \mathfrak{M}$, such that the mapping $f = \psi_b \circ g : \mathfrak{H} \to \mathfrak{Z}$ defined on some neighborhood of the point $a \in \mathfrak{H}$ is $m$-times differentiable at $a$ in the sense of Definition 2.1.

In this case it is also easy to show that the differentiability of $g$, in the indicated sense, does not depend on the choice of the local system of coordinates $\psi_b$. Moreover, we have the next obvious result.

**PROPOSITION 2.1.** *A mapping $g : \mathfrak{H} \to \mathfrak{M}$ is m-times differentiable at a point $a \in \mathfrak{H}$ — in the sense of Definition 2.2 — if there exist two local systems of coordinates, $\varphi_a$, at the point $a \in \mathfrak{H}$ and $\psi_b$ at the point $b = g(a) \in \mathfrak{M}$, such that the mapping $\psi_b \circ g \circ \varphi_a^{-1} : \mathfrak{Y} \to \mathfrak{Z}$ is m-times Fréchet differentiable at the point $\varphi_a(a) \in \mathfrak{Y}$ in the usual sense.*

Analogously we introduce the notion of continuously differentiable mapping and the notion of analytic mapping (if the manifold is analytic) at a point, or, on some open subset, of a manifold $\mathfrak{H}$.

**DEFINITION 2.3.** Let $\mathfrak{H}$ and $\mathfrak{M}$ be two manifolds modelled on the spaces $\mathfrak{Y}$ and $\mathfrak{Z}$, respectively, and let $g$ be an one-to-one mapping from $\mathfrak{H}$ onto $\mathfrak{M}$ such that both $g$ and $g^{-1}$ are smooth mappings at any point. Then $g$ is called a *diffeomorphism between the manifolds $\mathfrak{H}$ and $\mathfrak{M}$, and the manifolds $\mathfrak{H}$ and $\mathfrak{M}$ are said to be diffeomorphic.*

Let us note that the model spaces of some diffeomorphic manifolds coincide, up to a linear isomorphism. In particular, two finite-dimensional diffeomorphic manifolds have the same dimension.

## §3. Submanifolds

Let $\mathfrak{M}$ be a manifold modelled on the space $\mathfrak{X}$ and let $\mathfrak{H}$ be a subset of $\mathfrak{M}$.

**DEFINITION 3.1.** The set $\mathfrak{H}$ is called a *submanifold* of $\mathfrak{M}$ if for any $a \in \mathfrak{H}$ there exists a chart $(\mathfrak{D}_a, \Phi_a)$, $\mathfrak{D}_a \subseteq \mathfrak{M}$, such that $\Phi_a(a) = 0$ and $\Phi_a(\mathfrak{D}_a \cap \mathfrak{H}) = \mathfrak{V}_a \cap \mathfrak{E}_a$, where $\mathfrak{V}_a = \Phi_a(\mathfrak{D}_a)$ and $\mathfrak{E}_a$ is a linear subspace of $\mathfrak{X}$.

A submanifold $\mathfrak{H}$ is called a *direct submanifold* if each of the spaces $\mathfrak{E}_a$ has a direct complement in $\mathfrak{X}$, i.e., $\mathfrak{X} = \mathfrak{E}_a \dot{+} C\mathfrak{E}_a$, where $C\mathfrak{E}_a$ is a linear subspace of $\mathfrak{X}$.

It is clear that, at least locally, the set $\mathfrak{H}$ itself is a manifold in the sense of Definition 1.1, with the model space $\mathfrak{E}_a$. In addition, if the sets $\Phi_a(\mathfrak{D}_a \cap \mathfrak{H})$ are open subsets in a fixed subspace $\mathfrak{Y} \subseteq \mathfrak{X}$ , for all $a \in \mathfrak{H}$, then $\mathfrak{H}$ is a manifold modelled on $\mathfrak{Y}$. In this case, an atlas for $\mathfrak{H}$ is provided by the family of charts $\{(\mathfrak{U}_a, \varphi_a)\}_{a\in\mathfrak{H}}$, where $\mathfrak{U}_a = \mathfrak{D}_a \cap \mathfrak{H}$ and $\varphi_a$ is the restriction of the homeomorphism $\Phi_a$ to the set $\mathfrak{U}_a$.

The manifold $\mathfrak{M}$ appearing in Definition 3.1 above is often called the *ambient manifold* of the submanifold $\mathfrak{H}$.

The notion of submanifold plays an important role in a series of problems. The case when the ambient manifold is a Banach space is especially interesting.

However, not any manifold $\mathfrak{H}$ which is a subset of a manifold $\mathfrak{M}$ must be, at the same time, a submanifold of $\mathfrak{M}$ (see, for instance, J. Milnor [1]).

Now, we consider this problem in the case when the ambient manifold $\mathfrak{M}$ is a Banach space $\mathfrak{X}$.

**THEOREM 3.1.** *Let $\mathfrak{X}$ be a Banach space and let $\mathfrak{H} \subseteq \mathfrak{X}$ be a manifold modelled on $\mathfrak{Y}$. Assume that $\mathfrak{Y}$ is a subspace of $\mathfrak{X}$ with a direct complement in $\mathfrak{X}$, i.e.,*

$$\mathfrak{X} = \mathfrak{Y} \dot{+} \mathfrak{Z}, \tag{3.1}$$

*where $\mathfrak{Z}$ is a linear subspace of $\mathfrak{X}$. Then the following conditions are equivalent:*

*a) for any point $a \in \mathfrak{H}$ there exists a chart $(\mathfrak{U}_a, \varphi_a)$ such that the homeomorphism $\psi_a = \varphi_a^{-1} : \mathfrak{V}_a \to \mathfrak{U}_a$, $\mathfrak{V}_a = \varphi_a(\mathfrak{U}_a)$, is continuously Fréchet differentiable as a mapping from $\mathfrak{Y}$ into $\mathfrak{X}$ and the linear operator $P\psi_a'(b) : \mathfrak{Y} \to \mathfrak{Y}$ is continuously invertible, where $P$ is the projection from $\mathfrak{X}$ onto $\mathfrak{Y}$, and $\psi_a'(b)$ is the Fréchet derivative of $\psi_a$ at $b = \varphi_a(a) \in \mathfrak{Y}$;*

*b) for any point $a \in \mathfrak{H}$ there exist a neighborhood $\mathfrak{U}_a' \subseteq \mathfrak{H}$ of the point $a$, a neighborhood $\mathfrak{V}_a'$ of the point $c = Pa \in \mathfrak{Y}$, and a Fréchet differentiable mapping $f : \mathfrak{V}_a' \to \mathfrak{Z}$, such that*

$$\{x \in \mathfrak{X} \ : \ x = u + f(u), \ u \in \mathfrak{V}_a'\} = \mathfrak{U}_a'.$$

◁ It is sufficient to prove the implication a) ⇒ b) only, since the inverse implication b) ⇒ a) is obvious. Indeed, the chart $(\mathfrak{U}_a, \varphi_a)$ where $\mathfrak{U}_a = \mathfrak{U}_a'$ and $\varphi_a$ is the inverse of the mapping $\psi_a(u) = u + f(u)$ (i.e., $\varphi_a$ is the projection mapping from $\mathfrak{U}'_a$ into $\mathfrak{Y}$) satisfies all the requirements in condition a). More precisely we have $P\psi'_a(b) = I \mid \mathfrak{Y}$, where $I$ is the identity mapping on $\mathfrak{X}$ and $b = c = Pa$.

Assume that condition a) is fulfilled. By the Inverse Mapping Theorem it follows that $P\psi_a : \mathfrak{V}'_a \to \mathfrak{Y}$ is a diffeomorphism on some neighborhood $\mathfrak{V}'_a \subseteq \mathfrak{Y}$ of the point $b \in \mathfrak{Y}$, i.e., for all $x$ sufficiently close to the point $a$ and $y$ sufficiently close to the point $b$, the equation

$$Px = P\psi_a(y)$$

has the unique solution

$$y = g(Px) = g(u), \quad u = Px,$$

and the mapping $g$ is differentiable on a neighborhood of the point $c = Pa$.

Further, let $\mathfrak{U}'_a = \psi_a(\mathfrak{V}'_a)$. Then the points of the set $\mathfrak{U}'_a$ are exactly the points $x \in \mathfrak{X}$ for which the equality

$$Qx = Q\psi_a(g(Px)) \tag{3.2}$$

is fulfilled, where $Q = I - P$. Setting $x = u + v$, where $u = Px$, $v = Qx$ and $f(u) = Q\psi_a(g(u))$ the desired conclusion follows. ▶

**THEOREM 3.2.** *Assume that the hypotheses of Theorem* 3.1 *and one of the conditions* a) *or* b) *are fulfilled. Then* $\mathfrak{H}$ *is a direct submanifold of* $\mathfrak{X}$.

◁ Consider the mapping $\Phi : \mathfrak{X} \to \mathfrak{X}$ defined on some neighborhood $\mathfrak{D}_a \subset \mathfrak{X}$ of the point $a$ by the equality

$$h = \Phi x = P(x - a) + Qx - f(Px), \tag{3.3}$$

where $f$ is given by the right hand side of (3.2).

Setting $u = Px$, $v = Qx$, we have $\Phi'_u(a) = P - f'_u(Pa)$ and $\Phi'_v(a) = Q$. Since both the partial derivatives are continuous on some neighborhood of the point $a$, $\Phi$ is differentiable on that neighborhood. Moreover, $\Phi$ is a diffeomorphism, since its inverse mapping $\Psi = \Phi^{-1} : \mathfrak{X} \to \mathfrak{D}_a$, defined by the equality $x = h + f(P(h + a)) + Pa$, is also differentiable on some neighborhood of the point $h = 0$. Further, by (3.2) and (3.3) we obtain $\Phi(a) = 0$ and $\Phi(x) = P(x - a) \in \mathfrak{Y}$ for all $x \in \mathfrak{H} \cap \mathfrak{D}_a$, i.e., by (3.1), $\mathfrak{H}$ is a direct submanifold of $\mathfrak{X}$. ▶

The mapping $\Psi_a : \mathfrak{Y} \to \mathfrak{H}$ satisfying condition a) in Theorem 3.1 is called a *local parametrization* of the manifold $\mathfrak{H}$.

The mapping $\Psi_a(u) = u + f(u)$ appearing in condition b) of Theorem 3.1 gives also a parametrization of the manifold $\mathfrak{H}$, which, due to its special form, is called an *explicit* or a *direct parametrization.*

In the case of a finite-dimensional space $\mathfrak{X}$ any submanifold admits obviously a local parametrization given by a suitable permutation of the coordinates. Therefore, from Theorems 3.1 and 3.2 the next result follows.

**COROLLARY 3.1.** *Let $\mathfrak{X} = \mathbb{K}^N$ and let $\mathfrak{H} \subseteq \mathfrak{X}$ be a smooth manifold of dimension $n \leqslant N$. Then $\mathfrak{H}$ is a submanifold of $\mathfrak{X}$ if and only if one of the following conditions is fulfilled:*

a) *for any point $a \in \mathfrak{H}$ there exists a chart $(\mathfrak{U}_a, \varphi_a)$ such that the homeomorphism $\psi_a = \varphi_a^{-1}$ is continuously differentiable as a mapping from $\mathbb{K}^N$, and the linear mapping $\psi'_a(a) \in L(\mathbb{K}^n, \mathbb{K}^N)$ has rank $n$;*

b) *for any point $a \in \mathfrak{H}$ there exist a suitable permutation of the coordinates $(x_1, \dots, x_N) \in \mathbb{K}^N$ of the ambient space $\mathbb{K}^N$, and $N - n$ numerical functions $f_{n+1}(x_1, \dots, x_n), \dots, f_N(x_1, \dots, x_n)$ which are continuously differentiable on a neighborhood of the point $(a_1, \dots, a_n)$, such that the set of the points $x \in \mathfrak{H}$ sufficiently close to $a$ is given by the equations*

$$x_1 = x_1, x_2 = x_2, \dots, x_n = x_n,$$

$$x_{n+1} = f_{n+1}(x_1, \dots, x_n), \dots, x_N = f_N(x_1, \dots, x_n).$$

In many mathematical theories the notion of submanifold is more convenient and more useful than the abstract notion of manifold. Moreover, in the mathematical literature, the term "manifold" is often used to call a set which, according to our definition above, is a submanifold (see, for instance, H. Cartan [1] and J. Milnor [2]). Nevertheless, in applications, both notions are equally justified. To be more specific, let us notice that a mechanical system with a finite number of degrees of freedom can be considered as a finite-dimensional manifold, and for this description it is not necessary to use any ambient space. For example, a system with four degrees of freedom, which is given as a set of points in the three-dimensional space, can not be considered as a submanifold of that space. However, it is possible to "embed" such a system in a certain abstract space, or a manifold, of a greater dimension using some one-to-one mappings, and this enables us, sometimes, to obtain a series of new properties of the considered system.

**DEFINITION 3.2.** Let $\mathfrak{H}$ and $\mathfrak{M}$ be two manifolds. An *embedding* of $\mathfrak{H}$ into $\mathfrak{M}$ is any mapping $f : \mathfrak{H} \to \mathfrak{M}$ such that the image $f(\mathfrak{H})$ is a submanifold of $\mathfrak{M}$, and $f$ induces a diffeomorphism between $\mathfrak{H}$ and $f(\mathfrak{H})$.

For our purposes it is not necessary to develop here the theory of embeddings. We mention only, without proof, that any smooth finite-dimensional manifold can be smoothly embedded in some finite-dimensional Euclidean space.

## §4. Complex manifolds and Stein manifolds

**1.** As we have already mentioned, a manifold $\mathfrak{H}$ is called complex if its model space $\mathfrak{Y}$ is a complex Banach space.

Of course, any smooth complex manifold $\mathfrak{H}$ modelled on $\mathfrak{Y}$ can be considered as a real smooth manifold modelled on $\mathfrak{Y}_{\mathbb{R}}$ — the realification of the space $\mathfrak{Y}$ (in the finite-dimensional case, a complex manifold of dimmension $n$ is a real manifold of dimension $2n$).

Further, it is clear that if the mappings $\varphi_x \circ \varphi_y^{-1}$ and $\varphi_y \circ \varphi_x^{-1}$ are smooth in the complex sense (i.e., Fréchet differentiable in the complex sense on $\varphi_y(\mathfrak{U}_x \cap \mathfrak{U}_y)$, and on $\varphi_x(\mathfrak{U}_x \cap \mathfrak{U}_y)$, respectively), for any two charts $(\mathfrak{U}_x, \varphi_x)$ and $(\mathfrak{U}_y, \varphi_y)$ of $\mathfrak{H}$ such that $\mathfrak{U}_x \cap \mathfrak{U}_y \neq \emptyset$, then they are analytic, and, consequently, the manifold $\mathfrak{H}$ is also analytic.

Note that in the finite-dimensional case it is enough to require the analyticity of one of the indicated mappings only, since its inverse is also analytic, by a theorem of Osgood.

In what follows we will be concerned mainly with complex manifolds, i.e., manifolds of class $O$, modelled on complex Banach spaces.

The notion of a Fréchet differentiable mapping $f$ from a complex analytic manifold $\mathfrak{H}$ into a complex analytic manifold $\mathfrak{M}$ is introduced with respect to the previous Definitions 2.1 and 2.2. If such a mapping $f$ is Fréchet differentiable, in the complex sense, on neighborhoods of any point in $\mathfrak{H}$, then the mapping $f$ is called *analytic* (or *holomorphic*).

The term "holomorphic mapping" is mostly used in the theory of complex manifolds. In what follows we will use it in order to keep the tradition.

In particular, if $\mathfrak{M} = \mathbb{C}$ is the complex field, then $f$ is called a *complex-valued analytic function* on $\mathfrak{H}$. The set of all complex-valued analytic functions on $\mathfrak{H}$ will be denoted by $A(\mathfrak{H})$.

Let $F(\mathfrak{H})$ be an arbitrary set of functions defined on $\mathfrak{H}$ with values in $\mathbb{C}$.

**DEFINITION 4.1.** For an arbitrary compact set $\mathfrak{G} \subset \mathfrak{H}$ the set

$$\hat{\mathfrak{G}}_{F(\mathfrak{H})} = \left\{ x \in \mathfrak{H} \ : \ |f(x)| \leqslant \sup_{y \in \mathfrak{G}} |f(y)| \text{ for all } f \in F(\mathfrak{H}) \right\}$$

is called the *$F(\mathfrak{H})$-convex hull* of the set $\mathfrak{G}$.

In particular, the $A(\mathfrak{H})$-convex hull of a compact set $\mathfrak{G} \subseteq \mathfrak{H}$ is called the *holomorphic convex hull* of the set.

Obviously if $F_1(\mathfrak{H}) \subseteq F_2(\mathfrak{H})$ then

$$\hat{\mathfrak{G}}_{F_1(\mathfrak{H})} \supseteq \hat{\mathfrak{G}}_{F_2(\mathfrak{H})}. \tag{4.1}$$

**DEFINITION 4.2.** A manifold $\mathfrak{H}$ is called $F(\mathfrak{H})$-*convex* if $\hat{\mathfrak{G}}_{F(\mathfrak{H})}$ is a compact subset of $\mathfrak{H}$ for any compact subset $\mathfrak{G} \subseteq \mathfrak{H}$.

In particular, the $A(\mathfrak{H})$-convex manifolds are called *holomorphically convex.*

In order to give some examples of holomorphically convex manifolds let us consider first the case of a complex Banach space $\mathfrak{X}$. For any subset $\mathfrak{G} \subseteq \mathfrak{X}$ its geometric closed convex hull $\overline{\text{co}}\,\mathfrak{G}$ can be also defined as

$$\overline{\text{co}}\,\mathfrak{G} = \left\{ x \in \mathfrak{X} \;:\; |l(x)| \leqslant \sup_{y \in \mathfrak{G}} |l(y)| \text{ for all } l \in \mathfrak{X}^* \right\}.$$

If $\mathfrak{G}$ is compact, then by a theorem of Mazur, $\overline{\text{co}}\,\mathfrak{G}$ is compact, too. Since $\mathfrak{X}^* = L(\mathfrak{X}, \mathbb{C}) \subseteq A(\mathfrak{X})$, by (4.1) we obtain that $\hat{\mathfrak{G}}_{A(\mathfrak{H})} \subset \overline{\text{co}}\,\mathfrak{G}$, hence $\hat{\mathfrak{G}}_{A(\mathfrak{H})}$ is compact for all compact subsets of $\mathfrak{X}$, and, consequently, $\mathfrak{X}$ is a holomorphically convex manifold.

Let now $\mathfrak{H} \subseteq \mathfrak{X}$ be a complex-analytic manifold modelled on $\mathfrak{Y}$, where $\mathfrak{Y} \subseteq \mathfrak{X}$ is a subspace of $\mathfrak{X}$. Assume that the hypoteses in Theorem 3.1 and one of the conditions a) or b) are fulfilled. According to Theorem 3.2, $\mathfrak{H}$ is a submanifold of $\mathfrak{X}$. Since the restriction to $\mathfrak{H}$ of any analytic function on $\mathfrak{X}$ is an analytic function on $\mathfrak{H}$, then using again (4.1) we have

$$\hat{\mathfrak{G}}_{A(\mathfrak{H})} \subseteq \mathfrak{H} \cap \hat{\mathfrak{G}}_{A(\mathfrak{X})}, \tag{4.2}$$

for any compact subset $\mathfrak{G} \subseteq \mathfrak{H}$, where $\hat{\mathfrak{G}}_{A(\mathfrak{X})}$ denotes the $A(\mathfrak{X})$-convex hull of $\mathfrak{G}$ considered as a compact subset of $\mathfrak{X}$.

Clearly, by (4.2) and since $\hat{\mathfrak{G}}_{A(\mathfrak{X})}$ is compact, it follows that if $\mathfrak{H}$ is a closed subset of $\mathfrak{X}$, then $\mathfrak{H}$ is a holomorphically convex manifold. Actually this last assumption can be replaced by a weaker one, taking into account the fact that the geometric convex hull of a compact subset of the Banach space $\mathfrak{X}$ is closed. The easy proof of the next result is left to the reader.

**LEMMA 4.1.** *Let $\mathfrak{X}$ and $\mathfrak{H}$ be as above, with $\mathfrak{H}$ not necessarily closed in $\mathfrak{X}$. Let $\mathfrak{D} \subseteq \mathfrak{X}$ be a domain, convex in the geometric sense, such that $\mathfrak{H} \subseteq \mathfrak{D}$. If $\mathfrak{H}$ is closed in $\mathfrak{D}$ with respect to the induced topology on $\mathfrak{D}$, then $\mathfrak{H}$ is a holomorphically convex manifold.*

We will now define a new class of manifolds.

**DEFINITION 4.3.** A complex-analytic manifold $\mathfrak{H}$ of a finite dimension $n$, which is countable at infinity, is called a *Stein manifold* if

a) $\mathfrak{H}$ is holomorphically convex;

b) $\mathfrak{H}$ is holomorphically separate, i.e., for any $a, b \in \mathfrak{H}$ with $a \neq b$ there exists $f \in A(\mathfrak{H})$ such that $f(a) \neq f(b)$;

c) for any $a \in \mathfrak{H}$ there exists a collection of $n$ functions $f_1, \dots, f_n \in A(\mathfrak{H})$ which provide a local coordinate system at $a$.

Let us recall that a Hausdorff topological space $\mathfrak{H}$ is said to be countable at infinity if there exists a countable family $\mathfrak{G}_1, \mathfrak{G}_2, \dots$ of compact subsets of $\mathfrak{H}$ such that each compact subset of $\mathfrak{H}$ is contained in some $\mathfrak{G}_j$, $j = 1, 2, \dots$.

**THEOREM 4.1.** *Any complex-analytic manifold $\mathfrak{H}$ which is a closed submanifold of a Stein manifold $\mathfrak{M}$ is a Stein manifold.*

◁ Conditions a) and b) are obvious, since the restriction to $\mathfrak{H}$ of any function which is analytic in the whole manifold $\mathfrak{M}$ is an analytic function in $\mathfrak{H}$.

Let us prove condition c). Take an arbitrary point $a \in \mathfrak{H}$. Since $a \in \mathfrak{M}$, there exists a system of functions $f_1, \dots, f_m \in A(\mathfrak{M})$, $m = \dim \mathfrak{M}$, which provide a local coordinate system for $\mathfrak{M}$ at $a$. On the other hand, since $\mathfrak{H}$ is a submanifold of $\mathfrak{M}$ of dimension $n \leqslant m$, there exists a local chart $(\mathfrak{D}, \varphi)$ for $\mathfrak{M}$ with $\varphi = (\varphi_1, \dots, \varphi_m)$ and $a \in \mathfrak{D}$, such that

$$\mathfrak{D} \cap \mathfrak{H} = \{x \in \mathfrak{M} \ : \ \varphi_{n+1}(x) = \dots = \varphi_m(x) = 0\}.$$

Since the Jacobian

$$\det \left( \frac{\partial f_i}{\partial \varphi_j} \right)_{i,j=1,\dots,m} \neq 0,$$

then for a suitable choice of $1 \leqslant i_1, \dots, i_n \leqslant m$ we have

$$\det \left( \frac{\partial f_{i_\mu}}{\partial \varphi_\nu} \right)_{\mu,\nu=1,\dots,n} \neq 0.$$

The last relation shows that the restrictions of $f_{i_1}, \dots, f_{i_n}$ to $\mathfrak{H}$ provide a local system of coordinates for $\mathfrak{H}$ at $a$. This completes the proof. ▶

Since the space $\mathbb{C}^m$, as well as any convex domain in $\mathbb{C}^m$ are Stein manifolds we have:

**COROLLARY 4.2.** *Every complex-analytic manifold $\mathfrak{H}$ which is a closed submanifold of $\mathbb{C}^m$ is a Stein manifold of dimension $n \leqslant m$.*

Assume now that the ambient manifold for $\mathfrak{H}$ is a convex domain $\mathfrak{D}$ in an infinite dimensional Banach space $\mathfrak{X}$. In this case, we can not use Theorem 4.1 since $\mathfrak{D}$ is not a Stein manifold (in particular $\mathfrak{D}$ is not finite-dimensional). However the next generalization of Corollary 4.2 is true.

**THEOREM 4.2.** *Let $\mathfrak{D}$ be a convex domain in a Banach space $\mathfrak{X}$ (in particular, $\mathfrak{D} = \mathfrak{X}$), and let $\mathfrak{H}$ be a finite-dimensional countable at infinity complex manifold, which is a submanifold of $\mathfrak{D}$ and is closed with respect to the induced topology on $\mathfrak{D}$. Then $\mathfrak{H}$ is a Stein manifold.*

◁ The holomorphic convexity of the manifold $\mathfrak{H}$ follows directly from Lemma 4.1. For the holomorphic separation of $\mathfrak{H}$ it is enough to recall that $\mathfrak{X}^* = L(\mathfrak{X}, \mathbb{C}) \subseteq \subseteq A(\mathfrak{H})$ and to use the fact that for any $a, b \in \mathfrak{H}$ ($\subseteq \mathfrak{X}$) with $a \neq b$, there exists a linear functional $l \in \mathfrak{X}^*$ such that $\langle a, l\rangle \neq \langle b, l\rangle$.

Finally, let $a$ be an arbitrary point in $\mathfrak{H}$ and let $\mathfrak{E}_a$ be a finite-dimensional subspace of $\mathfrak{X}$ which is the model space for some neighborhood of the point $a$. Choose in $\mathfrak{X}^*$ a family of functionals $(f_1, \ldots, f_n)$ with $n = \dim \mathfrak{E}_a = \dim \mathfrak{H}$, such that

$$\langle e_i, f_j\rangle = \delta_{ij}, \quad i, j = 1, \ldots, n,$$

where $(e_1, \ldots, e_n)$ is a fixed basis in $\mathfrak{E}_a$, and consider the restrictions of these functionals to $\mathfrak{H}$. Since $\operatorname{rank}(f_1, \ldots, f_n) = \dim \mathfrak{H}$, then $\{f_i\}_{i=1}^n$ ($\subseteq A(\mathfrak{H})$) provides a local system of coordinates for $\mathfrak{H}$ at the point $a$. The theorem is proved. ▶

# Chapter VIII

# Non-regular solutions of nonlinear equations

## §1. Ramification of solutions. Statement of the problem

Let $\mathfrak{D}$ and $\Omega$ be domains in the Banach spaces $\mathfrak{X}$ and $\Lambda$, respectively, and let $F$ be an operator acting from the topological product $\mathfrak{H} = \mathfrak{D} \times \Omega$ into $\mathfrak{X}$. Assume that, for a given $\lambda_0 \in \Omega$, the equation

$$x = F(x, \lambda) \tag{1.1}$$

has the solution $x_0 \in \mathfrak{D}$, i.e., $x_0 = F(x_0, \lambda_0)$.

The point $(x_0, \lambda_0) \in \mathfrak{H}$ is called a *regular point* of equation (1.1) if, for any $\lambda$ in some neighborhood $\mathfrak{U} \subseteq \Omega$ of the point $\lambda_0$, there exists a unique solution $x(\lambda) \in \mathfrak{D}$ of equation (1.1) such that $x(\lambda_0) = x_0$. Otherwise the point $(x_0, \lambda_0)$ is said to be *non--regular*. There are two kinds of non-regular points. A non-regular point $(x_0, \lambda_0)$ of the first kind is characterized by the following property: there exists a neighborhood $\mathfrak{U}$ of the point $\lambda_0$ such that equation (1.1) has no solutions for any $\lambda \in \mathfrak{U}$ with $\lambda \neq \lambda_0$, and for $\lambda = \lambda_0$ it has the unique solution $x_0$. Such points are called *non-prolongation points.* A non-regular point $(x_0, \lambda_0)$ of the second kind has the property that in any neighborhood of the point $\lambda_0$ there exists $\lambda$ for which equation (1.1) has more than one solution. Such points are called *ramification points.* If on some neighborhood $\mathfrak{U}$ of the point $\lambda_0$ the identity

$$x_0 \equiv F(x_0, \lambda)$$

is true, and the point $(x_0, \lambda_0)$ is a ramification point, then $\lambda_0$ is called a *bifurcation point.*

Bifurcation points play an important role in different applications. For instance, in perturbation theory for elastic systems, bifurcation points define the so-called critical loadings; in wave theory the bifurcation points define characteristic flows in which the waves occur, and so on.

Let us consider some examples.

**EXAMPLE 1.1.** Let $\mathfrak{X} = \Lambda = \mathbb{C}$. Consider the equation

$$x = x^2 + \lambda,$$

whose solutions are

$$x(\lambda) = \frac{1}{2} \pm \frac{\sqrt{1-4\lambda}}{2}.$$

The point $(0,0)$ is a regular point for this equation, since for all $\lambda$ sufficiently close to $\lambda_0 = 0$ the equation has the unique solution $x(\lambda) = (1-\sqrt{1-4\lambda})/2$ close to $x_0 = 0$ and satisfying the condition $x(0) = 0$.

At the same time $\left(\frac{1}{2}, \frac{1}{4}\right)$ is a ramification point since for all $\lambda$ close to $\lambda_0 = \frac{1}{4}$ there exist two solutions

$$x_1(\lambda) = \frac{1}{2} - \frac{\sqrt{1-4\lambda}}{2} \quad \text{and} \quad x_2(\lambda) = \frac{1}{2} + \frac{\sqrt{1-4\lambda}}{2}$$

satisfying the condition $x_i\left(\frac{1}{4}\right) = \frac{1}{2}$, $i = 1, 2$.

**EXAMPLE 1.2.** Let $\mathfrak{X} = C[0,\,1]$ and $\Lambda = \mathbb{R}$. Consider on $\mathfrak{X}$ the equation

$$x(t) = \int_0^1 (\lambda x(s) + x^2(s))ds.$$

We can easily see that this equation has exactly two solutions: $x_1(t) \equiv 0$ and $x_2(t) \equiv$ $\equiv 1-\lambda$. If we consider as $(x_0, \lambda_0)$ the point $(0,0)$, then this point is not a ramification point since in the domain $\mathfrak{D}$ of all $x \in \mathfrak{X}$ such that $\|x\| = \max_{[0,\,1]} |x(t)| < \frac{1}{2}$, the considered equation has, for any $\lambda < \frac{1}{2}$, the unique solution $x(t) \equiv 0$. But if we consider the point $(0,1)$, then this is a ramification point: the solutions $x_1(t)$ and $x_2(t)$ "branch out" from it. Moreover, the point $\lambda_0 = 1$ is a bifurcation point; setting $F(x,\lambda) = \int_0^1 (\lambda x(s) + x^2(s)ds$ we obtain $F(0,\lambda) \equiv 0$ for all $\lambda \in \mathbb{R}$.

An important particular case of equation (1.1) is offered by an equation of the form

$$x = \lambda \Phi(x), \tag{1.2}$$

where the operator $\Phi$ is defined on a domain $\mathfrak{D}$ which contains $0 \in \mathfrak{X}$.

If $x = 0$ is a fixed point for the operator $\Phi$ then equation (1.2) has the trivial solution $x(\lambda) \equiv 0$, $\lambda \in \mathbb{K}$. If, in addition, $(0, \lambda_0)$ is a ramification point for this equation, then $\lambda_0$ is a bifurcation point.

When equation (1.2) has a non-trivial solution $x^* \neq 0$, $x^* \in \mathfrak{D}$, for some $\lambda \in \mathbb{K}$, then, by analogy with the theory of linear operators, the number $\lambda_0^{-1}$ is called an eigenvalue of the operator $\Phi$ and $x^*$ is said to be an eigenvector corresponding to the eigenvalue $\lambda_0^{-1}$.

According to the definition, if $\lambda_0$ is a bifurcation point, then $\lambda_0^{-1}$ is the limit of a sequence of eigenvalues for the operator $\Phi$. However, $\lambda_0^{-1}$ may not be an eigenvalue of the operator $\Phi$. If this is the case, then $x = 0$ is the unique — and, therefore, an isolated — fixed point for the operator $\Phi$.

In the general case the local study of equation (1.1) leads to the following problems:

1. Under what conditions the fixed point $x_0$ of $F(\cdot, \lambda_0)$ is isolated?

2. Under what conditions a point $(x_0, \lambda_0)$ is regular for equation (1.1)?

3. The study of conditions for ramification of the solutions of equation (1.1) and, correspondingly, the determination of the number of solutions (branches) for equation (1.1).

4. Under what conditions a fixed point $x_0$ for the operator $F(\cdot, \lambda_0)$ is not isolated and how can the set of all fixed points for this operator be described?

We know already some answers to these problems. To be more specific, let us assume that the operator $F$ is Fréchet differentiable in a neighborhood of the point $(x_0, \lambda_0)$ with $x_0 = F(x_0, \lambda_0)$.

Then, by Theorem V.5.1 about implicit operators, the condition for $\lambda = 1$ to be a regular point of the operator $A = F'_x(x_0, \lambda_0)$ gives a possible answer to Problems 1 and 2. This condition is but a sufficient one. In §6 of Chapter VI we gave some examples which showed that the above mentioned condition is not a necessary one; in those examples the point $(x_0, \lambda_0)$ was a regular point for equation (1.1), but the number $\lambda = 1$ was a point of the spectrum of the operator $A$. By the same Theorem V.5.1, this last condition is necessary for solving Problems 3 and 4.

## §2. Equations of ramification

**1.** THE GENERAL SETTING. Let $(x_0, \lambda_0)$ be a given point satisfying equation (1.1), i.e., $x_0 = F(x_0, \lambda_0)$. Without loosing the generality, we may assume, and we will, that $(x_0, \lambda_0) = (0, 0)$. (If this is not the case, then we can make the following changes: $y = x - x_0$, $\mu = \lambda - \lambda_0$, $\Phi(y, \mu) = F(y + x_0, \mu + \lambda_0) - x_0$, and can consider the equation $y = \Phi(y, \mu)$ which is equivalent to equation (1.1) and, moreover, satisfies $\Phi(0, 0) = 0$.)

Further, let us assume that the following conditions are satisfied:

i) the operator $F$ is Fréchet differentiable in the domain $\mathfrak{H} = \mathfrak{D} \times \Omega$, where $\mathfrak{D} \subseteq \mathfrak{X}$, $\Omega \subseteq \Lambda$;

ii) 1 is an eigenvalue of the operator $A = F'_x(0, 0)$;

iii) the operator $B = I - A$ is normally solvable and splitable (see §7 of Chapter IV).

Then the operator $B$ admits a regular matrix representation

$$B = \begin{pmatrix} B_{11} & B_{12} \\ B_{21} & B_{22} \end{pmatrix}$$

obtained using a quadruple of projections $P_1$, $P_2 = I - P_1$, $Q_1$, $Q_2 = I - Q_1$. We denote (as in Chapter IV) $\mathfrak{E}_1 = P_1\mathfrak{X}$, $\mathfrak{E}_2 = P_2\mathfrak{X}$, $\mathfrak{E}_1^0 = Q_1\mathfrak{X}$, $\mathfrak{E}_2^0 = Q_2\mathfrak{X}$.

Then equation (1.1) can be written as a system of equations

$$\begin{cases} B_{11}u + B_{12}v = Q_1\Phi(u + v, \lambda) \\ B_{21}u + B_{22}v = Q_2\Phi(u + v, \lambda), \end{cases} \tag{2.1}$$

where $u = P_1x$, $v = P_2x$, $\Phi(\cdot, \lambda) = F(\cdot, \lambda) - A$. Let us rewrite the second equation of (2.1) as

$$H(v, u, \lambda) = B_{21}u + B_{22}v - Q_2\Phi(u + v, \lambda) = 0,$$

and consider the point $(u, v) \in \mathfrak{E}_1 \times \Lambda$ as a parameter. Since the operator $H'_v(0, 0, 0)$ $(= B_{22} - Q_2\Phi'_v(0, 0, 0) = B_{22})$ is continuously invertible, by Theorem V.5.1 about implicit operators the last equation has a unique solution

$$v = R(u, \lambda) \tag{2.2}$$

defined on some neighborhood of the point $(0, 0) \in \mathfrak{E}_1 \times \Lambda$ and satisfying $R(0, 0) = 0$.

By substituing the right hand side of (2.2) in the first equation of (2.1) we obtain the equation

$$f(u, \lambda) = 0, \tag{2.3}$$

which defines the implicit operator $u = u(\lambda)$ on the subspace $P_1\mathfrak{X}$. If there exists a (single- or a multiple-valued) solution $u(\lambda)$ of equation (2.3), then $u(\lambda)$ provides a solution $x(\lambda)$ of the initial equation (1.1) given by the formula

$$x(\lambda) = u(\lambda) + R(u(\lambda), \lambda). \tag{2.4}$$

Equation (2.3) is called the *equation of ramification* for equation (1.1).

Using formula (IV.7.4) it is easy to see that the operator $f$ has the form

$$f(u,\lambda) = B_{12}B_{22}^{-1}Q_2\Phi(u+R(u,\lambda),\lambda) - Q_1\Phi(u+R(u,\lambda),\lambda),$$

and acts from the space $\mathfrak{E}_1$ into $\mathfrak{E}_1^0$.

Equation (2.3) like the initial equation (1.1) is degenerate, i.e., $f'_u(0,0) = 0$ (this follows easily by a direct computation). Therefore the Implicit Function Theorem can not be applied in this case. However, sometimes this equation can be studied using methods from finite-dimensional analysis. Namely, if the subspaces $\mathfrak{E}_1$ and $\mathfrak{E}_1^0$ are finite-dimensional, then equation (2.3) can be represented as a system with a finite number of equations and a finite number of unknowns. There are many well-known different methods for the search and the study of solutions for such equations: elimination methods, Newton diagrams, algebraic methods, and others (see, for instance, M. M. Vainberg and V. A. Trenogin [1]).

The above described general setting allows us to find different equations of ramification, depending on the properties of the operator $B$ and on the choice of the subspaces $\mathfrak{E}_1$, $\mathfrak{E}_1^0$, $\mathfrak{E}_2$, $\mathfrak{E}_2^0$, and of the projections $P_1$, $P_2$, $Q_1$, $Q_2$ relatively to which the regular matrix representation of the operator $B$ is obtained.

**2.** LIAPUNOV EQUATION OF RAMIFICATION. Let us consider the case when $\mathfrak{E}_1 = N(B)$ and $\mathfrak{E}_2^0 = R(B)$ (recall that $B$ is a normally solvable operator). Then the operator $B$ admits the matrix representation

$$B = \begin{pmatrix} 0 & 0 \\ 0 & B_{22} \end{pmatrix},$$

and system (2.1) becomes

$$\begin{cases} 0 = (I-Q)\Phi(u+v,\lambda) \\ B_{22}v = Q\Phi(u+v,\lambda), \end{cases}$$

where $Q$ is a projection onto $R(B)$, $P$ is a projection onto $N(B)$ and $u = Px$, $v = (I-P)x$.

Using the first of these equations, the second one can be written as

$$B_{22}v = \Phi(u+v,\lambda).$$

If $v = R(u,\lambda)$ is a solution of this equation, then the equation of ramification (2.3) is

$$f(u,\lambda) = (I-Q)\Phi(u+R(u,\lambda),\lambda) = 0. \tag{2.5}$$

Assume, in addition, that $B$ is a Fredholm operator (see §6 of Chapter IV), i.e., $\dim N(B) = \dim N(B^*) = n$. Let $\{\varphi_i\}_{i=1}^n$ and $\{\psi_i\}_{i=1}^n$ be bases in $N(B)$ and $N(B^*)$, respectively, and let $\{\gamma_i\}_{i=1}^n \subset \mathfrak{X}^*$ and $\{z_i\}_{i=1}^n \subset \mathfrak{X}$ be two systems of elements biorthogonal to these bases, respectively. Moreover, we consider that $\{\gamma_i\}_{i=1}^n$ was chosen such that $\langle v, \gamma_i \rangle = 0$, $i = 1, \ldots, n$, for any $v \in \mathfrak{X} \dot{-} N(B)$. This can be done by Hahn-Banach Theorem. Then every element $u \in N(B)$ can be represented as

$$u = \sum_{i=1}^{n} \zeta_i \varphi_i, \quad \zeta_i \in \mathbb{K}$$

($\mathbb{K}$ is, as usual, the field of scalars).

At the same time the equality

$$(I - Q)x = \sum_{i=1}^{n} \langle x, \psi_i \rangle z_i$$

holds for the projection $I - Q$. Taking into account the linear independence of the elements $\{z_i\}_{i=1}^n$, equation (2.5) can be rewritten as a system of $n$ equations with $n$ unknowns $\zeta_1, \ldots, \zeta_n$:

$$f_k(\zeta_1, \ldots, \zeta_n, \lambda) = \left\langle \Phi \left( \sum_{i=1}^{n} \zeta_i \varphi_i + R \left( \sum_{i=1}^{n} \zeta_i \varphi_i, \lambda \right), \lambda \right), \psi_k \right\rangle = 0, \tag{2.6}$$

$k = 1, 2, \ldots, n$.

Equations as in (2.6) above, corresponding to integral equations with analytic functions, have been considered for the first time by A. M. Liapunov in connection with the equilibrium figures of a fluid mass in a revolving motion. Therefore, equations (2.6) are usually called the *Liapunov equations of ramification*. The above described approach, which reduces the initial equation to the system (2.6), is called the *Liapunov procedure*. This procedure can be also effective in the study of the solutions for equations with fixed values of the parameter $\lambda$ (see §4 below).

**3.** SCHMIDT EQUATION OF RAMIFICATION. With the notations and under the assumptions of the previous subsection, we infer other equations of ramification using essentially Lemma IV.6.1 (Schmidt Lemma). According to this lemma, the Fredholm operator $B$ can be represented as

$$B = C - K,$$

where $C$ is a continuously invertible operator on $\mathfrak{X}$, and $K$ is a finite rank operator of the form

$$Kx = \sum_{i=1}^{n} \langle x, \gamma_i \rangle z_i.$$

Then equation (1.1) can be rewritten as

$$Cx = \Phi(x, \lambda) + \sum_{i=1}^{n} \omega_i z_i, \tag{2.7}$$

where

$$\omega_i = \langle x, \gamma_i \rangle, \quad i = 1, \dots, n. \tag{2.8}$$

If in equation (2.7) we consider $(\omega_1, \dots, \omega_n, \lambda) \in \mathbb{K}^n \times \Lambda$ as a parameter, then, using the Implicit Function Theorem, we obtain a solution for this equation of the form

$$x = S(\omega, \lambda), \tag{2.9}$$

where $\omega = (\omega_1, \dots, \omega_n)$. To determine $\omega$, we substitute (2.9) in (2.8). Thus we obtain the following system of equations

$$\omega_i = \langle S(\omega_1, \dots, \omega_n, \lambda), \gamma_i \rangle, \quad i = 1, 2, \dots, n, \tag{2.10}$$

which is called the *Schmidt system of equations of ramification.* This procedure, in contrast to the ones described in the previous Subsections 1 and 2, has the advantage that the considered operators are defined on the whole space $\mathfrak{X}$, and it is not necessary to split the operator $B$ or to use the projections $P_1$, $P_2$, $Q_1$, $Q_2$.

At the same time, there exists a direct connection between the Liapunov and the Schmidt equations of ramification.

◁ In order to establish this connection, we rewrite first equation (2.10) into a slightly different form. Set $x = u + v$, where $u = \sum_{i=1}^{n} \zeta_i \varphi_i \in N(B)$, and remark that by the equalities $\langle \varphi_i, \gamma_j \rangle = \delta_{ij}$, $i, j = 1, \dots, n$ and $\langle v, \gamma_j \rangle = 0$, $j = 1, \dots, n$, we have

$$\zeta_i = \langle u, \gamma_i \rangle = \langle x - v, \gamma_i \rangle = \omega_i.$$

Further, $C\varphi_i = B\varphi_i + K\varphi_i = z_i$ and, therefore

$$Cx = C\left(\sum_{I=1}^{n} \zeta_i \varphi_i + v\right) = \sum_{i=1}^{n} \zeta_i z_i + Cv.$$

Consequently, equation (2.7) can be written as

$$Cv = \Phi(u+v,\lambda) = \Phi\left(\sum_{i=1}^{n} \zeta_i\varphi_i + v, \lambda\right). \tag{2.11}$$

Let $v = \tilde{R}(\zeta_1,\dots,\zeta_n,\lambda)$ be a solution of equation (2.11), considered as an implicit function of the parameter $(\zeta_1,\dots,\zeta_n,\lambda)$. Substituting $x = \sum_{i=1}^{n} \zeta_i\varphi_i + \tilde{R}(\zeta_1,\dots,\zeta_n,\lambda)$ in (2.8) we obtain for $\zeta_j$ the equations

$$\zeta_j = \left\langle \sum_{i=1}^{n} \zeta_i\varphi_i + \tilde{R}(\zeta_1,\dots,\zeta_n,\lambda), \gamma_j \right\rangle, \quad j = 1,\dots,n,$$

which are equivalent to the equations in (2.10).

According to $\langle\varphi_i,\gamma_j\rangle = \delta_{ij}$, $i,j = 1,\dots,n$, we finally get

$$\left\langle \tilde{R}\left(\sum_{i=1}^{n} \zeta_i\varphi_i, \lambda\right), \gamma_j \right\rangle = 0, \quad j = 1,\dots,n. \tag{2.12}$$

Let us show now that system (2.12) represents nothing else but the Liapunov system of equations of ramification.

In the literature (see, for instance, V. A. Trenogin [1]) (2.12) is often called the Liapunov-Schmidt system of equations of ramification.

Indeed, since $C^* = B^* + K^*$, then $C^*\psi_i = \gamma_i$, $i = 1,\dots,n$, and, consequently, the system (2.12) can be written as

$$\left\langle \tilde{R}\left(\sum_{i=1}^{n} \zeta_i\varphi_i, \lambda\right), C^*\psi_j \right\rangle = \left\langle C\tilde{R}\left(\sum_{i=1}^{n} \zeta_i\varphi_i, \lambda\right), \psi_j \right\rangle = 0, \quad j = 1,\dots,n. \tag{2.13}$$

It is easy to see that the space $\mathfrak{X}\dot{-}N(B)$ is the kernel of the operator $K$, i.e., $Kv = 0$ for all $v \in \mathfrak{X}\dot{-}N(B)$. Therefore equation (2.11) is equivalent to the next equation

$$\hat{B}v = \Phi(u+v,\lambda),$$

where $\hat{B}$ is the restriction of the operator $B$ to the subspace $\mathfrak{X}\dot{-}N(B)$. Hence it follows that $\tilde{R}(u,\lambda) = B_{22}^{-1}\Phi(u+\tilde{R}(u,\lambda),\lambda) = R(u,\lambda)$, where $B_{22}$ and $R(\cdot,\cdot)$ are the same as in Subsection 2.

From (2.11) and (2.12) we get the system of equations

$$\left\langle \Phi\left(\sum_{i=1}^{n} \zeta_i\varphi_i + R\left(\sum_{i=1}^{n} \zeta_i\varphi_i, \lambda\right), \lambda\right), \psi_j \right\rangle = 0, \quad j = 1,\dots,n,$$

which coincides with the Liapunov system of equations of ramification (2.6). ▶

As we have noted above, the Liapunov system is degenerated, i.e.,

$$\det \left\| \frac{\partial f_i}{\partial \zeta_j}(0,0) \right\| = 0.$$

Therefore, if it has a solution $\zeta(\lambda) = (\zeta_1(\lambda), \dots, \zeta_n(\lambda))$, then the functions $\zeta_i(\lambda)$ can be either single-valued or multiple-valued. Their feature determines the feature of the vector-function $u(\lambda)$, and, consequently, the feature of the solution $x(\lambda)$ of the initial equation (2.1).

## §3. Equations of ramification for an analytic operator. The problem of coefficients

**1.** Let us assume that the operator $F$ in equation (1.1) is analytic with respect to the arguments $(x, \lambda)$ in some neighborhood of $(0,0) \in \mathfrak{X} \times \Lambda$. As before we will use the notations $A = F'_x(0,0)$, $B = I - A$, and we will suppose that the operator $B$ is splitable.

We consider again the system of equations

$$\hat{B}v = \Phi(u + v, \lambda) \tag{3.1}$$

$$f(u, \lambda) = (I - Q)\Phi(u + R(u(\lambda), \lambda)) = 0 \tag{3.2}$$

equivalent to equation (1.1), where $u = Px$, $v = (I - P)x$, $\hat{B}$ is the restriction of the operator $QB(I - P)$ to the subspace $\mathfrak{E}_2 = (I - P)\mathfrak{X}$, $P$ is a projection onto $N(B)$, $Q$ is a projection onto $R(B)$, and $\Phi = F - A$.

By the Implicit Function Theorem, equation (3.1) has a solution $v = R(u, \lambda)$ on some neighborhood $\mathfrak{U} \times \Omega$ of $(0,0)$, where $\mathfrak{U} \subseteq N(B)$, $\mathfrak{U} \ni 0$, $\Omega \subseteq \Lambda$, $\Omega \ni 0$. This solution is analytic with respect to the variables $(u, \lambda)$ in the whole neighborhood $\mathfrak{U} \times \Omega$. Consequently, the operator $f(\cdot, \cdot)$ that defines the left hand side of the equation of ramification (3.2), is also analytic in $\mathfrak{U} \times \Omega$. Moreover, from $\Phi'_x(0,0) \equiv 0$ it follows that $R'_u(0,0) \equiv 0$, hence $f'_u(0,0) \equiv 0$. This means that the operator $f$ admits on the domain $\mathfrak{U} \subseteq N(B)$ the representation

$$f(u, \lambda) = f(u, 0) + \sum_{i=0}^{\infty} L^i(u, \lambda), \tag{3.3}$$

where $L^i(u, \lambda)$ are homogeneous forms of degree $i = 0, 1, 2, \dots$ in $u$ for a fixed $\lambda \in \Omega$, and analytic operators in $\lambda$ for a fixed $u \in \mathfrak{U}$. In addition $L^i(u, 0) \equiv 0$.

On the other hand

$$f(u,0) = \sum_{k=2}^{\infty} H^k(u), \tag{3.4}$$

where $H^k$ are homogeneous forms of degree $k = 2, 3, \ldots$. The series in (3.3) and (3.4) have non-zero radii of absolute convergence. The operators $H^k$, $k = 2, 3, \ldots$, and $L^i$, $i = 0, 1, 2, \ldots$, are called the *coefficients of ramification* for equation (3.2).

If all the coefficients of ramification are zero, then Problem 3 in §1 about the number of "branches" has an obvious answer: this number is infinite. Indeed, equality (3.2) is identically fulfilled on $\mathfrak{U}$. Replacing $u(\lambda)$ in equality (2.4) by an operator $u : \Omega \to \mathfrak{U}$, we obtain different small solutions for equation (1.1). The features of these solutions are determined by the feature of the chosen operator $u(\cdot)$. In particular, for $\lambda = 0$, the point $x = 0$ is a non-isolated solution for equation $x = F(x, 0)$. Moreover, according to $R'_u(0,0) = 0$ we obtain that the operator $G = I \mid N(B) + R(\cdot, \cdot)$ is a bijection from the set $\mathfrak{U} \subseteq N(B)$ onto the set $\mathfrak{F} \cap \mathfrak{D}$, where $\mathfrak{D}$ is a neighborhood of 0 in $\mathfrak{X}$ and $\mathfrak{F}$ is the set of all fixed points for the operator $F(\cdot, 0)$ (see Problem 4 in §1).

If $B$ is a Fredholm operator, and not all the coefficients of ramification $H^k$, $k = 2, 3, \ldots$, are equal to zero, then it can be shown that the equation of ramification has a finite number of small solutions, determined by the first operator $H^k$ which is different from zero (see, for instance, M. M. Vainberg and V. A. Trenogin [1]).

In practice, it turns out frequently that both the construction of the equation of ramification and the computation of its coeficients are either rather cumbersome, or lead to difficult to solve problems. That is why the equation of ramification is not easy to handle. In the above cited book there are examples in which the first two or three coefficients of ramification are computed by reccurent methods, but these computations require a lot of work.

In §4 below we will show that if the operator $F(\cdot, 0)$ leaves invariant some neighborhood of the point $x = 0$, then all the coefficients $H^k$ in the decomposition of $f(u, 0)$ are equal to zero.

## §4. The description of the set of fixed points for an analytic operator

In this section we are mainly concerned with Problem 4 of §1, in the case of a complex Banach space $\mathfrak{X}$. More precisely, we will study the set of fixed points for the analytic operator $T = F(\cdot, 0)$ acting from $\mathfrak{X}$ into $\mathfrak{X}$, or, equivalently, the set of solutions of equation (1.1) for a fixed value of the parameter. We show that under some natural conditions this set has a quite complete description.

Let us assume that in the set where $T$ is defined there exists a bounded domain $\mathfrak{D} \subseteq \mathfrak{X}$ which is invariant for $T$, i.e.,

$$T(\mathfrak{D}) \subseteq \mathfrak{D}, \tag{4.1}$$

and let $\mathfrak{F}$ be the set of all fixed points for $T$ in $\mathfrak{D}$.

If $\mathfrak{D}$ is a ball in a strictly convex space, then the set $\mathfrak{F}$ has a very simple structure: either $\mathfrak{F}$ is void, or it is an affine variety in $\mathfrak{X}$ (see Rudin Theorem — Theorem 3.3. in §3 of Chapter VI).

We emphasize that the condition of strict convexity is essential in this respect. For example, the operator $F$ acting on the space $\mathfrak{X} = \mathbb{C}^2$ and defined by $F(x) = (x_1, 0.5(x_1 + x_2^2))$, $x = (x_1, x_2)$, leaves invariant the unit polydisk

$$\mathfrak{D} = \{x \in \mathbb{C}^2 \; : \; |x_i| < 1, \; i = 1, 2\},$$

but the set $\mathfrak{F}$ is not linear. Moreover, the intersection of $\mathfrak{N}$ and $\mathfrak{F}$, where $\mathfrak{N}$ is the set of all eigenvectors of the operator $F'(0)$ corresponding to the eigenvalue 1, has but one point, $x = 0$. Thus, in the absence of the strict convexity, Rudin Theorem is not true.

Nevertheless, as we will see below, the assertions analogous to some consequences of Rudin Theorem such as the regularity of an isolated point, the continuousness of the set of non-regular points, its connectivity, the approach to the boundary, and others, are also true for a larger class of spaces.

Let $T$ be an analytic operator on an arbitrary domain $\mathfrak{D}$ of a complex Banach space $\mathfrak{X}$ satisfying condition (1.1), and let $\mathfrak{F}$ be the set of all fixed points for $T$ in $\mathfrak{D}$.

**THEOREM 4.1.** *Let $\mathfrak{M}$ be a connected component of the set $\mathfrak{F}$ containing a given point $x^* \in \mathfrak{F}$. If the operator $\Phi = I - T$ is Fredholm on some neighborhood of the set $\mathfrak{M}$, then:*

1) *$\mathfrak{M}$ is a direct complex-analytic manifold of dimension $n$, where $n = \dim\ker(I - T'(x^*))$;*

2) *if the set $\mathfrak{D}$ is convex and $\mathfrak{F}$ contains more than one point, then $n > 0$.*

Recall that a nonlinear operator $\Phi$ is called Fredholm at a point $x^*$ if the linear operator $\Phi'(x^*)$ is Fredholm.

**REMARK 4.1.** By assertion 1) in Theorem 4.1 it follows, in particular, that whenever $x^*$ is an isolated point in $\mathfrak{F}$, then $n = 0 = \dim\ker(I - T'(x^*))$, a conclusion which —

by Banach Theorem — implies the continuous invertibility of the operator $I - T'(x^*)$. This means that an isolated fixed point is regular. Moreover, from assertion 2) it follows that, in this case, $x^*$ is the only fixed point for $T$ in $\mathfrak{D}$. Thus, for an operator $T$ which leaves invariant some convex domain $\mathfrak{D}$, and has a Fredholm complement $\Phi$ up to the identity operator, the next alternative is true: either a fixed point for $T$ (if it exists) is unique and regular, or there is an infinity of fixed points and all of them are non-regular and non-isolated.

**THEOREM 4.2.** *Let $\mathfrak{D}$ be a convex domain in $\mathfrak{X}$. Assume that the operator $T$ admits a continuous extension on $\overline{\mathfrak{D}}$ and the operator $\Phi = I - T$ is proper. Then:*

*1) the closure $\overline{\mathfrak{F}}$ of the set $\mathfrak{F}$ in $\mathfrak{D}$ is connected;*

*2) if $\Phi$ is Fredholm on some neighborhood of $\mathfrak{F}$, then $\mathfrak{F}$ is a Stein manifold, and, moreover, if $\mathfrak{F}$ contains more than one point, then $\overline{\mathfrak{F}} \cap \partial\mathfrak{D} \neq \emptyset$.*

The proofs of both theorems are based on the following preliminary results.

Let $\mathfrak{D}$ be an arbitrary bounded domain in $\mathfrak{X}$, $\mathfrak{F} \neq \emptyset$, $x^* \in \mathfrak{F}$, and $\mathfrak{N} = \ker(I - A)$, where $A = T'(x^*)$. Assume that for some subspace $\mathfrak{E} \subseteq \mathfrak{N}$ there exists a linear projection $P$ onto $\mathfrak{E}$, commuting with the operator $A$:

$$PA = AP. \tag{4.2}$$

**LEMMA 4.1.** *Let $\kappa$ be an analytic operator defined on some neighborhood $\mathfrak{U} \subseteq \mathfrak{E}$ of the point $0 \in \mathfrak{E}$, with values in $\ker P$ such that $\|\kappa(u)\| = o(\|u\|)$. Then the identity*

$$PT(x^* + u + \kappa(u)) \equiv Px^* + u \tag{4.3}$$

*is true, for all $u \in \mathfrak{U}$ such that $x^* + u + \kappa(u) \in \mathfrak{D}$.*

$\lhd$ Without loosing the generality we will consider that $x^* = 0 \in \mathfrak{X}$ and that $\mathfrak{U}$ is a ball centered at the point $0 \in \mathfrak{E}$, such that $u + \kappa(u) \in \mathfrak{D}$ for all $u \in \mathfrak{U}$. Let us fix an arbitrary $u \in \mathfrak{U}$ and consider the vector-functions

$$\varphi_n(\zeta) = PT^n(\zeta u + \kappa(\zeta u)),$$

where $\zeta \in \overline{\Delta} = \{\zeta \in \mathbb{C} \ : \ |\zeta| \leqslant 1\}$, $T^n = T \circ T^{n-1}$, $n = 1, 2, \dots$, and $T^0 = I$. Since the operator $T$ satisfies the condition $T(\mathfrak{D}) \subseteq \mathfrak{D}$ and the domain $\mathfrak{D}$ is bounded, then all the functions $\varphi_n$ are defined on $\overline{\Delta}$ and there exists $M < \infty$ such that

$$\|\varphi_n(\zeta)\| \leqslant M, \quad n = 1, 2, \dots .$$

The functions $\varphi_n$ have the power series representations

$$\varphi_n(\zeta) = \varphi_n'(0)\zeta + h_n\zeta^{m_n} + \cdots,$$

where $h_n \in \mathfrak{X}$ is the first non-zero coefficient — if such a coefficient exists — in the representation of the nonlinear part of $\varphi_n$, and $m_n \geqslant 2$.

By the Cauchy inequalities (see §2 of Chapter III) we have

$$\|h_n\| \leqslant M, \quad n = 1, 2, \dots . \tag{4.4}$$

From the conditions $APx = Px$ and (4.2) we get

$$PA^n x \equiv Px, \quad n = 1, 2, \dots .$$

By these relations, a direct computation of $\varphi_n(\zeta)$, based on an induction argument, gives

$$\varphi_1(\zeta) = PT(\zeta u + \kappa(\zeta u)) =$$
$$= PA(\zeta u) + PA\kappa(\zeta u) + PS(\zeta u + \kappa(\zeta u)) = \zeta u + h_1\zeta^{m_1} + \cdots ,$$

where $S = T - A$, and

$$\varphi_2(\zeta) = PT^2(\zeta u + \kappa(\zeta u)) = PT(A((\zeta u + \kappa(\zeta u)) + S(\zeta u + \kappa(\zeta u))) =$$
$$= PA^2(\zeta u) + PA\kappa(\zeta u) + PAS(\zeta u + \kappa(\zeta u)) + PS(\zeta u + A\kappa(\zeta u) + S(\zeta u + \kappa(\zeta u))) =$$
$$= \zeta u + 2h_1\zeta^{m_1} + \cdots$$
$$\cdots$$
$$\varphi_n(\zeta) = \zeta u + nh_1\zeta^{m_1} + \cdots .$$

Thus we conclude that $m_n = m_1$ and $h_n = n \cdot h_1$, $n = 1, 2, \dots$. By (4.4) this is possible if and only if $h_1 = 0$. Consequently

$$\varphi_1(\zeta u) \equiv \zeta u,$$

and setting $\zeta = 1$ we obtain (4.3). ►

**LEMMA 4.2.** *Let $\mathfrak{D}$ be a convex domain in $\mathfrak{X}$ and let $\mathfrak{M}$ be a connected component of the set $\mathfrak{F}$. Let us assume that for some neighborhood $\mathfrak{U} \subseteq \mathfrak{D}$ of $\mathfrak{M} \cap \mathfrak{D}$, such that $\mathfrak{U} \cap (\mathfrak{F} \setminus \mathfrak{M}) = \emptyset$ the operator $\Phi = I - T$ is uniformly non-degenerate on the set $\Gamma = \partial\mathfrak{U} \setminus \partial\mathfrak{D}$, i.e., there exists $\rho > 0$ such that*

$$\|\Phi(x)\| \geqslant \rho, \tag{4.5}$$

*for all* $x \in \Gamma$. *Then* $\mathfrak{M} = \mathfrak{F}$, *i.e.,* $\mathfrak{F}$ *is connected.*

◁ Let $y \in \mathfrak{X}$ be an element which runs through some compact set $\mathfrak{Y} \subseteq \mathfrak{D}$. For any $\varepsilon \in (0,\ 1]$ we consider the operator $\tilde{T}$ $(= \tilde{T}(\varepsilon, y))$ defined by the equality

$$\tilde{T}(x) = (1-\varepsilon)T(x) + \varepsilon y. \tag{4.6}$$

By the convexity of $\mathfrak{D}$ and by condition (4.1) the operator $\tilde{T}$ maps the domain $\mathfrak{D}$ strictly inside $\mathfrak{D}$. Consequently, the equation

$$x = \tilde{T}(x) \tag{4.7}$$

has a unique solution $x$ $(= x(\varepsilon, y))$ which depends continuously on $y \in \mathfrak{Y}$ and satisfies the relation

$$\|x(\varepsilon, y) - Tx(\varepsilon, y)\| < \varepsilon M, \tag{4.8}$$

where $M$ is the radius of a ball which contains completely the domain $\mathfrak{D}$.

Choose now two arbitrary points $a, b \in \mathfrak{F}$ and set $y(\alpha) = (1-\alpha)a + \alpha b$, where $\alpha \in [0,\ 1]$. Then the vector-function $x(\varepsilon, y(\alpha))$ defined by equation (4.7) is continuous in $\alpha$ on the interval $[0,\ 1]$ and satisfies the conditions

$$x(\varepsilon, y(0)) = a, \quad x(\varepsilon, y(1)) = b.$$

Now if we assume that $\mathfrak{F} \neq \mathfrak{M}$, then choosing $b \in \mathfrak{F} \setminus \mathfrak{M}$ and $a \in \mathfrak{M} \cap \mathfrak{D}$ we obtain that there exists $\alpha_0 \in [0,\ 1]$ for which $x(\varepsilon, y(\alpha_0)) \in \Gamma$. Setting $\varepsilon = \rho M^{-1}$, by (4.8) we get the inequality $\|\Phi x(\varepsilon, y(\alpha_0))\| < \rho$, which contradicts (4.5). Consequently our assumption is false, hence $\mathfrak{F} = \mathfrak{M}$. ▶

*Proof of Theorem 4.1.* We will consider that $x^* = 0 \in \mathfrak{X}$. Let $A = T'(0)$. From the relations $A^m = (T^m)'(0)$, by the boundedness of the domain $\mathfrak{D}$ and from Cauchy inequalities it follows that

$$\|A^m\| \leqslant M, \tag{4.9}$$

for a suitable constant $M < \infty$. From Corollary V.8.2 we obtain the decomposition

$$\mathfrak{X} = \mathfrak{N} \dot{+} \mathfrak{R}, \tag{4.10}$$

where $\mathfrak{N} = \ker(I - A)$ and $\mathfrak{R} = \operatorname{Im}(I - A)$. We use now the Liapunov procedure. Let $P$ and $Q$ be linear projections from $\mathfrak{X}$ onto the subspaces $\mathfrak{N}$ and $\mathfrak{R}$, respectively. Then the equation

$$x = T(x),$$

for $x$ of a sufficiently small norm, is equivalent to the system of equations

$$u = PT(u+v) \tag{4.11}$$

$$v = QT(u+v), \tag{4.12}$$

where $u = Px$ and $v = Qx$. By the Implicit Function Theorem equation (4.12) has a unique solution $v = \kappa(u) \in \mathfrak{R}$ $(= \ker P)$, which is analytic in some neighborhood $\mathfrak{U}$ of the point $0 \in \mathfrak{N}$. A direct computation shows that $\|\kappa(u)\| = o(\|u\|)$. By substituing this solution in (4.11), according to Lemma 4.1 we obtain an identity which is satisfied for all $u \in \mathfrak{U}$. In other words, the point

$$x = u + \kappa(u) \tag{4.13}$$

is a fixed point for the operator $T$, for all $u \in \mathfrak{U} \subseteq \mathfrak{N}$. Now, if we choose a neighborhood $\Omega \subseteq \mathfrak{X}$ of the point $0 \in \mathfrak{X}$ such that $P\Omega \subseteq \mathfrak{U} \subseteq \mathfrak{N}$, then the homeomorphism $\psi : \Omega \to \mathfrak{X}$ defined by the equality $\Psi(x) = x - \kappa(Px)$ gives a local chart of the domain $\mathfrak{D}$ as a manifold, which satisfies the condition $\Psi(0) = 0$ and, by (4.13), the condition $\Psi(\mathfrak{M} \cap \Omega) \subseteq \mathfrak{N}$, too.

Thus, by (4.10) it follows that $\mathfrak{M}$ is a direct submanifold of $\mathfrak{X}$ of dimension $n = \dim \mathfrak{N}$. The assertion 1) in Theorem 4.1 is proved.

We prove now assertion 2). Assume that $0 \in \mathfrak{F}$ is an isolated point in this set $(n = 0)$. Then the operator $\Phi = I - T$ is locally invertible, i.e., we can find the numbers $r_1, r_2 > 0$ such that for all $y$ with $\|y\| \leqslant r_2$ the equation $\Phi(x) = y$ has a unique solution $x = \Phi^{-1}(y)$ in the ball $\|x\| \leqslant r_1$, which depends analyticaly on $y$ and satisfies the condition $x(0) = 0$. Then for all $x$ with $\|x\| = r_1$ inequality (4.5) is fulfilled for a suitable $\rho$. By Lemma 4.2, we obtain that $\mathfrak{F} = \{0\}$, which contradicts the assumption in 2). ▶

*Proof of Theorem 4.2.* 1) Let $\mathfrak{M}$ be a connected component of the set $\mathfrak{F}$. Choose a neighborhood $\mathfrak{U} \subset \mathfrak{D}$ which contains $\mathfrak{M} \cap \mathfrak{D}$ and such that $\mathfrak{U} \cap (\mathfrak{F} \setminus \mathfrak{M}) = \emptyset$. Then $\Phi(x) \neq 0$, whence, by the fact that $\Phi$ is proper, inequality (4.5) follows with a suitable $\rho$. Our assertion is a consequence of Lemma 4.2.

2) If we assume in addition that $\Phi$ is a Fredholm operator, then, according to Theorem 4.1, it follows that $\mathfrak{M}$ is a finite-dimensional complex-analytic submanifold of $\mathfrak{D}$. Again by the fact that $\Phi$ is proper we conclude that $\mathfrak{F}$ is countable at infinity and thus, by Theorem VII.4.2, it is a Stein manifold.

Finally, let us suppose $\overline{\mathfrak{F}} \cap \partial\mathfrak{D} = \emptyset$. Then there exists a neighborhood $\mathfrak{D}_1$ $(\subseteq \mathfrak{D})$ of the set $\mathfrak{F}$ such that all the values of the operator $\Phi$ $(= I - T)$ on $\partial\mathfrak{D}_1$ are different from zero. Consequently, for $\Phi_1 = \Phi$ considered as the vector field $I - T$ — due to the fact that it is Fredholm and proper — we can define the quasi-winding number $\gamma^*(\Phi_1, \partial\mathfrak{D}_1)$ (see, for instance, M. A. Krasnoselskii and P. P. Zabreiko [1]). Since $\mathfrak{D}_1 \supset \overline{\mathfrak{F}}$, then $\gamma^*(\Phi_1, \partial\mathfrak{D}_1) \neq 0$. At the same time the field $\Phi_\varepsilon = I - \tilde{T}$, where $\tilde{T}$ is defined by formula (4.6) for sufficiently small $\varepsilon$, does not vanish on $\partial\mathfrak{D}_1$ and is quasi-homotopic with the field $\Phi_1 = I - T$. Hence $\gamma^*(\Phi_\varepsilon, \partial\mathfrak{D}_1) = \gamma^*(\Phi_1, \partial\mathfrak{D}_1) \neq 0$. Consequently, equation (4.7) has at least one solution $x_\varepsilon$ in $\mathfrak{D}_1$. But, as we have already mentioned, this solution is unique and regular, i.e., $\gamma^*(\Phi_\varepsilon, \partial\mathfrak{D}_1) = 1$ (recall that the considered fields are analytic). Therefore $\gamma^*(\Phi, \partial\mathfrak{D}_1) = 1$ and $\mathfrak{F}$ consists of a single point, in contradiction with our assumption. ▶

In Theorem 4.1 we used that $\Phi$ is a Fredholm operator only to decompose the space $\mathfrak{X}$ into the direct sum of subspaces $\mathfrak{N}$ and $\mathfrak{R}$.

In §4 of Chapter IV we pointed out a few different conditions which guarantee the decomposition $\mathfrak{X} = \mathfrak{N} \dot{+} \mathfrak{R}$ (∗). During the translation of this book the paper of P. Mazet and J.-P. Vigue [1] appeared in which it is also proved that under condition (∗) the set $\mathfrak{F}$ of the fixed points of the operator $T$ can be represented locally as a complex-analytic manifold. Some versions of this theorem have been proved earlier by E. Vesentini [2], [3] and J.-P. Vigue [2] in the $n$-dimensional case, by M. Abd-Alla [1] for the product of Hilbert balls, and by D. Shoikhet [2] in the general case of Banach spaces (see also M. Herve [2] and D. Shoikhet [3]). In addition let us note that P. Mazet and J.-P. Vigue established also some other conditions which are equivalent to condition (∗).

By Theorem IV.8.2 such a decomposition holds too in the case when the operator $\Phi'(x^*)$ is normally solvable and the space $\mathfrak{X}$ is reflexive. Therefore the next analogue of Theorem 4.1 is true.

**THEOREM 4.3.** *Let $\mathfrak{X}$ be a reflexive complex Banach space, $\mathfrak{D}$ a bounded domain in $\mathfrak{X}$, and $\mathfrak{F}$ the set of all fixed points for an operator $T : \mathfrak{D} \to \mathfrak{D}$, analytic in $\mathfrak{D}$. If the operator $I - T'(x)$ is normally solvable for any $x \in \mathfrak{F}$, then:*

*1) $\mathfrak{F}$ is a direct complex-analytic submanifold in $\mathfrak{D}$ modelled on the space $\ker(I - T'(x))$ with $x \in \mathfrak{F}$;*

*2) if $\mathfrak{D}$ is convex and $\mathfrak{F}$ contains an isolated point then this point is regular and is the unique fixed point for $T$ in $\mathfrak{D}$.*

# Chapter IX

# Operators on spaces with indefinite metric

## §1. Spaces with indefinite metric

Let us consider a complex Banach space $\mathfrak{B}$ endowed with a norm $\|\cdot\|$, and decomposed in a topological direct sum

$$\mathfrak{B} = \mathfrak{B}_+ \dot{+} \mathfrak{B}_- .$$

The bounded mutually complementar projections generated by this decomposition are denoted by $P_\pm$, such that $\mathfrak{B}_\pm = P_\pm \mathfrak{B}$, $P_\pm^2 = P_\pm$ and $P_+ + P_- = I$. We define a functional $J_\nu$ on $\mathfrak{B}$ by the formula

$$J_\nu(x) = \|x_+\|^\nu - \|x_-\|^\nu, \quad \nu > 0,\ x \in \mathfrak{B},\ x_\pm = P_\pm x. \tag{1.1}$$

We say that the functional $J_\nu$ introduces an *indefinite metric* on the space $\mathfrak{B}$, which is also called a *$J_\nu$-metric*. The space $\mathfrak{B}$ endowed with a $J_\nu$-metric is called, for brevity, a *$J_\nu$-space*.

$J_\nu$-spaces appeared as a natural generalization of Hilbert spaces with indefinite metric, which we will briefly recall in what follows. Let us consider a complex Hilbert space $\mathfrak{H}$ with the inner product $\langle\cdot,\cdot\rangle$ (see §11 of Chapter 0). Let $G$ be a bounded self-adjoint invertible operator on $\mathfrak{H}$ (i.e., the operator $G$ is defined on the whole space $\mathfrak{H}$ and the inverse operator $G^{-1}$ exists). We define a sesquilinear form $[\cdot,\cdot]$ on $\mathfrak{H}$, called the *G-metric*, by the formula $[x,y] = \langle Gx, y\rangle$, $x, y \in \mathfrak{H}$. If the operator $G$ is indefinite, i.e., its spectrum on the real line lies on both sides of zero, then the $G$-metric is also indefinite: for a non-zero vector $x \in \mathfrak{H}$ the "scalar square" $[x,x]$ can be positive, negative or equal to zero. Assume now that the operator $G^{-1}$

is bounded. Then the space $\mathfrak{H}$ endowed with the $G$-metric is called a *Krein space.* Denoting by $E_\lambda$ the spectral resolution of the operator $G$ we set

$$P_- = \int_{-\infty}^{0} \mathrm{d}E_\lambda, \quad P_+ = \int_{0}^{+\infty} \mathrm{d}E_\lambda.$$

Then $P_-$ and $P_+$ are mutually orthogonal projections: $P_\pm^2 = P_\pm = P_\pm^*$, $P_+ + P_- = I$. Setting $\mathfrak{H}_\pm = P_\pm \mathfrak{H}$ we obtain a decomposition of the space $\mathfrak{H}$ in the orthogonal direct sum $\mathfrak{H} = \mathfrak{H}_+ \oplus \mathfrak{H}_-$. Let us introduce on $\mathfrak{H}$ a new inner product $(\cdot,\cdot)$ by the formula

$$(x,y) = [P_+x, y] - [P_-x, y], \quad x, y \in \mathfrak{H}.$$

This new inner product is equivalent to the initial one. Indeed, if $x \in \mathfrak{H}$, then

$$(x,x) = \langle G(P_+ - P_-)x, x\rangle \leqslant \|G\| \langle x, x\rangle,$$

and, conversely,

$$(x,x) = \langle GP_+x, x\rangle - \langle GP_-x, x\rangle \geqslant$$
$$\geqslant \lambda_1 \langle P_+x, x\rangle + \lambda_2 \langle P_-x, x\rangle \geqslant \min\{\lambda_1, \lambda_2\} \langle x, x\rangle,$$

where $\lambda_1 = \inf\{\lambda \;:\; \lambda \in \sigma(G \mid \mathfrak{H}_+)\} > 0$ and $-\lambda_2 = \sup\{\lambda \;:\; \lambda \in \sigma(G \mid \mathfrak{H}_-)\} < 0$. Relatively to the inner product $(\cdot,\cdot)$, the form $[\cdot,\cdot]$ can be written as

$$[x,y] = (Jx, y), \quad J = P_+ - P_-. \tag{1.2}$$

The operator $J$ satisfies the properties $J^* = J^{-1} = J$ and is called the *canonical symmetry* of the considered Krein space. The form $[\cdot,\cdot]$ given by (1.2) is called the *J-metric* of the space, and the space $\mathfrak{H}$ with the inner product $(\cdot,\cdot)$ and the $J$-metric $[\cdot,\cdot]$ is called a *J-space.*

Setting $\|x\| = (x,x)^{\frac{1}{2}}$ for each $x \in \mathfrak{H}$, we have

$$[x,x] = \|x_+\|^2 - \|x_-\|^2, \quad x_\pm = P_\pm x, \tag{1.3}$$

hence a Krein space is a Hilbert $J_2$-space, in which $P_+$, $P_-$ are orthogonal projections.

Let us return now to an arbitrary $J_\nu$-space $\mathfrak{B}$. A vector $x \in \mathfrak{B}$ is said to be *positive*, *negative* or *neutral* if it satisfies the condition $J_\nu(x) > 0$, $J_\nu(x) < 0$ or $J_\nu(x) = 0$, respectively. A vector $x$ such that $J_\nu(x) \geqslant 0$ (respectively, $J_\nu(x) \leqslant 0$) is called *non-negative* (respectively, *non-positive*). The set of all non-negative (respectively, non-positive) vectors in $\mathfrak{B}$ will be denoted by $\mathfrak{K}_+$ (respectively, $\mathfrak{K}_-$). By $\mathfrak{K}_{++}$ ($\mathfrak{K}_{--}$) we will denote the set of all positive (respectively, negative) vectors in $\mathfrak{B}$, and by $\mathfrak{K}_0$ we will denote the set of all neutral vectors.

Before introducing a new notion, let us recall that by a lineal in $\mathfrak{B}$ we mean any subset $\mathcal{L} \subseteq \mathfrak{B}$ such that $\lambda x + \mu y \in \mathcal{L}$, for all $x, y \in \mathcal{L}$ and $\lambda, \mu \in \mathbb{C}$. The term subspace is constantly used to call a lineal that is closed in the norm topology. A lineal $\mathcal{L}$ is said to be *non-negative* if the inequality $J_\nu(x) \geqslant 0$ is true for all $x \in \mathcal{L}$. Analogously we define the non-positive, positive, negative or neutral lineals. The non-negative or non-positive lineals are also called *semi-definite* lineals, whereas the negative or positive lineals are said to be *definite*. Among the definite lineals we distinguish the *uniformly definite* ones, i.e., those definite lineals $\mathcal{L}$ with the property

$$|J_\nu(x)| \geqslant c(\mathcal{L})\|x\|^\nu,$$

for all $x \in \mathcal{L}$, where $c(\mathcal{L}) > 0$ is a suitable constant. By the continuity of the functional $J_\nu$, the closure in $\mathfrak{B}$ of any semi-definite (uniformly definite) lineal is still semi-definite (uniformly definite). For definite but not uniformly definite lineals a similar property is no longer true, as the next example shows.

We need a simple preliminary result.

**LEMMA 1.1.** *Let $\mathfrak{X}$ be a Banach space with the norm $\|\cdot\|$, and let $\mathcal{M}$ be a lineal in $\mathfrak{X}$ on which a bounded linear operator $T$ is defined. If there exists a projection operator $P_\mathcal{M}$ from $\mathfrak{X}$ onto the lineal $\mathcal{M}$, with the norm $\|P_\mathcal{M}\| = 1$, then the operator $T$ can be extended on the whole space $\mathfrak{X}$, without increasing the norm, by setting the extension equal to 0 on $(I - P_\mathcal{M})\mathfrak{X}$.*

Recall (see Chapter 0) that a linear operator $\tilde{T}$ defined on a lineal $\mathfrak{D}(\tilde{T})$ is said to be an extension of the operator $T$ defined on a lineal $\mathfrak{D}(T)$ if $\mathfrak{D}(T) \subset \mathfrak{D}(\tilde{T})$ and $\tilde{T}x = Tx$, for all $x \in \mathfrak{D}(T)$.

◁ Let $\tilde{T}$ be the operator defined by the equalities $\tilde{T}x = Tx$, $x \in \mathcal{M}$, and $\tilde{T}y = 0$, $y \in (I - P_\mathcal{M})\mathfrak{X}$. Then $\tilde{T}$ is an extension of $T$ on the whole space $\mathfrak{X}$ and for all $z = x + y \in \mathcal{M} \dot{+} (I - P_\mathcal{M})\mathfrak{X} = \mathfrak{X}$ we have

$$\|\tilde{T}z\| = \|Tx\| = \|TP_\mathcal{M}z\| \leqslant \|T\| \cdot \|z\|,$$

i.e., $\|\tilde{T}\| \leqslant \|T\|$. Since the inequality $\|T\| \leqslant \|\tilde{T}\|$ is obvious, we get $\|\tilde{T}\| = \|T\|$. ►

**EXAMPLE 1.1.** Let $\mathfrak{B}$ be a $J_\nu$-space with $\mathfrak{B}_+$ infinite dimensional, and let $x_+ \in \mathfrak{B}_+$, $x_- \in \mathfrak{B}_-$ be two fixed elements such that $\|x_+\| = \|x_-\| = 1$. We define the operator

$K_+ : \text{lin}\{x_+\} \to \text{lin}\{x_-\}$ by the equality $K_+ x_+ = x_-$. Then $\|K_+\| = 1$. It is well-known (see V. P. Odinec [1]) that for the one-dimensional subspace $\mathcal{M} = \text{lin}\{x_+\}$ there exists a projection operator $P_{\mathcal{M}}$ from $\mathfrak{B}$ onto $\mathcal{M}$, with the norm $\|P_{\mathcal{M}}\| = 1$. We extend $K_+$ to an operator $\tilde{K}_+$ defined on the whole space $\mathfrak{B}_+$ by $\tilde{K}_+ \mid (I - P_{\mathcal{M}})\mathfrak{B}_+ = 0$. According to Lemma 1.1 we have $\|\tilde{K}_+\| = 1$. By a direct verification we obtain that the lineal $\mathcal{L}_+ = (P_+ + \tilde{K}_+)\mathfrak{B}_+$ is a non-negative subspace of $\mathfrak{B}$. Moreover, $\mathcal{L}_+$ is infinite dimensional. Let $\mathcal{E} \subset \mathcal{L}_+$ be a lineal which is dense in $\mathcal{L}_+$ and such that $x_+ + x_- \notin \mathcal{E}$ (for example we can choose $\mathcal{E} = \ker \varphi$, where $\varphi$ is a discontinuous linear functional on $\mathcal{L}_+$, with $\varphi(x_+ + x_-) \neq 0$). Then the lineal $\mathcal{E}$ is positive, whereas its closure $\overline{\mathcal{E}} = \mathcal{L}_+$ contains a neutral vector $x_+ + x_- \neq 0$. ▶

## §2. Angle operators

We introduce now the angle operator of a semi-definite lineal, one of the most important objects of study in this chapter.

Let $\mathcal{L}_+$ be a non-negative lineal of $\mathfrak{B}$. Since $\mathcal{L}_+ \subset \mathfrak{K}_+$, it follows that $\|P_- x\| \leqslant \|P_+ x\|$, for all $x \in \mathcal{L}_+$. Consequently, there exists a well-defined linear operator $K_+ : P_+\mathcal{L}_+ \to P_-\mathcal{L}_+$, given by $K_+(P_+ x) = P_- x$, $x \in \mathcal{L}_+$. Equivalently, for all $x = x_+ + x_- \in \mathcal{L}_+$ with $x_\pm = P_\pm x$, we have $x_- = K_+ x_+$. It is clear that $\|K_+\| \leqslant 1$. The operator $K_+$ is called the *angle operator* of the lineal $\mathcal{L}_+$. We define analogously the angle operator $K_- : P_-\mathcal{L}_- \to P_+\mathcal{L}_-$ of a non-positive lineal $\mathcal{L}_-$.

Further we will study mainly the non-negative lineals. The corresponding results for non-positive lineals can be obtained analogously.

If the lineal $\mathcal{L}_+$ is positive (i.e., $\mathcal{L}_+ \setminus \{0\} \subset \mathfrak{K}_{++}$), then $\|K_+ x_+\| < \|x_+\|$, for all $0 \neq x_+ \in P_+\mathcal{L}_+$. If, in addition, $\mathcal{L}_+$ is a uniformly positive lineal, then $\|K_+\| \leqslant 1 - \gamma(\mathcal{L}_+)$, where $0 < \gamma(\mathcal{L}_+) \leqslant 1$ is a suitable constant. Indeed, by the definition of the uniform positivity of the lineal $\mathcal{L}_+$ we have $\|x_+\|^\nu - \|x_-\|^\nu \geqslant c(\mathcal{L}_+)\|x\|^\nu$, for all $x = x_+ + x_- \in \mathcal{L}_+$, where $c(\mathcal{L}_+) > 0$. Hence, and by the obvious relation $\|x_+\| = \|P_+ x\| \leqslant \|P_+\|\,\|x\|$, it follows that

$$\|x_-\| \leqslant \left(1 - \frac{c(\mathcal{L}_+)}{\|P_+\|^\nu}\right)^{\frac{1}{\nu}} \|x_+\|,$$

thus $\|K_+\| \leqslant 1 - \gamma(\mathcal{L}_+)$, where

$$\gamma(\mathcal{L}_+) = 1 - \left(1 - \frac{c(\mathcal{L}_+)}{\|P_+\|^\nu}\right)^{\frac{1}{\nu}}.$$

A semi-definite lineal is said to be *maximal semi-definite* provided that it is not a proper subset of another semi-definite lineal. The maximal definite, maximal uniformly definite and maximal neutral lineals are introduced analogously. Considering on the set of all semi-definite lineals the order relation corresponding to the inclusion, and using Zorn Lemma (see §1 of Chapter 0), we conclude easily that any semi-definite lineal is contained in a maximal semi-definite lineal. The analogous conclusions for definite or neutral lineals are also true.

Let us consider the class $\mathfrak{M}_+$ of all non-negative lineals $\mathcal{L}_+$ such that $P_+\mathcal{L}_+ = \mathfrak{B}_+$. Analogously, we define the class $\mathfrak{M}_-$ of non-positive lineals. It is obvious that all lineals $\mathcal{L}_+ \in \mathfrak{M}_+$ are maximal non-negative subspaces. The converse is, generally speaking, false. Namely, as Lemma 1.1 shows, the question whether a maximal non-negative subspace $\mathcal{L}_+$ belongs to the class $\mathfrak{M}_+$ is related to the question about the existence of a projection operator of norm 1 from the space $\mathfrak{B}_+$ onto the subspace $P_+\mathcal{L}_+$. It is well-known (see §11 of Chapter 0) that in the class of all Banach spaces, the Hilbert spaces are the only ones which have the property that any subspace is the range of a projection operator of norm 1. Therefore, when $\mathfrak{B}_+$ is a Hilbert space, the set of all maximal non-negative subspaces coincides with $\mathfrak{M}_+$. In the general case of a $J_\nu$-space $\mathfrak{B}$, this is no longer true. (For more information about projections of norm 1 on a Banach space see V. P. Odinec [1].) However, as Lemma 1.1 shows, if $\dim \mathfrak{B}_+ \geqslant 2$, then we have $\dim \mathcal{L}_+ \geqslant 2$ for any maximal non-negative subspace $\mathcal{L}_+$ of $\mathfrak{B}$. In what follows we will denote by $\mathfrak{M}_+^0$ ($\mathfrak{M}_-^0$) the set of all uniformly positive (respectively, negative) subspaces $\mathcal{L}_+ \in \mathfrak{M}_+$ ($\mathcal{L}_- \in \mathfrak{M}_-$).

Later on we will use the next result.

**PROPOSITION 2.1.** *Every vector $x \in \mathfrak{K}_+$ ($\mathfrak{K}_{++}$) is contained in a subspace $\mathcal{L}_+ \in \mathfrak{M}_+$ ($\mathfrak{M}_+^0$).*

◁ Let $K_+$ be an angle operator corresponding to the one-dimensional subspace $\operatorname{lin}\{x_+\} \subset \mathfrak{K}_+$, where $x_+ = P_+x$. If $x \in \mathfrak{K}_{++}$ then $\|K_+\| < 1$. As we have noted above, any one-dimensional subspace is the range of a projection of norm 1. Therefore the operator $K_+$ can be extended, preserving the norm, to an operator $\tilde{K}_+$ defined on the whole space $\mathfrak{B}_+$. The subspace $\mathcal{L}_+ = (P_+ + \tilde{K}_+)\mathfrak{B}_+$ satisfies the desired condition. ▶

Let us consider a subspace $\mathcal{L}_+$ in the class $\mathfrak{M}_+$. For any such subspace $\mathcal{L}_+$ we have an angle operator $K_+ : \mathfrak{B}_+ \to \mathfrak{B}_-$, with $\|K_+\| \leqslant 1$, defined on the whole space $\mathfrak{B}_+$, such that

$$\mathcal{L}_+ = \{x : x = x_+ + K_+x_+,\ x_+ \in \mathfrak{B}_+\}, \tag{2.1}$$

or $\mathcal{L}_+ = (P_+ + K_+)\mathfrak{B}_+$. In order to determine the operator $K_+$ we observe first the next simple fact.

**LEMMA 2.1.** *If $\mathcal{L}$ is a non-negative lineal then the projection $P_+$ maps $\mathcal{L}$ homeomorphically onto $P_+\mathcal{L}$.*

◁ For any $x \in \mathcal{L}$ we have $\|P_+x\| \geqslant \|P_-x\|$. Therefore $\|x\| \leqslant \|P_+x\| + \|P_-x\| \leqslant 2\|P_+x\|$. On the other hand, $\|P_+x\| \leqslant \|P_+\|\ \|x\|$. ▶

Now let $\mathcal{L}_+ \in \mathfrak{M}_+$ be fixed. By Lemma 2.1 the operator $P_+ \mid \mathcal{L}_+$, i.e., the restriction of the projection $P_+$ to $\mathcal{L}_+$, has a bounded inverse $(P_+ \mid \mathcal{L}_+)^{-1}$; for all $x$ in $\mathcal{L}_+$ the equality $x = (P_+ \mid \mathcal{L}_+)^{-1}x_+$ is fulfilled. Then $x_- = P_-(P_+ \mid \mathcal{L}_+)^{-1}x_+$. Consequently, the operator $K_+$ in (2.1) has the form

$$K_+ = P_-(P_+ \mid \mathcal{L}_+)^{-1}. \tag{2.2}$$

We denote by $\mathcal{K}_+$ the closed unit ball of the space $\mathcal{L}(\mathfrak{B}_+, \mathfrak{B}_-)$ of all bounded linear operators acting from $\mathfrak{B}_+$ into $\mathfrak{B}_-$, and by $\mathcal{K}_+^0$ its interior. By (2.1) any $K_+ \in \mathcal{K}_+$ determines a subspace $\mathcal{L}_+$ in the class $\mathfrak{M}_+$.

Summing up all the considerations above we obtain the next result.

**THEOREM 2.1.** *There exists a bijective correspondence between the sets $\mathfrak{M}_+$ and $\mathcal{K}_+$ given by the relations* (2.1) *and* (2.2).

## §3. Plus-operators

We consider a linear operator $A$ defined on a lineal $\mathfrak{D}(A)$ of a $J_\nu$-space $\mathfrak{B}$.

**DEFINITION 3.1.** The operator $A$ is called a *plus-operator* if $A(\mathfrak{K}_+ \cap \mathfrak{D}(A)) \subset \mathfrak{K}_+$.

**DEFINITION 3.2.** A plus-operator $A$ is said to be *strict* if $J_\nu(Ax) \geqslant \mu(A)J_\nu(x)$, for all $x \in \mathfrak{K}_+ \cap \mathfrak{D}(A)$, where $\mu(A) > 0$ is a constant.

**DEFINITON 3.3.** The operator $A$ is called $J_\nu^+$*-expansive* if

$$J_\nu(Ax) \geqslant J_\nu(x), \tag{3.1}$$

for all $x \in \mathfrak{D}(A) \cap \mathfrak{K}_+$. If the inequality (3.1) is fulfilled for all $x \in \mathfrak{D}(A)$, then the operator $A$ is said to be $J_\nu$*-expansive.*

**DEFINITION 3.4.** The operator $A$ is called *uniformly* $J_\nu^+$*-expansive* if

$$J_\nu(Ax) \geqslant J_\nu(x) + \gamma(A)\|x\|^\nu, \tag{3.2}$$

for all $x \in \mathfrak{D}(A) \cap \mathfrak{K}_+$, where $\gamma(A) > 0$ is a constant. If the inequality (3.2) is fulfilled for all $x \in \mathfrak{D}(A)$, then $A$ is said to be *uniformly $J_\nu$-expansive.*

**DEFINITION 3.5.** The operator $A$ is said to be a *$J_\nu$-isometry* if

$$J_\nu(Ax) = J_\nu(x),$$

for all $x \in \mathfrak{D}(A)$. A $J_\nu$-isometry is called a *$J_\nu$-semi-unitary operator* if $\mathfrak{D}(A) = \mathfrak{B}$, and a *$J_\nu$-unitary operator* if $\mathfrak{D}(A) = \mathfrak{B}$ and, in addition, $R(A)\ (= A\mathfrak{D}(A)) = \mathfrak{B}$.

In the general case, a plus-operator on a $J_\nu$-space $\mathfrak{B}$ is unbounded. We give next an example of an unbounded $J_\nu$-expansive plus-operator $A$ with $\mathfrak{D}(A) = \mathfrak{B}$.

**EXAMPLE 3.1.** Let $\mathfrak{B}$ be a $J_\nu$-space with $\mathfrak{B}_+$ infinite dimensional. Relatively to the decomposition $\mathfrak{B} = \mathfrak{B}_+ \dot{+} \mathfrak{B}_-$ we define the linear operator $A$ on $\mathfrak{B}$ by the matrix

$$A = \begin{pmatrix} A_{11} & 0 \\ 0 & I \end{pmatrix},$$

where $A_{11}$ is an unbounded expansive operator on $\mathfrak{B}_+$, i.e., $\|A_{11}x_+\| \geqslant \|x_+\|$ for all $x_+ \in \mathfrak{B}_+$. Then the operator $A$ is unbounded and

$$J_\nu(Ax) = \|A_{11}x_+\|^\nu - \|x_-\|^\nu \geqslant \|x_+\|^\nu - \|x_-\|^\nu = J_\nu(x), \quad x \in \mathfrak{B}.$$

Our next aim is to establish a criterion for the continuity of a plus-operator. We need first a preliminary result.

**LEMMA 3.1.** *Let $A$ be a plus-operator such that $\mathfrak{D}(A) \cap \mathfrak{K}_{++} \neq \emptyset$. Then the boundedness of the operator $A$ is equivalent to the boundedness of the operator $P_+A$.*

$\triangleleft$ The necesity of the condition is obvious. In order to prove the sufficiency we assume, on the contrary, that $A$ is unbounded. Then there exists a sequence $\{x_n : n \in \mathbb{N}\}$, $x_n \in \mathfrak{D}(A)$, such that $x_n \to 0$ and $\|Ax_n\| \to \infty$ for $n \to \infty$. Since $P_+A$ is bounded we have $P_+Ax_n \to 0$, hence $\|P_-Ax_n\| \to \infty$. For any sequence $\{z_m : m \in \mathbb{N}\}$, $z_m \in \mathfrak{B}$, with the property $z_m \to 0$ for $m \to \infty$, we clearly obtain that $\|P_-Ax_n + P_-z_m\| \to \infty$, for $n, m \to \infty$. Consequently, $J_\nu(Ax_n + z_m) \to -\infty$ for $n, m \to \infty$. Set $z_m = m^{-1}Ay$, where $y \in \mathfrak{D}(A) \cap \mathfrak{K}_{++}$ is fixed. Then there exists $p \in \mathbb{N}$ such that $J_\nu(Ax_n + z_m) < 0$, for all $n, m \geqslant p$. But

$$J_\nu\left(\frac{1}{p}y\right) = \frac{1}{p^\nu} J_\nu(y) > 0.$$

By the continuity of $J_\nu$ there exists $l \in \mathbb{N}$, $l \geqslant p$, such that $J_\nu(x_n + (1/p)y) > 0$, for all $n \geqslant l$. Put $n = l$. Then $J_\nu(x_l + (1/p)y) > 0$, and $J_\nu(Ax_l + A(1/p)y) = J_\nu(A(x_l + (1/p)y)) < 0$, which contradicts the assumption that $A$ is a plus-operator. Thus the boundedness of $P_+A$ implies the boundedness of $A$. ▶

**THEOREM 3.1.** *Let $A$ be a plus-operator with the property that $A(\mathfrak{D}(A) \cap \mathfrak{K}_{++}) \subset \mathfrak{K}_+ \setminus \{0\}$. If $\mathfrak{D}(A)$ contains a uniformly positive lineal $\mathcal{L}$ such that $P_+A\mathcal{L} = P_+A\mathfrak{D}(A)$, then the boundedness of the operator $A$ is equivalent to the boundedness of the operator $P_+A \mid \mathcal{L} : \mathcal{L} \to \mathfrak{B}_+$.*

◁ The boundedness of $P_+A \mid \mathcal{L}$ follows obviously from the boundedness of $A$. Conversely, let us assume that $P_+A \mid \mathcal{L} : \mathcal{L} \to \mathfrak{B}_+$ is a bounded operator. By Lemma 3.1 it is sufficient to establish the boundedness of $P_+A$. Let $d = \|P_+A \mid \mathcal{L}\|$. Since $P_+A\mathcal{L} = P_+A\mathfrak{D}(A)$ it follows that for any $z \in P_+A\mathfrak{D}(A)$ with $\|z\| > d$ there exists $y \in \mathcal{L}$ with $\|y\| > 1$ such that $P_+Ay = z$. By the uniform positivity of $\mathcal{L}$ we find $\varepsilon > 0$ such that $J_\nu(u) > 0$ for any $u$ in the $\varepsilon$-neighborhood of the set $\mathcal{M} = \{x : x \in \mathcal{L}, \|x\| \geqslant 1\}$. Let $w \in \mathfrak{D}(A)$ be an arbitrary element with $\|w\| < \varepsilon$. Then $J_\nu(y - w) > 0$ for any $y \in \mathcal{M}$. Since $A(\mathfrak{D}(A) \cap \mathfrak{K}_{++}) \subset \mathfrak{K}_+ \setminus \{0\}$ we obtain $0 \neq J_\nu(Ay - Aw) = J_\nu(A(y - w)) \geqslant 0$, hence $P_+Ay \neq P_+Aw$, for all elements $y \in \mathcal{M}$. Consequently, there exists a number $s > 0$ such that $\|P_+Aw\| \leqslant s$, for all $w \in \mathfrak{D}(A)$ with $\|w\| < \varepsilon$, a condition which clearly implies the boundedness of $P_+A$. As we have already noted, by Lemma 3.1 the proof of Theorem 3.1 is complete. ▶

Actually, under our assumptions, we can avoid the use of Lemma 3.1. Indeed, take as above $w \in \mathfrak{D}(A)$ with $\|w\| < \varepsilon$, and $y \in \mathcal{M}$. Since $J_\nu(y + w) > 0$ we have $y + w \in \mathfrak{D}(A) \cap \mathfrak{K}_{++}$, hence $0 \neq J_\nu(Ay + Aw) = J_\nu(A(y + w)) \geqslant 0$. It follows that

$$\|P_+Ay + P_+Aw\| \geqslant \|P_-Ay + P_-Aw\|.$$

On the other hand we have

$$\|P_-Aw\| = \|(P_-Ay + P_-Aw) - P_-Ay\| \leqslant \|P_-Ay + P_-Aw\| + \|P_-Ay\| \leqslant$$
$$\leqslant \|P_+Ay + P_+Aw\| + \|P_-Ay\| \leqslant \|P_+Ay\| + \|P_+Aw\| + \|P_-Ay\|.$$

Therefore, taking into account that $\|P_+Aw\| \leqslant s$, we obtain

$$\|Aw\| \leqslant \|P_+Aw\| + \|P_-Aw\| \leqslant$$
$$\leqslant 2\|P_+Aw\| + \|P_+Ay\| + \|P_-Ay\| \leqslant 2s + \|P_+Ay\| + \|P_-Ay\|,$$

for all $w \in \mathfrak{D}(A)$ with $\|w\| < \varepsilon$, and the boundedness of $A$ follows.

Theorem 3.1 enables one to obtain a simple and natural condition for the boundedness of a strict plus-operator defined on the whole space $\mathfrak{B}$. This very class of operators will be, mainly, the object of our further investigations.

**THEOREM 3.2.** *Let $A$ be a strict plus-operator defined on $\mathfrak{B}$. If there exists $\mathcal{L}_+^0 \in \mathfrak{M}_+^0$ such that $A\mathcal{L}_+^0 \in \mathfrak{M}_+$, then $A$ is bounded.*

◁ We show the boundedness of the operator $P_+A \mid \mathcal{L}_+^0 : \mathcal{L}_+^0 \to \mathfrak{B}_+$. Since $\mathcal{L}_+^0 \in \mathfrak{M}_+^0$ we have $J_\nu(x) \geqslant c(\mathcal{L}_+^0)\|x\|^\nu$, for all $x \in \mathcal{L}_+^0$, where $c(\mathcal{L}_+^0) > 0$. Therefore

$$\|P_+Ax\|^\nu = J_\nu(P_+Ax) \geqslant J_\nu(Ax) \geqslant \mu(A)J_\nu(x) \geqslant \mu(A)c(\mathcal{L}_+^0)\|x\|^\nu,$$

for all $x \in \mathcal{L}_+^0$, where $\mu(A) > 0$. Hence it follows that $P_+A(\mathcal{L}_+^0)$ is closed in $\mathfrak{B}_+$, $P_+A \mid \mathcal{L}_+^0$ is one-to-one, and $(P_+A \mid \mathcal{L}_+^0)^{-1}$ maps continuously the Banach space $P_+A(\mathcal{L}_+^0)$ onto the Banach space $\mathcal{L}_+^0$. By Theorem 0.8.5 the operator $P_+A \mid \mathcal{L}_+^0$ is continuous, too. Applying Theorem 3.1 we obtain the boundedness of $A$. ▶

We will state next an obvious consequence of Theorem 3.2.

**COROLLARY 3.1.** *Let $A$ be a strict plus-operator defined on the whole space $\mathfrak{B}$. If $A_{11}$ $(= P_+AP_+)$ maps $\mathfrak{B}_+$ onto $\mathfrak{B}_+$, then the operator $A$ is bounded.*

Further we will consider bounded plus-operators defined on the whole space $\mathfrak{B}$. We need the following result.

**LEMMA 3.2.** *A strict plus-operator $A$ maps a uniformly positive subspace onto a uniformly positive subspace.*

◁ Let $\mathcal{L}_+$ be a uniformly positive subspace. Then

$$\|Ax\|^\nu \geqslant J_\nu(Ax) \geqslant \mu(A)J_\nu(x) \geqslant \mu(A)c(\mathcal{L}_+)\|x\|^\nu = c\|x\|^\nu,$$

where $x \in \mathcal{L}_+$ and $c = \mu(A)c(\mathcal{L}_+) > 0$. If $z \in \overline{A\mathcal{L}_+}$, then $z = \lim_{n\to\infty} Ax_n$, where $x_n \in \mathcal{L}_+$. Since $\|Ax_n - Ax_m\| \geqslant c^{\frac{1}{\nu}}\|x_n - x_m\|$, the sequence $\{x_n : n \in \mathbb{N}\}$ is fundamental in $\mathcal{L}_+$ and therefore $x_n \to x$ for $n \to \infty$, where $x \in \mathcal{L}_+$. This means that $z$ $(= \lim_{n\to\infty} Ax_n) = Ax \in A\mathcal{L}_+$, hence $A\mathcal{L}_+$ is a subspace. Hence and from $J_\nu(Ax) \geqslant c\|x\|^\nu$ the uniform positivity of the subspace $A\mathcal{L}_+$ follows. ▶

We consider now the action of a strict plus-operator on the subspaces belonging to $\mathfrak{M}_+$. By the *defect number* of a subspace $\mathcal{L}$ $(\subset \mathfrak{K}_+)$ we mean the cardinal number $\operatorname{def} \mathcal{L} = \dim(\mathfrak{B}_+/P_+\mathcal{L})$ (note that by Lemma 2.1 $P_+\mathcal{L}$ is a subspace of $\mathcal{L}_+$; $\mathfrak{B}_+/P_+\mathcal{L}$ is the quotient of $\mathfrak{B}_+$ by $P_+\mathcal{L}$).

**THEOREM 3.3.** *If $A$ is a strict plus-operator, then the defect number $\operatorname{def} A\mathcal{L}_+$ is the same for all $\mathcal{L}_+ \in \mathfrak{M}_+^0$.*

◁ By Lemma 3.2 we obtain $\overline{A\mathcal{L}_+} = \mathcal{L}_+$, for all $\mathcal{L}_+ \in \mathfrak{M}_+^0$. Let $\mathcal{L}_+^1, \mathcal{L}_+^2 \in \mathfrak{M}_+^0$ and let $K_+^1$, $K_+^2$ be their angle operators. By the convexity of $K_+^0$ the operator $K_+^t = (1-t)K_+^1 + tK_+^2$ is the angle operator of some $\mathcal{L}_+^t \in \mathfrak{M}_+^0$ for any $t \in [0, 1]$. We associate to the family $\{K_+^t\}$ a new family of operators $\{W_A(K_+^t) : W_A(K_+^t) = P_+A(P_+ + K_+^t),\ t \in [0, 1]\}$. Since $A$ takes non-zero values on the set of positive vectors, all the operators $W_A(K_+^t)$ are $\Phi_+$-operators (we are using here the terminology in I. C. Gohberg and M. G. Krein [1]). By Theorem 7.1 in that book, for any $t \in [0, 1]$ we can find $\tau > 0$ such that for all $s \in [0, 1]$ with $|s - t| < \tau$ the operators $W_A(K_+^s)$ have the same defect index. We use the Heine-Borel Lemma and cover the interval $[0, 1]$ with a finite number of open intervals, each of them corresponding to a constant value of the defect index. Therefore $\operatorname{def} A\mathcal{L}_+^1 = \operatorname{def} A\mathcal{L}_+^2$. ▶

The possibility of extending Lemma 3.2 and Theorem 3.3 to the whole class $\mathfrak{M}_+$ is related to the values of the constant $\nu$. Namely, — as the next example shows — for $\nu \leqslant 1$ this is impossible.

**EXAMPLE 3.2.** Let $\mathfrak{B} = \mathfrak{B}_+ \dot{+} \mathfrak{B}_-$ be a two-dimensional $J_\nu$-space, where $\nu \leqslant 1$ and $\mathfrak{B}_\pm = \operatorname{lin}\{e_\pm : \|e_\pm\| = 1\}$. With respect to the basis $\{e_+, e_-\}$ we define a linear operator $A$ by the matrix

$$A = \begin{pmatrix} 1 & 1 \\ 0 & 0 \end{pmatrix}.$$

Then

$$J_\nu(A(\alpha e_+ + \beta e_-)) = |\alpha + \beta|^\nu \geqslant |\alpha|^\nu - |\beta|^\nu = J_\nu(\alpha e_+ + \beta e_-),$$

i.e., the operator $A$ is $J_\nu$-expansive. But $A(e_+ - e_-) = 0$ and the vector $e_+ - e_- \neq 0$ is neutral.

When $\nu > 1$ it is possible to extend Theorem 3.3 to the whole class $\mathfrak{M}_+$. We need the next result.

**LEMMA 3.3.** *Let $A$ be a strict plus-operator acting on a $J_\nu$-space $\mathfrak{B}$, with $\nu > 1$. Then there exists $\delta(A) > 0$ such that $\|Ax\| \geqslant \delta(A)\|x\|$, for all $x \in \mathfrak{K}_+$.*

◁ Assume the conclusion is false. Then there exists a sequence $\{x^n : x^n \in \mathfrak{K}_+, \|x^n\| = 1\}$ such that $Ax^n \to 0$ for $n \to \infty$. We have $\|x_+^n\| = \|P_+x^n\| \geqslant \frac{1}{2}\|x^n\| = \frac{1}{2}$. Set $y^n = (1 + \lambda_n)^{\frac{1}{\nu}} x_+^n + \lambda_n^{\frac{1}{\nu}} x_-^n$, where $\lambda_n^{\frac{1}{\nu}} = \|Ax^n\|^{-\frac{1}{2}}$, for $Ax^n \neq 0$, and $\lambda_n = n$, otherwise. Then $J_\nu(y^n) \geqslant 2^{-\nu}$ and

$$J_\nu(Ay^n) = J_\nu\left(A\left((1 + \lambda_n)^{\frac{1}{\nu}} - \lambda_n^{\frac{1}{\nu}}\right) x_+^n + \lambda_n^{\frac{1}{\nu}} Ax^n\right) \to 0,$$

for $n \to \infty$, which contradicts the fact that $A$ is a strict plus-operator. ▶

Now using Lemma 3.3 we can extend Theorem 3.3 on the whole $\mathfrak{M}_+$.

**THEOREM 3.4.** *Let $A$ be a strict plus-operator acting on a $J_\nu$-space $\mathfrak{B}$, with $\nu > 1$. Then the defect number $\operatorname{def} A\mathcal{L}_+$ is the same for all $\mathcal{L}_+ \in \mathfrak{M}_+$.*

◁ The proof is almost the same as the one of Theorem 3.3. ▶

**COROLLARY 3.2.** *If a strict plus-operator $A$ on a $J_\nu$-space $\mathfrak{B}$, with $\nu > 1$, maps at least one subspace from $\mathfrak{M}_+$ onto a subspace in $\mathfrak{M}_+$, then $A\mathcal{L}_+ \in \mathfrak{M}_+$, for all $\mathcal{L}_+ \in \mathfrak{M}_+$.*

## §4. Symmetric properties of a plus-operator and its adjoint

In the dual space $\mathfrak{B}^*$ of the $J_\nu$-space $\mathfrak{B}$ we define the sets $\mathfrak{K}^*_\pm$ by

$$\mathfrak{K}^*_\pm = \left\{x^* \ : \ x^* \in \mathfrak{B}^*, \ \left\|P^*_\pm x^*\right\| \geqslant \left\|P^*_\mp x^*\right\|\right\},$$

where $P^*_\pm$ are the adjoint operators of the projections $P_\pm$. Clearly $P^*_\pm$ are projections on $\mathfrak{B}^*$ and $\mathfrak{B}^* = P^*_+\mathfrak{B}^* \dot{+} P^*_-\mathfrak{B}^* = \mathfrak{B}^*_+ \dot{+} \mathfrak{B}^*_-$. We introduce also $\mathfrak{K}^*_0 = \mathfrak{K}^*_+ \cap \mathfrak{K}^*_-$. As in the case of the space $\mathfrak{B}$, for every lineal $\mathcal{M}_\pm \subset \mathfrak{K}^*_\pm$ there exists the angle operator $Q_\pm : P^*_\pm\mathcal{M}_\pm \to P^*_\mp\mathcal{M}_\pm$ such that $\|Q_\pm\| \leqslant 1$ and $\mathcal{M}_\pm = \left\{x^*_\pm + Q_\pm x^*_\pm \ : \ x^*_\pm \in P^*_\pm\mathcal{M}_\pm\right\}$. By $\mathfrak{M}^*_\pm$ we denote the class of all subspaces $\mathcal{M}_\pm \subset \mathfrak{K}^*_\pm$ such that $P^*_\pm\mathcal{M}_\pm = \mathfrak{B}^*_\pm$. We single out also the class $\mathfrak{M}^{*0}_\pm$ of all uniform subspaces in $\mathfrak{M}^*_\pm$. As in §2 above (Proposition 2.1) we establish easily that any vector $x^* \in \mathfrak{K}^*_+$ ($x^* \in \mathfrak{K}^*_-$) is contained in some subspace $\mathcal{M}_+ \in \mathfrak{M}^*_+$ ($\mathcal{M}_- \in \mathfrak{M}^*_-$), and any positive (negative) vector $x^*$ is contained in a subspace from $\mathfrak{M}^{*0}_+$ ($\mathfrak{M}^{*0}_-$).

We are interested to describe the structure of the orthogonal complements of the subspaces in the classes $\mathfrak{M}_+$ and $\mathfrak{M}_-$. All the considerations below will be explicitly developed for subspaces in the class $\mathfrak{M}_+$. The corresponding results for subspaces in the class $\mathfrak{M}_-$ can be establish analogously.

**THEOREM 4.1.** *If $\mathcal{L}_+ \in \mathfrak{M}_+$ ($\mathfrak{M}^0_+$), then $\mathcal{L}^\perp_+ \in \mathfrak{M}^*_-$ ($\mathfrak{M}^{*0}_-$). When the space $\mathfrak{B}_+$ is reflexive, then for any positive subspace $\mathcal{L}_+ \in \mathfrak{M}_+$ its othogonal complement $\mathcal{L}^\perp_+$ is negative.*

◁ If $\mathcal{L}_+ \in \mathfrak{M}_+$, then $\mathcal{L}_+ = (P_+ + K_+)\mathfrak{B}_+$, where $K_+ \in \mathcal{K}_+$. Further, we easily get that $\mathcal{L}^\perp_+ = (P^*_- + Q_-)\mathfrak{B}^*_-$, where $Q_- = -K^*_+$, $Q_- : \mathfrak{B}^*_- \to \mathfrak{B}^*_+$, and $\|Q_-\| \leqslant 1$. Hence $\mathcal{L}^\perp_+ \in \mathfrak{M}^*_-$. If $\mathcal{L}_+ \in \mathfrak{M}^0_+$, i.e., $\|K_+\| < 1$, then $\|Q_-\|$ $(= \|K_+\|) < 1$, therefore $\mathcal{L}^\perp_+ \in \mathfrak{M}^{*0}_-$. Assume now that $\mathfrak{B}_+$ is reflexive and let $\mathcal{L}_+ \in \mathfrak{M}_+$ be a positive subspace, i.e., $\|K_+x_+\| < \|x_+\|$ for all $0 \neq x_+ \in \mathfrak{B}_+$. Suppose that there exists $x_-^* \in \mathfrak{B}_-^*$, with $\|x_-^*\| = 1$, and such that $\|Q_-x_-^*\| = 1$. Then $\|K_+^*x_-^*\| = 1$. Since the space $\mathfrak{B}_+$ is reflexive, we find a vector $x_+ \in \mathfrak{B}_+$ with $\|x_+\| = 1$ and such that $\langle x_+, K_+^*x_-^*\rangle = 1$, that is $\langle K_+x_+, x_-^*\rangle = 1$. From this relation and the inequality $|\langle K_+x_+, x_-^*\rangle| \leqslant \|K_+x_+\|$ it follows that $\|K_+x_+\| = 1$, which contradicts the assumption that $\mathcal{L}_+$ is positive. ▶

**REMARK 4.1.** The condition in Theorem 4.1 for $\mathfrak{B}_+$ to be a reflexive space can not be dropped, as the following example shows.

◁ Let $\ell_1$ be the space of all summable sequences of the form $x = (\ldots, x_{-k}, \ldots, \ldots, x_{-1}, x_1, \ldots, x_k, \ldots)$, with the standard basis $\{e_n : n \in \mathbb{Z}, n \neq 0\}$, where

$$e_n = (\ldots, 0, \ldots, \underbrace{1}_{n}, \ldots, 0, \ldots).$$

We consider $\mathfrak{B} = \ell_1 \dot{+} \ell_1$ such that $\mathfrak{B}_\pm = \ell_1$, and define an angle operator $K_+$ by the equalities $K_+e_n^+ = \xi_n e_{n+1}^-$, where $\xi_n = 1 - \dfrac{1}{|n|}$, $n \in \mathbb{Z}$, $n \neq 0$, and $e_n^\pm$ are the elements in the standard bases of the spaces $\mathfrak{B}_\pm$. We easily see that $\|K_+x_+\| < \|x_+\|$ for all $0 \neq x_+ \in \mathfrak{B}_+$, hence the subspace $\mathcal{L}_+ = (P_+ + K_+)\mathfrak{B}_+$ is positive. Further we have $\mathfrak{B}^* = \boldsymbol{m} \dot{+} \boldsymbol{m}$, where $\boldsymbol{m} = \ell_1^*$, and $\mathfrak{B}_\pm^* = \boldsymbol{m}$. Let $z \in \mathfrak{B}_-^*$ be the vector with the coordinates $z_k = 1$, $k \in \mathbb{Z}$, $k \neq 0$. Then, using the equalities $K_+^*e_n^- = \xi_n e_{n-1}^+$, we obtain $\|K_+^*z\| = \|z\|$. Therefore the subspace $\mathcal{M}_- = \mathcal{L}_+^\perp = (P_-^* - K_+^*)\mathfrak{B}_-^*$ contains the neutral vector $z - K_+^*z \neq 0$. ▶

The next result follows from the equality ${}^\perp\mathcal{E}^\perp = \mathcal{E}$, which holds for any subspace $\mathcal{E} \subseteq \mathfrak{B}^*$.

**COROLLARY 4.1.** *If in Theorem 4.1 we replace the space $\mathfrak{B}$ by the space $\mathfrak{B}^*$, and instead of the orthogonal complements we consider the $*$-orthogonal complements, then the conclusions remain true.*

We return now to the study of plus-operators. Recall that we are dealing only with bounded plus-operators defined on the whole space $\mathfrak{B}$.

**DEFINITION 4.1.** A plus-operator $A$ is said to be a *double plus-operator* if $A^*\mathfrak{K}_+^* \subset \subset \mathfrak{K}_+^*$.

Of course, not every plus-operator is a double plus-operator. It is sufficient to consider on the two-dimensional $J_\nu$-space $\mathfrak{B} = \mathfrak{B}_+ \dot{+} \mathfrak{B}_-$, where $\mathfrak{B}_\pm = = \text{lin}\{e_\pm : \|e_\pm\| = 1\}$, the plus-operator $A$ given — with respect to the basis

$\{e_+, e_-\}$ — by the matrix

$$A = \begin{pmatrix} \alpha & \beta \\ 0 & 0 \end{pmatrix}, \quad \alpha, \beta \in \mathbb{C}.$$

For $|\alpha| < |\beta|$ the operator $A^*$ is not a plus-operator.

In the considered example the closure of $A\mathcal{L}_+$ for some subspaces $\mathcal{L}_+ \in \mathfrak{M}_+^0$ does not belong to $\mathfrak{M}_+$.

Our next aim is to formulate conditions under which a plus-operator is a double plus-operator. We need a preliminary result.

**LEMMA 4.1.** *Let $x^*$ be a negative element of $\mathfrak{B}^*$, i.e., $\|P_+^* x^*\| < \|P_-^* x^*\|$. Then there exists a subspace $\mathcal{L}_+ \in \mathfrak{M}_+^0$ such that $\langle \mathcal{L}_+, x^* \rangle = \{0\}$.*

◁ If $P_+^* x^* = 0$, then we can take $\mathcal{L}_+ = \mathfrak{B}_+$. Assume that $P_+^* x^* \neq 0$ and consider a vector $y = y_+ + y_-$ in $\mathfrak{B}$ such that $\langle y_+, x^* \rangle \neq 0$, $\|y_-\| = 1$, and $\langle y_-, x^* \rangle > \|P_+^* x^*\|$. Set

$$\mathcal{L}_+ = N(x^*) \cap \mathfrak{B}_+ \dot{+} \operatorname{lin}\left\{ y_+ - \frac{\langle y_+, x^* \rangle}{\langle y_-, x^* \rangle} y_- \right\}.$$

Then $P_+ \mathcal{L}_+ = \mathfrak{B}_+$ and $\langle \mathcal{L}_+, x^* \rangle = \{0\}$. We show that $\mathcal{L}_+ \in \mathfrak{M}_+^0$.

Let $x = x_+ + x_- = z + \alpha \left( y_+ - \dfrac{\langle y_+, x^* \rangle}{\langle y_-, x^* \rangle} y_- \right)$, where $\|x_+\| = 1$, $z \in N(x^*) \cap \cap \mathfrak{B}_+$, $\alpha \in \mathbb{C}$. We have $|\langle x_+, x^* \rangle| = |\alpha| \, |\langle y_+, x^* \rangle| \leqslant \|P_+^* x^*\|$. Thus

$$\|x_-\| = |\alpha| \left| \frac{\langle y_+, x^* \rangle}{\langle y_-, x^* \rangle} \right| \leqslant \|P_+^* x^*\| \cdot \langle y_-, x^* \rangle^{-1} < 1. \qquad ▶$$

**THEOREM 4.2.** *If a plus-operator $A$ on $\mathfrak{B}$ satisfies the condition*

$$\overline{A\mathcal{L}_+} \in \mathfrak{M}_+ \quad \textit{for all } \mathcal{L}_+ \in \mathfrak{M}_+^0, \tag{4.1}$$

*then $A$ is a double plus-operator.*

◁ Let us consider first a positive element $x^* \in \mathfrak{K}_+^*$ and assume that $A^* x^* \notin \notin \mathfrak{K}_+^*$. Then $\|P_+^* A^* x^*\| < \|P_-^* A^* x^*\|$, i.e., $A^* x^*$ is a negative element in $\mathfrak{B}^*$. By Lemma 4.1 there exists $\mathcal{L}_+ \in \mathfrak{M}_+^0$ such that $\langle \mathcal{L}_+, A^* x^* \rangle = \{0\}$. Then $\langle A\mathcal{L}_+, x^* \rangle = = \{0\}$, a relation which, by (4.1), contradicts a conclusion of Theorem 4.1. Therefore our assumption is false, hence $A^* x^* \in \mathfrak{K}_+^*$ for all positive elements of $\mathfrak{B}^*$.

Take now an arbitrary vector $0 \neq y^* \in \mathfrak{K}_+^*$. Then for all $y_n^* = \left(1 + \dfrac{1}{n}\right) P_+^* y^* + +P_-^* y^*$ we have $\|P_+^* y_n^*\| > \|P_-^* y_n^*\|$, i.e., each $y_n^*$ is a positive element. From the first part of the proof it follows that $A^*y_n^* \in \mathfrak{K}_+^*$. Taking the limit as $n \to \infty$, we obtain $A^*y^* \in \mathfrak{K}_+^*$. ▶

If a plus-operator $A$ has the property that $\overline{A\mathcal{L}_+} \in \mathfrak{M}_+$ for all $\mathcal{L}_+ \in \mathfrak{M}_+$, then, generally speaking, the adjoint operator $A^*$ fails to have the analogous property.

**EXAMPLE 4.1.** Let $\mathfrak{B} = \mathfrak{B}_+ \dot{+} \mathfrak{B}_-$, where the space $\mathfrak{B}_+$ is reflexive and has a Schauder basis $\{e_i \ : \ i \in \mathbb{N}\}$. We define a plus-operator $A$ on $\mathfrak{B}$ in the following way: on $\mathfrak{B}_+$ we put $Ae_1 = 0$, $Ae_i = e_{i-1}$, $i = 2, 3, \ldots$, and on $\mathfrak{B}_-$ we set $A \equiv 0$. Then the operator $A$ is bounded, and $\overline{A\mathcal{L}_+} \in \mathfrak{M}_+$ for all $\mathcal{L}_+ \in \mathfrak{M}_+$. For $A^*$ we have $A^*e_i = e_{i+1}$, $i = 1, 2, \ldots$, i.e., $\overline{A^*\mathcal{M}_+} \notin \mathfrak{M}_+^*$ for $\mathcal{M}_+ \in \mathfrak{M}_+^*$

However, if we add the next condition: "$A$ maps the non-zero vectors from $\mathfrak{B}_+$ into positive vectors", then the property "$\overline{A\mathcal{L}_+} \in \mathfrak{M}_+$ for all $\mathcal{L}_+ \in \mathfrak{M}_+$" is symmetric. More precisely, we have:

**THEOREM 4.3.** *Let $\mathfrak{B}_+$ be a reflexive space and assume that $A(\mathfrak{B}_+ \setminus \{0\}) \subset \mathfrak{K}_{++}$. If $\overline{A\mathcal{L}_+} \in \mathfrak{M}_+$ for all $\mathcal{L}_+ \in \mathfrak{M}_+^0$, then $\overline{A^*\mathcal{M}_+} \in \mathfrak{M}_+^*$ for all $\mathcal{M}_+ \in \mathfrak{M}_+^*$.*

◁ Assume that $\overline{A^*\mathcal{M}_+} \notin \mathfrak{M}_+^*$ for some $\mathcal{M}_+ \in \mathfrak{M}_+^*$. Then, by the reflexivity of $\mathfrak{B}_+$, there exists $0 \neq x_+ \in \mathfrak{B}_+$ such that $\langle x_+, A^*\mathcal{M}_+\rangle = \{0\}$, or, equivalently, $\langle Ax_+, \mathcal{M}_+\rangle = \{0\}$. By Theorem 4.1 we get that $Ax_+ \in \mathfrak{K}_-$, but, according to our assumptions, we must have $Ax_+ \in \mathfrak{K}_{++}$. The obtained contradiction proves the theorem. ▶

Let us consider a $J_p$-space $\mathfrak{B}$, with $p > 1$. We introduce on $\mathfrak{B}$ a new norm, equivalent to the initial one, by

$$\|x\|_p = (\|x_+\|^p + \|x_-\|^p)^{\frac{1}{p}}.$$

Correspondingly, we introduce on $\mathfrak{B}^*$ the conjugate norm $\|x^*\|_q = \left(\|x_+^*\|^q + \|x_-^*\|^q\right)^{\frac{1}{q}}$, where $q = \dfrac{p}{p-1}$, and define a $J_q$-metric on $\mathfrak{B}^*$ by the relation

$$J_q(x^*) = \|x_+^*\|^q - \|x_-^*\|^q, \quad x^* = x_+^* + x_-^* \in \mathfrak{B}^*.$$

**DEFINITION 4.2.** A strict plus-operator $A$ on the $J_p$-space $\mathfrak{B}$ is called *double-strict* if $A^*$ is a strict plus-operator on the $J_q$-space $\mathfrak{B}^*$.

**DEFINITION 4.3.** A strict plus-operator $A$ is said to be *focusing*, if

$$\|P_-Ax\| \leqslant (1 - \tilde{\gamma}(A))\|P_+Ax\|,$$

for all $x \in \mathfrak{K}_+$, where $\tilde{\gamma}(A) > 0$. (Compare with Definition 3.4.)

A focusing operator $A$ is said to be *double-focusing*, if $A^*$ is also a focusing plus-operator in the $J_q$-space $\mathfrak{B}^*$.

In the case of a Hilbert $J$-space $\mathfrak{H}$ any strict plus-operator $A$ with the property

$$A\mathcal{L}_+^0 \in \mathfrak{M}_+ \quad \text{for some } \mathcal{L}_+^0 \in \mathfrak{M}_+ \tag{4.2}$$

is double-strict and $\mu(A) = \mu(A^*)$ (see M. G. Krein and Y. A. Shmulian [1]).

**LEMMA 4.2.** *In a $J_p$-space $\mathfrak{B}$ every strict plus-operator $A$ with property (4.2) satisfies the condition*

$$\|A_{11}^{-1}A_{12}\| < 1. \tag{4.3}$$

◁ Assume that the conclusion is false. Then we find a sequence $\{x_-^n : n \in \mathbb{N}\}$ such that $x_-^n \in \mathfrak{B}_-$, $\|x_-^n\| = 1$ and $\|A_{11}^{-1}A_{12}x_-^n\| \to 1$ for $n \to \infty$. Set $\alpha_n = \|A_{11}^{-1}A_{12}x_-^n\|$ and $x_n = \alpha_n^{-1}A_{11}^{-1}A_{12}x_-^n - x_-^n$. Then $x_n \in \mathfrak{K}_0$ ($\subset \mathfrak{K}_+$) and

$$P_+Ax_n = \alpha_n^{-1}A_{12}x_-^n - A_{12}x_-^n = \frac{1-\alpha_n}{\alpha_n}A_{12}x_-^n \to 0,$$

for $n \to \infty$.

On the other hand, since the operator $A$ is strict, by Lemma 3.3 it follows that $\|Ax\| \geqslant \delta(A)\|x\|$ for all $x \in \mathfrak{K}_+$. Hence

$$\|P_+Ax_n\| \geqslant \frac{1}{2}\|Ax_n\| \geqslant \frac{\delta(A)}{2}\|x_n\| \geqslant \frac{\delta(A)}{2\|P_+\|} > 0.$$

So we obtained a contradiction. ▶

**THEOREM 4.4.** *Every focusing strict plus-operator $A$ with property (4.1) on a reflexive $J_p$-space $\mathfrak{B}$ is double-focusing and double-strict.*

◁ With respect to the basis $\{\mathfrak{B}_+, \mathfrak{B}_-\}$ the operator $A$ has the matrix representation

$$\begin{pmatrix} A_{11} & A_{12} \\ A_{21} & A_{22} \end{pmatrix},$$

where $A_{11} = P_+AP_+$, $A_{12} = P_+AP_-$, $A_{21} = P_-AP_+$, $A_{22} = P_-AP_-$. By Theorem 3.4, from condition (4.2) it follows that $A_{11} : \mathfrak{B}_+ \to \mathfrak{B}_+$ is a homeomorphism. Relatively to the basis $\{\mathfrak{B}_+^*, \mathfrak{B}_-^*\}$ the operator $A^*$ has the matrix representation

$$\begin{pmatrix} A_{11}^* & A_{12}^* \\ A_{21}^* & A_{22}^* \end{pmatrix}.$$

Let us show that for the operator $A^*$ the condition

$$\left\|(A_{11}^*)^{-1} A_{21}^*\right\| < 1, \tag{4.4}$$

analogous to (4.3), is fulfilled. Indeed, since $A$ is focusing it follows that $\|P_- Ax\| \leqslant \beta\|P_+ Ax\|$, for all $x \in \mathfrak{K}_+$, where $\beta < 1$. Hence we have

$$\|(A_{21} + A_{22}K_+)x_+\| \leqslant \beta\|(A_{11} + A_{12}K_+)x_+\|,$$

for all $x_+ \in \mathfrak{B}_+$ and any angle operator $K_+ \in \mathcal{K}_+$.

Setting $K_+ = 0$ and considering the adjoint operators we obtain $\|A_{21}^*\| \leqslant \beta\|A_{11}^*\|$, a relation equivalent to (4.4).

By the reflexivity of $\mathfrak{B}$, the angle operators of the subspaces in $\mathfrak{M}_+^*$ are exactly the adjoints of the angle operators of the subspaces in $\mathfrak{M}_-$. Next we show that

$$\left\|\left(A_{12}^* + A_{22}^*K_-^*\right)\left(A_{11}^* + A_{21}^*K_-^*\right)^{-1}\right\| \leqslant \beta^* < 1, \tag{4.5}$$

where $K_- \in \mathcal{K}_-$. Let us suppose, on the contrary, the existence of a sequence $\{K_-^n : n \in \mathbb{N}\}$, such that $K_-^n \in \mathcal{K}_-$ and $\|\hat{K}_-^{n*}\| \to 1$ as $n \to \infty$, where

$$\hat{K}_-^{n*} = \left(A_{12}^* + A_{22}^*K_-^{n*}\right)\left(A_{11}^* + A_{21}^*K_-^{n*}\right)^{-1}.$$

Then we find elements $y_-^n \in \mathfrak{B}_-$ such that $\|y_-^n\| = 1$ and $\delta_n = \|\hat{K}_-^n y_-^n\| \to 1$ as $n \to \infty$. Setting $x^n = \hat{K}_-^n y_-^n - \delta_n y_-^n$ we obtain $\|x_+^n\| = \|x_-^n\|$, i.e., $x^n \in \mathfrak{K}_0$ $(\subset \mathfrak{K}_+)$. Hence — using the fact that $A$ is focusing — and from (4.3) it follows that $\|P_- Ax^n\| \leqslant (1 - \tilde{\gamma}(A))\|P_+ Ax^n\|$ and $\|P_+ Ax^n\| \geqslant \varepsilon > 0$, for $n \geqslant n_0$. Setting $y^n = \hat{K}_-^n y_-^n - y_-^n$, we obtain $x^n - y^n \to 0$, whence

$$\|P_- Ay^n\| < \|P_+ Ay^n\|, \quad n \geqslant m. \tag{4.6}$$

Let $\mathcal{M}_+^n = \left(P_+^* + K_-^{n*}\right)\mathfrak{B}_+^*$. Then $A^*\mathcal{M}_+^n = \left(P_+^* + K_-^{n*}\right)\mathfrak{B}_+^*$. Further

$$\langle Ay^n, \mathcal{M}_+^n\rangle = \langle y^n, A^*\mathcal{M}_+^n\rangle = \left\langle\left(P_- + \hat{K}_-^n\right)y_-^n, \left(P_+^* + \hat{K}_-^{n*}\right)\mathfrak{B}_+^*\right\rangle = \{0\},$$

which, by (4.6), contradicts condition $\mathcal{M}_+^n \in \mathfrak{M}_+^*$; so (4.5) is proved. It remains to show that the plus-operator $A^*$ is strict. For any $x^* \in \mathfrak{K}_+^*$ we have:

$$J_q(A^*x^*) = \|P_+^*A^*x^*\|^q - \|P_-^*A^*x^*\|^q \geqslant (1 - \beta^{*q})\left\|\left(A_{11}^* + A_{21}^*K_-^*\right)x_+^*\right\|^q =$$
$$= (1 - \beta^*)^q\left\|A_{11}^*\left(I + (A_{11}^*)^{-1}A_{21}^*K_-^*\right)x_+^*\right\|^q \geqslant \xi\left\|x_+^*\right\|^q \geqslant \xi J_q(x^*),$$

where $\xi = (1-\beta^*)^q \left\|A_{11}^{-1}\right\|^{-q} (1-\beta)^q > 0$. ▶

Theorem 4.4 assures that any focusing strict plus-operator on a reflexive space, which satisfies condition (4.2), is collinear to a double uniformly $J_p^+$-expansive operator, i.e., to an operator which is uniformly $J_p^+$-expansive simultaneously with its adjoint.

To conclude this section we take into consideration some conditions under which a $J_p$-expansive operator $A$ is double $J_p$-expansive, i.e., both $A$ and $A^*$ are $J_p$-expansive.

**LEMMA 4.3.** *Let $A$ be a strict plus-operator on the $J_p$-space $\mathfrak{B}$. Then the operators $P_- \pm AP_+$ and $P_- \pm P_+A$ are continuously invertible.*

◁ We consider the operators $P_- \pm P_+A$. Assume that $\|x^n\| = 1$ and $(P_- \pm \pm P_+A)x^n \to 0$ as $n \to \infty$. Then $x_-^n \to 0$ and $P_+Ax^n \to 0$. Since $\|x^n\| = 1$ we have $\|x_+^n\| \geqslant a > 0$ for $n \geqslant n_0$. Further, $P_+Ax^n = A_{11}x_+^n + A_{12}x_-^n$ and, since $A_{12}x_-^n \to 0$, then $A_{11}x_+^n \to 0$, too. On the other hand, by Lemma 3.3, we have $\|A_{11}x_+^n\| \geqslant \delta\|x_+^n\| \geqslant \delta a > 0$, for $n \geqslant n_0$, a contradiction. Hence it follows that $\|(P_- \pm \pm P_+A)x\| \geqslant \xi_\pm \|x\|$, for all $x \in \mathfrak{B}$, where $\xi_\pm > 0$. By Theorem 0.4.5 we conclude that the operators $(P_- \pm P_+A)^{-1}$ exist and $\|(P_- \pm P_+A)^{-1}\| \leqslant \xi_\pm^{-1}$.

Assume now that $(P_- \pm AP_+)y^n \to 0$ as $n \to \infty$, where $\|y^n\| = 1$. Then $y_-^n + P_-Ay_+^n \to 0$ and $P_+Ay_+^n \to 0$. As above, from Lemma 3.3 it follows that $y_+^n \to 0$. This means that $P_-Ay_+^n \to 0$, whence $y_-^n \to 0$, too. Thus we obtain again that $\|(P_- + AP_+)y\| \geqslant \eta_\pm \|y\|$ for all $y \in \mathfrak{B}$, where $\eta_\pm > 0$. Hence the operators $(P_- \pm AP_+)^{-1}$ exist and $\|(P_- \pm AP_+)^{-1}\| \leqslant \eta_\pm^{-1}$. ▶

**LEMMA 4.4.** *If $A$ is a double-strict plus-operator on a $J_p$-space $\mathfrak{B}$, then*

$$P_+AP_+\mathfrak{B}_+ = \mathfrak{B}_+. \tag{4.7}$$

We notice that for a strict plus-operator on a $J_p$-space $\mathfrak{B}$ condition (4.7) is equivalent to condition (4.2).

◁ By Lemma 4.3 the operator $P_- + AP_+$ is a homeomorphism and therefore the lineal $(P_- + AP_+)\mathfrak{B}$ is closed. If $(P_- + AP_+)\mathfrak{B} \neq \mathfrak{B}$, then we find an element $x^* \neq 0$ in $\mathfrak{B}^*$ such that $\langle (P_- + AP_+)\mathfrak{B}, x^* \rangle = \{0\}$, that is $\left(P_-^* + P_+^*A^*\right)x^* = 0$. But $A^*$ is a strict plus-operator on $\mathfrak{B}^*$, hence, by Lemma 4.3, the operator $P_-^* + P_+^*A^*$ is invertible. It follows that $x^* = 0$ — a contradiction. Thus $(P_- + AP_+)\mathfrak{B} = \mathfrak{B}$, a relation which clearly leads to (4.7). ▶

**LEMMA 4.5.** *Let $A$ be a $J_p$-expansive operator on $\mathfrak{B}$. Then the operator*

$$U = (P_+ - P_-A)(P_- - P_+A)^{-1} \tag{4.8}$$

*is well-defined and* $\|Ux\|_p \leqslant \|x\|_p$ *for all* $x \in \mathfrak{D}(U)$.

◁ By Lemma 4.3 the operator $U$ is well-defined. Let $x \in \mathfrak{D}(U)$. Then $x = (P_- - P_+A)y$, where $y \in \mathfrak{B}$, and $Ux = (P_+ - P_-A)y$. Hence and by $J_p(Ay) \geqslant J_p(y)$ it follows that

$$\|Ux\|_p^p = \|P_+y\|_p^p + \|P_-Ay\|_p^p \leqslant \|P_-y\|_p^p + \|P_+Ay\|_p^p =$$
$$= \|(P_- - P_+A)y\|_p^p = \|x\|_p^p. \quad ▶$$

**LEMMA 4.6.** *If a strict plus-operator $A$ on the $J_p$-space $\mathfrak{B}$ satisfies condition (4.7), then the operator $U$ given by (4.8) is defined on the whole $\mathfrak{B}$ and the following inversion formulas are true:*

$$U = (P_+ - P_-A)(P_- - P_+A)^{-1} = -(P_- + AP_+)^{-1}(P_+ + AP_-), \tag{4.9}$$
$$A = -(P_+ + P_-U)(P_- + P_+U)^{-1} = (P_- - UP_+)^{-1}(P_+ - UP_-). \tag{4.10}$$

◁ We prove first that $(P_- - P_+A)\mathfrak{B} = \mathfrak{B}$. Let $y = y_+ + y_- \in \mathfrak{B}$ be fixed. The equation $y = (P_- - P_+A)x$ is equivalent to $y_+ = -P_+Ax$ and $y_- = x_-$, whence $-y_+ = A_{11}x_+ + A_{12}y_-$. Therefore, using (4.7) it follows that $x_+ = -A_{11}^{-1}y_+ - A_{11}^{-1}A_{12}y_-$.

The equality $(P_- + AP_+)\mathfrak{B} = \mathfrak{B}$ can be proved analogously. For any $y = y_+ + y_- \in \mathfrak{B}$ we find $x \in \mathfrak{B}$ such that $y = (P_- + AP_+)x$. We have $y_+ = A_{11}x_+$ and $y_- = x_- + A_{21}x_+$. From (4.7) it follows that $x_+ = A_{11}^{-1}y_+$, whence $x_- = y_- - A_{21}A_{11}^{-1}y_+$. By Lemma 4.3 the operators $P_- - P_+A$ and $P_- + AP_+$ map homeomorphically $\mathfrak{B}$ onto $\mathfrak{B}$.

Let us prove equalities (4.9). The first equality is the definition of $U$. For the second equality we have

$$(P_- + AP_+)^{-1}(P_+ + AP_-) + (P_+ - P_-A)(P_- - P_+A)^{-1} =$$
$$= (P_- + AP_+)^{-1}((P_+ + AP_-)(P_- - P_+A) + (P_- + AP_+)(P_+ - P_-A))(P_- - P_+A)^{-1} =$$
$$= (P_- + AP_+)^{-1}(AP_- - P_+A - P_-A + AP_+)(P_- - P_+A)^{-1} = 0.$$

We establish now the first equality in (4.10), i.e., $A = -(P_+ + P_-U)(P_- + P_+U)^{-1}$. By (4.8) we have

$$Ux = (P_+ - P_-A)(P_- - P_+A)^{-1}x,$$

i.e., $x = (P_- - P_+A)z$, $z \in \mathfrak{B}$ and $Ux = (P_+ - P_-A)z$. Hence we get the following relations

$$U(P_- - P_+A)z = (P_+ - P_-A)z,$$
$$-P_+U(P_- - P_+A)z = -P_+z,$$

and

$$(P_- - P_+U)(P_- - P_+A)z = P_-z - P_+z = -Jz,$$

where $J = P_+ - P_-$. As it was already proved above $(P_- - P_+A)\mathfrak{B} = \mathfrak{B}$. By Lemma 4.3 the operator $(P_- - P_+A)^{-1}$ exists and is bounded. Therefore, by (4.8), $P_+(P_- - P_+A)^{-1}P_+\mathfrak{B}_+ = P_+UP_+\mathfrak{B}_+ = \mathfrak{B}_+$. We establish, as above, that $(P_- - P_+U)\mathfrak{B} = \mathfrak{B}$ and that the operator $P_- - P_+U$ is continuously invertible. Therefore $(P_- - P_+A)z = -(P_- - P_+U)Jz$ and $-P_+Az = -(P_- - P_+U)^{-1}Jz - P_-z$. On the other hand $-P_-Az = U(P_- - P_+A)z - P_+z = -U(P_- - P_+U)^{-1}Jz - P_+z$. Then

$$Az = (P_- - P_+U)^{-1}Jz + P_-z + P_+z + U(P_- - P_+U)^{-1}Jz = z + (U+I)(P_- - P_+U)^{-1}Jz.$$

But

$$\begin{aligned} I + (U+I)(P_- - P_+U)^{-1}J &= I - (U+I)(P_- + P_+U)^{-1} = \\ = ((P_- + P_+U) - (U+I))(P_- + P_+U)^{-1} &= -(P_+ + P_-U)(P_- + P_+U)^{-1}. \end{aligned}$$

So the equality $A = -(P_+ + P_-U)(P_- + P_+U)^{-1}$ is established.

Finally, the second equality in (4.10) can be proved analogously with the second equality in (4.9). ▶

**THEOREM 4.5.** *A $J_p$-expansive operator $A$ is double $J_p$-expansive if and only if it satisfies condition* (4.7).

◁ The necessity of condition (4.7) was established in Lemma 4.4. We prove its sufficiency. By Lemma 4.6 the inversion formulas (4.9)–(4.10) are true. By Lemma 4.5 it follows that $\|Ux\|_p \leqslant \|x\|_p$, $x \in \mathfrak{B}$. This means that $\|U^*x^*\|_q \leqslant \|x^*\|_q$, $x^* \in \mathfrak{B}^*$. Taking the adjoint operators in (4.9)–(4.10) we obtain

$$\begin{aligned} U^* &= \left(P_-^* - A^*P_+^*\right)^{-1}\left(P_+^* - A^*P_-^*\right) = \\ &= -\left(P_+^* + P_-^*A^*\right)\left(P_-^* + P_+^*A^*\right)^{-1}, \end{aligned} \tag{4.11}$$

$$\begin{aligned} A^* &= \left(P_+^* - P_-^*U^*\right)\left(P_-^* - P_+^*U^*\right)^{-1} = \\ &= -\left(P_-^* + U^*P_+^*\right)^{-1}\left(P_+^* + U^*P_-^*\right). \end{aligned} \tag{4.12}$$

From (4.11) we get $x^* = \left(P_-^* + P_+^*A^*\right)y^*$, $y^* \in \mathfrak{B}^*$, $U^*x^* = -\left(P_+^* + P_-^*A^*\right)y^*$. Therefore

$$\left\|P_+^*y^*\right\|_q^q + \left\|P_-^*A^*y^*\right\|_q^q = \|U^*x^*\|_q^q \leqslant \|x^*\|_q^q = \left\|P_-^*y^*\right\|_q^q + \left\|P_+^*A^*y^*\right\|_q^q,$$

and, consequently, $J_q(A^*y^*) \geqslant J_q(y^*)$. ▶

## §5. The problem of invariant semi-definite subspaces

In 1943 S. L. Sobolev concluded his work [1] "About the motion of a symmetrical top whose cavity is filled with a liquid".

The motion of the top is described by an equation of the form

$$\frac{\mathrm{d}R}{\mathrm{d}t} = \mathrm{i}BR + R_0, \quad R(t=0) = R^{(0)}, \tag{5.1}$$

where $B$ is a linear operator and $R$ is a vector in the phase (complex infinite dimensional) space. The solution of this equation is given by

$$R = \mathrm{e}^{\mathrm{i}Bt}R^{(0)} + \int_0^t \mathrm{e}^{\mathrm{i}B(t-t_1)}R_0(t_1)\mathrm{d}t_1,$$

since $R(t=0) = R^{(0)}$. In this case the operator $B$ is considered to be bounded, so that, for sufficiently large absolute values of $\lambda \in \mathbb{C}$, there exists the resolvent $\Gamma_\lambda = (\lambda I - B)^{-1}$. Then

$$\mathrm{e}^{\mathrm{i}Bt} = \frac{1}{2\pi\mathrm{i}} \int_\gamma \mathrm{e}^{\mathrm{i}\lambda t}\Gamma_\lambda \mathrm{d}\lambda,$$

where $\gamma$ is a circle of a sufficiently large radius centered at $0 \in \mathbb{C}$. A Hermitean form $Q(R_1, R_2)$ can be constructed on the phase space $\{R\}$. Its behaviour is determined by the quantity

$$L = C_1 + C_2 - A_1 - A_2 - \frac{k}{\omega_2} \quad (k = g(l_1\mathcal{M}_1 + l_2\mathcal{M}_2)),$$

where: a) $k$ is the rotational momentum of the gravitational force and depends on the angle which expresses the deviation of the $z$-axis relatively to the vertical direction;

b) $\omega$ is the angular speed of the rotation of the top around its axis (the $z$-axis coincide with the symmetry axis of the top and the origin of the coordinate axes is considered at the fixed point);

c) $A_1$ is the innertial momentum of the shell relatively to the $x$- and $y$-axis associated to the top, and $A_2$ is the innertial momentum of the liquid relatively to the same axes;

d) $C_1$ and $C_2$ are the innertial momenta of the shell and of the liquid relatively to the $z$-axis;

e) $l_1$ and $l_2$ are the distances between the fixed point and the centers of mass of the shell and of the liquid, respectively;

f) $g$ is the gravitational acceleration;

g) $\mathcal{M}_1$ and $\mathcal{M}_2$ are the masses of the shell and of the liquid, respectively.

If $L > 0$, then the form $Q$ is positive definite, and if $L < 0$, then $Q$ is an indefinite form with one negative square.

The operator $B$ is self-adjoint relatively to the form $Q$. Therefore, for $L > 0$ its spectrum is real and $e^{iBt}$ is a bounded operator. If $L < 0$ then the operator $B$ can have a non-real spectrum. In this last case there exists a pair of complex-conjugate eigenvalues, coresponding to a pair of neutral eigenvalues. The presence — or the absence — of these non-real eigenvalues has a determining influence in the stability of the solution for equation (5.1).

Extending Sobolev's investigations, L. S. Pontryagin published in 1944 his work [1], devoted to the study of self-adjoint operators relatively to a Hermitean form with $\kappa < \infty$ negative squares. Let $A$ be such an operator. Using some subtle algebraic methods the existence of a pair $(\mathcal{L}_+, \mathcal{L}_-)$ of invariant subspaces for $A$ can be established, where $\mathcal{L}_+$ is a maximal non-negative subspace, and $\mathcal{L}_-$ is a maximal non-positive subspace, with the following properties concerning the spectra of $A \mid \mathcal{L}_+$ and $A \mid \mathcal{L}_-$: the eigenvalues in $\sigma(A \mid \mathcal{L}_+)$ and $\sigma(A \mid \mathcal{L}_-)$ are two by two complex-conjugate, and the imaginary parts of these eigenvalues are non-negative for $\mathcal{L}_+$ and non-positive for $\mathcal{L}_-$. As a recognition of Pontriagyn's contributions, the Hilbert spaces with an indefinite metric having a finite number $\kappa$ of positive or negative squares are called *Pontriagyn spaces*. Usually they are denoted by $\Pi_\kappa$.

By the end of the 40's and the beginning of the 50's, I. S. Iohvidov [1] using the Caley-Neumann transform, established the relationship between $\pi$-self-adjoint and $\pi$-unitary operators and proved that for a $\pi$-unitary operator $U$ there exists a pair of invariant maximal semi-definite subspaces $(\mathcal{L}_+, \mathcal{L}_-)$ such that the spectra of $U \mid \mathcal{L}_+$ and $U \mid \mathcal{L}_-$ lie outside and inside the unit disk, respectively.

In 1950 M. G. Krein's paper [1] appeared, where a theorem on the existence of an invariant subspace for a plus-operator on a space $\Pi_\kappa$ was proved by an ingenious use of Schauder's fixed point principle. Later on the Iohvidov-Krein Theorem was extended by M. L. Brodskii [1] to the case of a plus-operator acting on a Banach $J_\nu$-space $\mathfrak{B}$ with $\dim \mathfrak{B}_+ < \infty$

Afterwards, the invariant subspace problem for the general case of an operator acting on a space with indefinite metric, which has both the number of positive and negative squares infinite, was considered. The axioms of such spaces were elaborated by M. G. Krein [3], and, in his honour, they are called *Krein spaces*. M. G. Krein achieved also an important step in solving the invariant subspace problem. In 1964 his paper [2] was published, in which, using once again a fixed point principle, he established the existence of an invariant maximal non-negative subspace $\mathcal{L}_+$ for a plus-operator $A$ with a completely continuous "corner" $P_+AP_-$, and studied the spectrum of the restriction $A \mid \mathcal{L}_+$.

On the other hand, in 1960 R. S. Phillips [1] considered the problem of extending a pair of semi-definite subspaces which are invariant relatively to some commutative operator algebra $\mathfrak{A}$, to a pair of maximal semi-definite subspaces, by preserving the invariance relatively to $\mathfrak{A}$. To be more specific, this problem occurred in connection with the Cauchy problem for a dissipative system

$$\frac{\mathrm{d}x^i}{\mathrm{d}t} = \sum_{j=1}^{\infty} a_{ij}x^j, \quad x^i(0) = c^i, \ i = 1, 2, \ldots,$$

under the condition that the real part of the matrix $(a_{n_i n_j})$ is non-positive for any finite collection $\{n_i\}$ of natural numbers and where

$$\sum_{j=1}^{\infty} |a_{ij}|^2 < \infty \quad \text{for every } i \in \mathbb{N},$$

$$\sum_{i=1}^{\infty} |a_{ij}|^2 < \infty \quad \text{for every } j \in \mathbb{N}.$$

This type of problems arise in the theory of small perturbations of the dissipative systems with infinitely many degrees of freedom. The precise statement of Phillips' problem is the following. Let $\mathfrak{H}$ be a Krein space with the $J$-metric $[\cdot, \cdot]$, and let $\mathfrak{A}$ be a commutative algebra of operators acting on $\mathfrak{H}$, closed in the weak operator topology, and such that, together with any $A \in \mathfrak{A}$ the algebra $\mathfrak{A}$ contains the operator $A^c$, the adjoint of $A$ relatively to the $J$-metric. Let $(\mathcal{L}_+, \mathcal{L}_-)$ be a pair of subspaces which are invariant relatively to all operators in $\mathfrak{A}$, such that $\mathcal{L}_\pm \subset \mathfrak{K}_\pm$ and $[x, y] = 0$ for all $x \in \mathcal{L}_+$, $y \in \mathcal{L}_-$ (i.e., $\mathcal{L}_+$ and $\mathcal{L}_-$ are $J$-orthogonal). The problem is to find out whether there exists a pair $(\mathcal{L}_+^{\max}, \mathcal{L}_-^{\max})$ of invariant relatively to $\mathfrak{A}$ subspaces such that $\mathcal{L}_\pm^{\max} \in \mathfrak{M}_\pm$, $\mathcal{L}_\pm \subseteq \mathcal{L}_\pm^{\max}$, and $[\mathcal{L}_+^{\max}, \mathcal{L}_-^{\max}] = \{0\}$. Phillips reformulated this problem for a commutative group of $J$-unitary operators. In his paper [1] he stated the conjecture that the indicated extension of the subspaces is always possible. He proved that the conjecture is true only in the case of a finite-dimensional space $\mathfrak{H}$.

Soon after that, M. A. Naimark [1] proved that Phillips conjecture is true for commutative groups of $J$-unitary operators on a Pontriagyn space $\Pi_\kappa$. Afterwards, some advances were reached in a series of works (as, for example, V. S. Shulman [1], E. A. Larionov [1], [2], H. Langer [1], [2], V. A. Khatskevich [1]). On one hand the extension problem for invariant subspaces was solved for larger (even non-commutative) families of operators acting, as above, on a Pontriagyn space $\Pi_\kappa$ (and, sometimes, on $\Pi_1$). On the other hand, some restrictions were found for commutative groups of $J$-unitary operators on an arbitrary Krein space for which Phillips conjecture has a positive answer.

In 1970 the work of J. W. Helton [1] appeared. There the existence of a maximal semi-definite invariant subspace was proved for any commutative group $\mathfrak{U}$ of $J$-unitary operators on a Krein space which satisfies the next condition: there exists an operator $U$ in $\mathfrak{U}$ represented — relatively to the decomposition $\mathfrak{H} = \mathfrak{H}_+ \oplus \mathfrak{H}_-$ — by a matrix of the form

$$U = \begin{pmatrix} U_{11} & 0 \\ 0 & U_{22} \end{pmatrix} + C,$$

where $C$ is a completely continuous operator, and the spectra of $U_{11}$ and $U_{22}$ are disjoint, i.e., $\sigma(U_{11}) \cap \sigma(U_{22}) = \emptyset$.

Every group of $\pi$-unitary operators on a Pontriagyn space $\Pi_\kappa$ satisfies this condition since it contains the identity operator

$$I = \begin{pmatrix} I_+ & 0 \\ 0 & I_- \end{pmatrix} = \begin{pmatrix} I_+ & 0 \\ 0 & -I_- \end{pmatrix} + 2P_- = \begin{pmatrix} -I_+ & 0 \\ 0 & I_- \end{pmatrix} + 2P_+,$$

and either $P_+$ or $P_-$ is completely continuous.

Based on Helton's result, V. A. Khatskevich [2] proved that Phillips conjecture is true for such a group $\mathfrak{U}$. Improving Helton's ideas, T. Ya. Azizov considered the so-called $\mathcal{H}$ and $\mathcal{K}(\mathcal{H})$ classes of double $J$-expansive operators and solved the invariant subspaces extension problem. His results in this direction are summed up in T. Ya. Azizov and I. S. Iohvidov's book [1].

Other papers, known to us, in which the mentioned problem was solved for some families of operators acting this time on a $J_\nu$-space are: I. S. Iohvidov [1], V. A. Khatskevich [3], H. Langer [3].

In its general form Phillips' problem is not yet solved. Moreover, the problem of the existence of a maximal semi-definite invariant subspace for an arbitrary $J$-unitary operator on a Krein space $\mathfrak{H}$ with infinite dimensional subspaces $\mathfrak{H}_+$ and $\mathfrak{H}_-$ is unsolved too.

In the next section we will apply the results of Chapters V and VI about fixed points for holomorphic operators to the problem of invariant subspaces for a plus-operator on a $J_\nu$-space.

## §6. An application of fixed point principles for holomorphic operators to the invariant semi-definite subspace problem

Let $\mathfrak{B} = \mathfrak{B}_+ \dot{+} \mathfrak{B}_-$ be a Banach space with the $J_p$-norm

$$J_p(x) = \|x_+\|^p - \|x_-\|^p,$$

where $x = x_+ + x_- \in \mathfrak{B}$, $x_\pm \in \mathfrak{B}_\pm$ and $p > 1$. We consider a bounded strict plus-operator $A$ defined on the whole space $\mathfrak{B}$ and satisfying condition (4.7): $P_+AP_+\mathfrak{B}_+ = = \mathfrak{B}_+$. By Corollary 3.2 the operator $A$ maps any subspace $\mathcal{L}_+ \in \mathfrak{M}_+$ onto the subspace $A\mathcal{L}_+$ which belongs to $\mathfrak{M}_+$, too.

Let now $K_+ \in \mathcal{K}_+$. Then the subspace $\mathcal{L}_+ = (P_+ + K_+)\mathfrak{B}_+$ belongs to $\mathfrak{M}_+$, therefore $A\mathcal{L}_+ \in \mathfrak{M}_+$. By Theorem 2.1 we know that $A\mathcal{L}_+ = (P_+ + \hat{K}_+)\mathfrak{B}_+$, where $\hat{K}_+ = P_-(P_+ \mid A\mathcal{L}_+)^{-1} \in \mathcal{K}_+$. We have:

$$P_+A\mathcal{L}_+ = (A_{11} + A_{12}K_+)\mathfrak{B}_+,$$
$$P_-A\mathfrak{B}_+ = (A_{21} + A_{22}K_+)\mathfrak{B}_+.$$

Therefore, setting $\hat{K}_+ = F_A(K_+)$ we obtain:

$$F_A(K_+) = (A_{21} + A_{22}K_+)(A_{11} + A_{12}K_+)^{-1}. \tag{6.1}$$

Thus to each strict plus-operator $A$ with property (4.7) it corresponds the linear fractional transformation $F_A : \mathcal{K}_+ \to \mathcal{K}_+$ given by (6.1). From Lemma 4.2 it follows that $\|A_{11}^{-1}A_{12}\| < 1$. Therefore

$$\begin{aligned} F_A(K_+) &= (A_{21} + A_{22}K_+)\left(I + A_{11}^{-1}A_{12}K_+\right)^{-1} A_{11}^{-1} = \\ &= (A_{21} + A_{22}K_+)\left(\sum_{n=0}^{\infty}(-1)^n \left(A_{11}^{-1}A_{12}K_+\right)^n\right) A_{11}^{-1}, \end{aligned} \tag{6.2}$$

i.e., the mapping $F_A$ is holomorphic on $\mathcal{K}_+$.

From (6.1) it follows that any fixed point $K_+ \in \mathcal{K}_+$ for the transformation $F_A$ is the angle operator for a subspace $\mathcal{L}_+ = (P_+ + K_+)\mathfrak{B}_+ \in \mathfrak{M}_+$, invariant relatively to $A$, and, conversely, if $A\mathcal{L}_+ \subset \mathcal{L}_+$ for some $\mathcal{L}_+ \in \mathfrak{M}_+$ then the angle operator of the subspace $\mathcal{L}_+$ is a fixed point for $F_A$. Hence, based on Corollary VI.2.1, we obtain:

**THEOREM 6.1.** *Let $A$ be a focusing plus-operator with property (4.7). Then $A$ has in $\mathfrak{M}_+$ a unique invariant subspace $\mathcal{L}_+^0 \in \mathfrak{M}_+^0$ satisfying $A\mathcal{L}_+^0 = \mathcal{L}_+^0$, and there exists a number $r < 1$ such that $|\lambda| \leqslant r|\mu|$ for all $\mu \in \sigma(A \mid \mathcal{L}_+^0)$ and $\lambda \in \sigma(A) \setminus \sigma(A \mid \mathcal{L}_+^0)$.*

◁ Consider the transformation $F_A$ given by (6.1). From Definition 4.3 of a focusing plus-operator, it follows that

$$\|F_A(K_+)\| \leqslant r,$$

where $r = 1 - \dfrac{\gamma}{\delta(A)}$, $\delta(A)$ is the number appearing in Lemma 3.3, and $\gamma \leqslant \min\{\delta(A), \gamma(A)\}$. Thus $0 \leqslant r < 1$. By Corollary VI.2.1, the transformation $F_A$ has an s-fixed point $K_+^0 \in \mathcal{K}_+^0$. Therefore $\mathcal{L}_+^0 = (P_+ + K_+^0)\mathfrak{B}_+$ ($\in \mathfrak{M}_+^0$) is the unique invariant subspace of the operator $A$ in $\mathfrak{M}_+$.

The result on the spectrum of the operator $A$ was established by A. V. Sobolev and V. A. Khatskevich in [1]. Their proof uses other methods than those in the present book. ▶

According to Corollary 3.2 we have:

**COROLLARY 6.1.** *Let $\mathfrak{B}$ be a $J_p$-space, such that $\mathfrak{B}_-$ is the dual space of a normed space $\mathfrak{X}$, and let $A$ be a strict plus-operator on $\mathfrak{B}$ satisfying* (4.7), *such that its "corner" $A_{12}$ is completely continuous.*

*Then $A$ has an invariant subspace $\mathcal{L}_+ \in \mathfrak{M}_+$, such that $A\mathcal{L}_+ = \mathcal{L}_+$ and*

$$|\lambda| \leqslant |\mu| \quad \textit{for all } \mu = \sigma(A \mid \mathcal{L}_+),\ \lambda = \sigma(A) \setminus \sigma(A \mid \mathcal{L}_+). \tag{6.3}$$

◁ Let $\Phi_A(z, K_+) = zF_A(K_+)$, where $z \in \overline{\Delta}$ ($\Delta$ is the unit disk in the complex plane $\mathbb{C}$) and $K_+ \in \mathcal{K}_+$. Obviously, $\Phi_A(z, \mathcal{K}_+) \subseteq \mathcal{K}_+$ for any $z \in \overline{\Delta}$. If $z \in \Delta$ then $\Phi_A(z, \cdot)$ satisfies all the conditions in Corollary VI.2.1. Therefore there exists a unique fixed point $K_+^z$ in $\mathcal{K}_+$, i.e., $\Phi_A(z, K_+^z) = K_+^z$.

By Alaoglu-Bourbaki Theorem and by Theorem 0.2.19, the property of $\mathfrak{B}_-$ to be the dual space of a normed space implies that the set $\mathcal{K}_+$ is compact in the ultra-weak operator topology. The complete continuity of the operator $A_{12}$ implies — by Theorem 0.4.15 — the convergence in the strong operator topology of the sequence $\left(I + A_{11}^{-1}A_{12}K_+^n\right)^{-1}$ (see (6.2)) for any sequence $\{K_+^n \ : \ n \in \mathbb{N}\} \subset \mathcal{K}_+$ which is convergent in the ultra-weak operator topology. Therefore $\Phi_A(z, \cdot)$ is continuous in the ultra-weak operator topology.

Let $\{z_n \ : \ n \in \mathbb{N}\}$ be a sequence of complex numbers which converges to $z = 1$ and such that the sequence $\left\{K_+^{z_n} \ : \ n \in \mathbb{N}\right\}$ is convergent in the ultra-weak operator topology to $K_+ \in \mathcal{K}_+$. Then $\Phi_A\left(z_n, K_+^{z_n}\right) = K_+^{z_n} \to K_+$ for $n \to \infty$. But $\Phi_A\left(z_n, K_+^{z_n}\right) \to F_A(K_+)$, whence $K_+ = F_A(K_+)$.

By Theorem 6.1, for any $n \in \mathbb{N}$ there exists $r_n < 1$ such that $|\lambda| \leqslant r_n|\mu|$ for all $\mu \in \sigma(A_n \mid \mathcal{L}_n)$, $\lambda \in \sigma(A_n) \setminus \sigma(A_n \mid \mathcal{L}_n)$, where $\mathcal{L}_n = (P_+ + K_+^{z_n})\mathfrak{B}_+$ and $A_n = P_+A + z_nP_-$. Hence, taking the limit for $n \to \infty$, we obtain (6.3). ▶

**DEFINITION 6.1.** We say that a subspace $\mathcal{L}_\pm \subset \mathfrak{K}_\pm$ belongs to the class $\mathcal{T}_\pm$ if there exists $\mathcal{L}_\pm^{\max} \in \mathfrak{M}_\pm$ such that $\mathcal{L}_\pm \subseteq \mathcal{L}_\pm^{\max}$.

**COROLLARY 6.2.** *If an operator $A$ satisfies all the conditions in Corollary 6.1, then any subspace $\mathcal{L} \in \mathcal{T}_+$, such that $A\mathcal{L} = \mathcal{L}$, is contained in some subspace $\hat{\mathcal{L}}_+ \in \mathfrak{M}_+$ which is invariant for $A$.*

◁ Let us denote by $\hat{\mathcal{K}}_+$ the set of all those $K_+ \in \mathcal{K}_+$ for which $\mathcal{L} \subset (P_+ + +K_+)\mathfrak{B}_+$. Under our assumptions we have $\hat{\mathcal{K}}_+ \neq \emptyset$. We can easily prove that $\hat{\mathcal{K}}_+$ is convex and closed in the weak operator topology. From $\mathcal{L} = A\mathcal{L}$ it follows that $F_A(\hat{\mathcal{K}}_+) \subset \hat{\mathcal{K}}_+$. Repeating — for the set $\hat{\mathcal{K}}_+$ — the corresponding arguments in the proof of Theorem 6.1 and Corollary 6.1, we obtain the existence in $\hat{\mathcal{K}}_+$ of a fixed point $\hat{K}_+$ for the transformation $F_A$. Then $\hat{\mathcal{L}}_+ = (P_+ + \hat{K}_+)\mathfrak{B}_+$ is the desired invariant for $A$ subspace which extends $\mathcal{L}$. ▶

**DEFINITION 6.2.** A pair $(\mathcal{L}_+, \mathcal{L}_-)$ where $\mathcal{L}_\pm \subset \mathfrak{K}_\pm$ is called a *dual pair* (d.p.). If $\mathcal{L}_\pm \in \mathfrak{M}_\pm$ then $(\mathcal{L}_+, \mathcal{L}_-)$ is called a *maximal dual pair* (m.d.p.).

A dual pair is said to be *invariant* for an operator $A$ if $A\mathcal{L}_\pm \subseteq \mathcal{L}_\pm$.

**THEOREM 6.2.** *Any focusing strict plus-operator $A$ on a reflexive $J_p$-space $\mathfrak{B}$ which satisfies property (4.7) has a unique invariant m.d.p. $(\mathcal{L}_+^0, \mathcal{L}_-^0)$. Moreover, $\mathcal{L}_\pm^0 \in \in \mathfrak{M}_\pm^0$, $A\mathcal{L}_+^0 = \mathcal{L}_+^0$, $\mathcal{L}_+^0 \dot{+} \mathcal{L}_-^0 = \mathfrak{B}$, there exists a number $r < 1$ such that $|\lambda| \leqslant r|\mu|$ for all $\lambda \in \sigma(A \mid \mathcal{L}_-^0)$, $\mu \in \sigma(A \mid \mathcal{L}_+^0)$, and $\sigma(A) = \sigma(A \mid \mathcal{L}_+^0) \cup \sigma(A \mid \mathcal{L}_-^0)$.*

◁ By Theorem 4.4, the operator $A^*$ is a focusing strict plus-operator satisfying condition (4.7). By Theorem 6.1, the operators $A$ and $A^*$ have in $\mathfrak{M}_+$ and $\mathfrak{M}_+^*$ the unique invariant subspaces $\mathcal{L}_+^0 \in \mathfrak{M}_+^0$ and $\mathcal{L}_+^{0*} \in \mathfrak{M}_+^{0*}$, respectively, such that $A\mathcal{L}_+^0 = \mathcal{L}_+^0$ and $A^*\mathcal{L}_+^{0*} = \mathcal{L}_+^{0*}$. Setting $\mathcal{L}_-^0 = {}^\perp\mathcal{L}_+^{0*}$ we obtain $A\mathcal{L}_-^0 \subset \mathcal{L}_-^0$. By Theorem 4.1, we know that $\mathcal{L}_-^0 \in \mathfrak{M}_-^0$. We show now that $\mathcal{L}_+^0 \dot{+} \mathcal{L}_-^0 = \mathfrak{B}$. Let $K_\pm$ be the angle operators of $\mathcal{L}_\pm^0$. We have $P_+ + K_+^0 + P_- + K_-^0 = I + K$, where $K = K_+^0 + K_-^0$ and $\|K\|_p < 1$. By Theorem 0.4.6, the operator $I + K$ is a homeomorphism from $\mathfrak{B}$ onto $\mathfrak{B}$, whence $(I+K)\mathfrak{B} = \mathfrak{B}$, i.e., $\mathcal{L}_+^0 + \mathcal{L}_-^0 = \mathfrak{B}$. From the last equality it follows that $\sigma(A) = \sigma(A \mid \mathcal{L}_+^0) \cup \sigma(A \mid \mathcal{L}_-^0)$. Hence, by Theorem 6.1, the conclusions in Theorem 6.2 concerning the spectrum $\sigma(A)$ are true. ▶

**COROLLARY 6.3.** *Let $A$ be a double strict plus-operator on the reflexive $J_p$-space $\mathfrak{B}$, such that $A_{12}$ and $A_{21}$ are completely continuous. Then any invariant d.p. $(\mathcal{L}_+, \mathcal{L}_-)$ for which $A\mathcal{L}_\pm = \mathcal{L}_\pm$ and $\mathcal{L}_\pm \in \mathcal{T}_\pm$ can be extended to an invariant m.d.p. $(\mathcal{L}_+^{\max}, \mathcal{L}_-^{\max})$.*

◁ By Lemma 4.4, the operator $A$ satisfies condition (4.7). By Corollary 6.2, the subspace $\mathcal{L}_+$ can be extended to an invariant for $A$ subspace $\mathcal{L}_+^{\max} \in \mathfrak{M}_+$. Consider the subset $\tilde{\mathcal{K}}_+^*$ of those $Q_+ \in \mathcal{K}_+^*$ for which $(P_+ + Q_+^*)\mathfrak{B}_- \supset \mathcal{L}_-$. Since $\mathcal{L}_- \in \mathcal{T}_-$, the set $\tilde{\mathcal{K}}_+^*$ is non-void, and from $A\mathcal{L}_- = \mathcal{L}_-$ it follows that $F_{A^*}\left(\tilde{\mathcal{K}}_+^*\right) \subset \subset \tilde{\mathcal{K}}_+^*$. Hence, as in the proof of Corollary 6.2, we obtain the existence of a fixed point $Q_+, i.e., F_{A^*}(Q_+) = Q_+$. Setting $-K_- = Q_+^*$ and $\mathcal{L}_-^{\max} = (P_- + K_-)\mathfrak{B}_-$ we get that $A\mathcal{L}_-^{\max} \subset \mathcal{L}_-^{\max}$ and $\mathcal{L}_- \subset \mathcal{L}_-^{\max}$. ▶

To conclude this section we establish a characteristic spectral property of a plus-operator with a focusig power.

**THEOREM 6.3.** *Let $A$ be a strict plus-operator on a reflexive $J_p$-space $\mathfrak{B}$, satisfying property (4.7). Then a power $A^m$ of $A$ is a focusing operator if and only if there exists a m.d.p. $(\mathcal{L}_+, \mathcal{L}_-)$ like the one in Theorem 6.2.*

◁ Necessity. Let $A^m$ be a focusing operator for some $m \geqslant 1$. It is easy to see that $F_{A^m} = F_A^m$. By Theorem VI.2.2, the transformation $F_A$ has an s-fixed point $K_+^0 \in \mathcal{K}_+^0$. Therefore the subspace $\mathcal{L}_+^0 = (P_+ + K_+^0)\mathfrak{B}_+$ is invariant for $A$. By Theorem 4.4, the operator $A^{*m}$ is focusing, too. Arguing as above we find an invariant for $A^*$ subspace $\mathcal{L}_+^{0*} \in \mathfrak{M}_+^{*0}$. Setting $\mathcal{L}_-^0 = {}^{\perp}\mathcal{L}_+^{0*}$ we obtain the invariant m.d.p. $(\mathcal{L}_+^0, \mathcal{L}_-^0)$. By Theorem 6.2, the spectrum $\sigma(A)$ satisfies all the mentioned conditions.

Sufficiency. Assume the existence of an invariant m.d.p. $(\mathcal{L}_+^0, \mathcal{L}_-^0)$ such that $\mathcal{L}_+^0 \in \mathfrak{M}_+^0$ and $|\lambda| \leqslant r|\mu|$ for all $\lambda \in \sigma(A \mid \mathcal{L}_-^0)$, $\mu \in \sigma(A \mid \mathcal{L}_+^0)$, where $0 \leqslant r < 1$. Then $\mathfrak{B} = \mathcal{L}_+^0 \dot{+} \mathcal{L}_-^0$ and therefore any vector $x \in \mathfrak{B}$ can be represented as $x = y + z$, where $y \in \mathcal{L}_+^0$, $z \in \mathcal{L}_-^0$. Put

$$l_n(x) = \frac{\|P_- A^n x\|}{\|P_+ A^n x\|}, \quad 0 \neq x \in \mathfrak{K}_+.$$

We have

$$l_n(x) \leqslant \frac{\|P_- A^n y\| + \|P_- A^n z\|}{\|P_+ A^n y\| - \|P_+ A^n z\|} \leqslant \frac{\dfrac{\|P_- A^n y\|}{\|P_+ A^n y\|} + 2\dfrac{\|P_-\| \, \|A^n z\|}{\|A^n y\|}}{1 - 2\dfrac{\|P_+\| \, \|A^n z\|}{\|A^n y\|}}.$$

By Theorem IV.2.2, we find $m_1 \in \mathbb{N}$ such that

$$\left\| \left(A^{-m_1} \mid \mathcal{L}_+^0\right) \left(A^{m_1} \mid \mathcal{L}_-^0\right) \right\| \leqslant r_1,$$

where $r \leqslant r_1 < 1$. Therefore

$$\frac{\sup\limits_{\substack{z \in \mathcal{L}^0_-,\\ \|z\|=1}} \|A^n z\|}{\inf\limits_{\substack{y \in \mathcal{L}^0_-,\\ \|y\|=1}} \|A^n y\|} \to 0, \quad \text{for } n \to \infty.$$

Hence there exists a number $m \geqslant m_1$ such that $l_m(x) \leqslant \gamma < 1$, i.e., the operator $A^m$ is focusing. ▶

# References


ABD-ALLA, M.

[1] L'ensemble des pointes fixes d'une application holomorphe dans un produit fini de boules-unité d'espaces de Hilbert est une sous-variété banachique complexe, *Ann. Mat. Pura Appl. (4)*, **153**(1988), 63–76.

ABST, D.

[1] On injective holomorphic Fredholm mappings of index 0 in complex Banach spaces, *Comment. Math. Univ. Carolin.*, **21:3**(1980), 513–525.

AHMEDOV, K. T.

[1] The analytic method of Nekrasov-Nazarov in non-linear analysis (Russian), *Uspehi Mat. Nauk,* **12:4**(1957), 135–153.

AIZENBERG, L. A.; YUZHAKOV, A. P.

[1] *Integral representations and residues in multidimensional complex analysis* (Russian), Nauka, Novosibirsk, 1979.

ANDO, T.

[1] *Linear operators in Krein spaces*, Sapporo, Japan, 1979.

AZIZOV, T. YA.; IOHVIDOV, I. S.

[1] *Foundations of the theory of linear operators in spaces with indefinite metric* (Russian), Nauka, Moscow, 1986.

BESSAGE, C.

[1] On the converse of the Banach fixed point principle, *Colloq. Math.*, **7:1**(1959), 41–43.

BRODSKII, M. L.

[1] On properties of an operator mapping the non-negative part of a space with indefinite metric into itself (Russian), *Uspehi Mat. Nauk,* **14:1**(1959), 147–152.

CARTAN, H.

[1] *Differential calculus. Differential forms* (Russian transl.), Mir, Moscow, 1971.

DUNFORD, N.

[1] Uniformity in linear spaces, *Trans. Amer. Math. Soc.*, **44**(1938), 304–356.

DUNFORD, N.; SCHWARTZ, J.

[1] *Linear operators. I,* Interscience, New York, 1958.

EARLY, C.; HAMILTON, R.

[1] A fixed point theorem for holomorphic mappings, *Proc. Sympos. Pure Math.*, **16**(1970), 61–65.

EDELSTEIN, M.

[1] An extension of Banach contraction principle, *Proc. Amer. Math. Soc.*, **12**(1961), 7–10.

GELFAND, I. M.; RAIKOV, D. A.; SHILOV, G. E.

[1] *Commutative normed rings* (Russian), Fizmatgiz, Moscow, 1960.

GELMAN, A. E.

[1] Theorems on implicit abstract functions (Russian), *Dokl. Akad. Nauk SSSR,* **132:3**(1960), 501–503.

GOEBEL, K.; REICH, S.

[1] *Uniform convexity, hyperbolic geometry and nonexpansive mappings*, in Pure and Appl. Math. Series of Monographs and Textbooks, Marcel Dekker Inc.,1984.

GOHBERG, I. C.; KREIN, M. G.

[1] Fundamental aspects on defect numbers, root numbers and indexes of linear operators, *Uspehi Mat. Nauk,* **12:2**(1957), 43–48.

GOLUZIN, G. M.

[1] *Geometric theory of functions of a complex variable* (Russian), Nauka, Moscow, 1966.

HARRIS, L.

[1] Schwarz's lemma in normed linear spaces, *Prod. Nat. Acad. Sci. USA,* **62**(1969), 1014–1017.

[2] A continuous form of Schwarz's lemma in normed linear spaces, *Pacific J. Math.*, **38**(1971), 635–639.

[3] The numerical range of holomorphic functions in Banach spaces, *Amer. J. Math.*, **93**(1971), 1005–1019.

[4] Schwarz-Pick systems of pseudometrics for domains in normed linear spaces, *Advances in Holomorphy*, North Holland, 1979, pp. 345–406.

HAYDEN, T.; SUFFRIDGE, T.

[1] Fixed points of holomorphic maps in Banach spaces, *Proc. Amer. Math. Soc.*, **60**(1976), 95–105.

[2] Biholomorphic maps in Hilbert spaces have a fixed point, *Pacific J. Math.*, **38**(1971), 419–422.

HELTON, J. W.

[1] Operators unitary in an indefinite metric and linear fractional transformations, *Acta Sci. Math.*, **32**(1971), 261–266.

HERVE, M.

[1] *Several complex variables*, Tata Institute of Fundamental Research, Bombay and Oxford Univ. Press, 1963.

[2] *Analyticity in infinite dimensional spaces*, in Studies in Math., **10**, De Gruyter, 1989.

HILLE, E.; PHILLIPS, R. S.

[1] *Functional analysis and semigroups*, A.M.S. Providence, 1957.

IOHVIDOV, I. S.

[1] On spectra of hermitean and unitary operators on spaces with indefinite metric (Russian), *Dokl. Akad. Nauk SSSR*, **71:2**(1950), 225–228.

[2] Banach spaces with a $J$-metric, $J$-non-negative operators (Russian), *Dokl. Akad. Nauk SSSR*, **169:2**(1966), 259–261.

[3] Banach spaces with a $J$-metric and certain classes of linear operators in these spaces (Russian), *Bul. Acad. Stiince RSS Moldoven.*, **1**(1968), 60–80.

[4] Studies in the theory of linear operators on spaces with a $J$-metric and the theory of Toeplitz forms (Russian), Ph. D. Thesis, Odessa, 1976.

IVANOV, A. A.

[1] Fixed points for mappings in metric spaces (Russian), in *Studies in topology*, **66**(1976), Nauka, Leningrad, pp. 5–102.

KADEC, M. I.; MITYAGIN, B. S.

[1] Complementable subspaces in Banach spaces (Russian), *Uspehi Mat. Nauk*, **28:6** (1973), 77–94.

KANTOROVIC, L. V.; AKILOV, G. P.

[1] *Functional analysis* (Russian), Nauka, Moscow, 1977.

KATO, T.

[1] *Perturbation theory for linear operators,* Springer-Verlag, Berlin—Heidelberg—New York, 1966.

KHATSKEVICH, V. A.

[1] On an application of the contraction principle in the operator theory on spaces with indefinite metric (Russian), *Funct. Anal. i Pril.*, **12:1**(1978), 88–89.

[2] On $J$-semi-unitary operators on Hilbert $J$-spaces (Russian), *Mat. Zametki,* **17:4** (1975), 639–647.

[3] On invariant subspaces for some classes of operators on normed spaces with indefinite metric (Russian), *Mat. Issled.*, **6:3**(1971), 133–147.

KHATSKEVICH, V. A.; SHOIKHET, D. M.

[1] Fixed points for analytic operators on Banach spaces and applications (Russian), *Sib. Mat. J.*, **XXV:1**(1984), 188–200.

KRASNOSELSKII, M. A.

[1] Some problems of non-linear analysis (Russian), *Uspehi Mat. Nauk,* **9:3**(1954), 57–114.

[2] On some new fixed point principles (Russian), *Dokl. Akad. Nauk SSSR,* **208:6** (1973), 1280–1281.

[3] *Topological methods in the theory of non-linear integral equations* (Russian), Gostehizdat, Moscow, 1956.

[4] *Integral operators on spaces of summable functions* (Russian), Nauka, Moscow, 1966.

KRASNOSELSKII, M. A.; SOBOLEV, A. V.

[1] On finite rank cones (Russian), *Dokl. Akad. Nauk SSSR,* **225:6**(1975), 1256–1259.

KRASNOSELSKII, M. A.; VAINIKKO, G. M.; ZABREIKO, P. P.; (et al)

[1] *Approximative solutions of operatorial equations* (Russian), Nauka, Moscow, 1969.

KRASNOSELSKII, M. A.; ZABREIKO, P. P.;

[1] *Geometric methods in nonlinear analysis* (Russian), Nauka, Moscow, 1975.

KREIN, M. G.

[1] On an application of a fixed point principle in the theory of linear mappings in spaces with an indefinite metric (Russian), *Dokl. Akad. Nauk SSSR,* **5:2**(1950), 180–190.

[2] An introduction in the geometry of indefinite $J$-spaces and the theory of operators on these spaces (Russian), in *Second Summer School in Mathematics. I*, Kiev, 1965, pp. 15–92.

[3] On a new application of a fixed point principle in the theory of operators on spaces with indefinite metrics (Russian), *Dokl. Acad. Nauk SSSR,* **154:5**(1964), 1023–1026.

KREIN, M. G.; RUTMAN, M. A.

[1] Linear operators having an invariant cone in a Banach space (Russian), *Uspehi Mat. Nauk,* **3:1**(1948), 3–95.

KREIN, M. G.; SHMULIAN, YU. L.

[1] On plus-operators in spaces with indefinite metric (Russian), *Mat. Issled.,* **1:2** (1966), 172–210.

[2] On functional-linear transformations with operatorial coefficients (Russian), *Mat. Issled.,* **2:3**(1967), 64–96.

LANGER, H.

[1] Invariant subspaces for a class of operators in spaces with indefinite metric, *J. Funct. Anal.,* **19**(1975), 232–241.

[2] On invariant subspaces for linear operators acting in spaces with indefinite metric (Russian), *Dokl. Akad. Nauk SSSR,* **169**(1966), 12–15.

[3] Remarks on invariant subspaces for linear operators in Banach spaces with indefinite metric (Russian), *Mat. Issled.,* **4:1**(1969), 27–31.

LARIONOV, E. A.

[1] Extension of dual subspaces (Russian), *Dokl. Akad. Nauk SSSR,* **176:3**(1967), 515–517.

[2] Extensions of dual subspaces invariant relatively to an algebra (Russian), *Mat. Zametki,* **3:3**(1968), 253–260.

LEBEDEV, N. A.

[1] *Area principle in the theory of single-valued functions* (Russian), Nauka, Moscow, 1975.

LEVIN, A. JU.; LIFSIC, E. A.

[1] On the generalized construction principle of M. A. Krasnoselskii (Russian), in *Problems of the mathematical analysis of compound systems,* **1,** Voronets State University, 1967.

LIAPUNOV, A. M.

[1] On the equilibrium figures, slightly different from ellipsoids, of a homogeneous

liquid mass in a revolving motion, in *Collected Works*, vol. **4**, Izdat. Akad. Nauk SSSR, Moscow, 1959

[2] Sur les figures d'equilibre peudifférentes des ellipsoides d'une masse liquide homogéne donée d'un mouvement de rotation. Premiére partie, in *Zap. Acad. Nauk*, **1**, St. Petersburg, 1906.

MARKUS, A. S.; MACAEV, V. I.

[1] On spectral properties of holomorphic operator-functions in Hilbert spaces (Russian), *Mat. Issled.*, **9:4**(1974), 79–91.

MAZET, P.; VIGUE, J.-P.

[1] Points fixes d'une application holomorphe d'un domain borné dans lui-même, *Acta Math.*, **166**(1991), 1–26.

MEYERS, P. R.

[1] A converse to Banach's contraction theorem, *J. Res. Nat. Bur. Standarts*, **716** (1967), 73–76.

MILNOR, J.

[1] *Critical points of complex hypersurfaces* (Russian transl.), Mir, Moscow, 1971.

2] *Topology from the differentiable viewpoint*, University Press, Virginia, 1965.

NACHBIN, L.R

[1] Recent developments in infinite dimensional holomorphy, *Notas fis. Cent. brasil. pesquisas fis.*, **20:13**(1973), 243–275.

NAIMARK, M. A.

[1] On commuting unitary operators in the space $\Pi_\kappa$, *Dokl. Akad. Nauk SSSR*, **146:6**(1963), 1261–1263.

NAZAROV, N. N.

[1] *Integral equations of the Hammerstein type* (Russian), Trudy SAGU, **33**, Tashkent, 1941.

NEKRASOV, A. I.

[1] On steady waves, *Izv. Ivanobo-Voznesenskovo Polytech. Inst.*, **6**(1922), 155–171.

ODINEC, V. P.

[1] *Minimal projections in Banach spaces. Uniqueness and existence problems and applications* (Russian), WSP, Warsaw, 1985.

OPIAL, Z.

[1] *Lecture notes on nonexpansive and monotone mappings in Banach spaces*, Center for Dinamical Systems, Brown University, Providence, USA, 1967.

ORAVA, J.; HALME, A.

[1] Inversion of generalized power series representations, *J. Math. Appl.*, **45**(1974), 136–141.

PETTIS, B

[1] On integration in vector spaces, *Trans. Amer. Math. Soc.*, **44**(1938), 277-304.

PHILLIPS, R.

[1] The extension of dual subspaces invariant under an algebra, in *Proc. Internat. Sympos. Linear Spaces, Ierusalim, 1960,* Ierusalim Acad. Press, Oxford — London — New York — Paris, 1961, pp. 366–398.

PONTRYAGIN, L. S.

[1] Hermitean operators in spaces with indefinite metric (Russian), *Izv.Akad. Nauk SSSR, Ser. Mat.*, **8**(1944), 243–280.

RIESZ, F.

[1] Some mean ergodic theorems, *J. London Math. Soc.*, **13**(1938), 274–278.

RUDIN, W.

[1] *Function theory in the unit ball of* $\mathbb{C}^n$ (Russian transl.), Mir, Moscow, 1984.

[2] The fixed-point sets of some holomorphic maps, *Bull. Malaysian Math. Soc.*, **1**(1978), 25–28.

[3] *Functional analysis* (Russian transl.), Mir, Moscow, 1975.

RYBIN, P. P.

[1] Analytic study of a perturbed linear integral equation, *Uchenye Zapiskii Kazansk. Univ.*, **117:2**(1957), 75–78.

SENDEROV, V. A.; KHATSKEVICH, V. A.

[1] On normed $J$-spaces and some classes of linear operators in these spaces (Russian), *Mat. Issled.*, **8**(1973), 56–75.

SCHMIDT, E.

[1] Zur Theorie linearen und nichtlinearen Integralgleichungen. 3. Teil: Über die Auflösung der nichtlinearen Integralgleichungen und die Verzweigung iher Losungen, *Math. Ann.*, **65**(1908), 370–399.

SCHWARTZ, L.

[1] *Analysis. II* (Russian transl.), Mir, Moscow, 1972.

SHABAT, B. V.

[1] *Introduction in complex analysis* (Russian), Nauka, Moskow, 1969.

[2] *Introduction in complex analysis (second edition)* (Russian), Nauka, Moscow, 1976.

SHOIKHET, D. M.

[1] Remarks on fixed points for non-expansive analytic operators, in *Complex analysis and mathematical physics,* Krasnoiarsk, 1988, pp. 145–150.

[2] Invariance principle in the theory of fixed points for analytic operators, preprint, Institute of Physics, Siberian Branch, Akademy of Sciences of USSR, no. 33M, Krasnoiarsk, 1986.

[3] Some properties of Fredholm mappings in Banach analytic manifolds, *Integral Equations Operator Theory,* **16**(1993), 430–450.

SHULMAN, V. C.

[1] On the fixed points of fractional-linear transformations (Russian), *Funct. Anal. i Priloj.,* **14:2**(1980), 93–94.

SOBOLEV, A. V.; KHATSKEVICH, V. A.

[1] On definite invariant subspaces and the structure of the spectrum of a focusing plus-operator (Russian), *Funct. Anal. i Priloj.,* **15:1**(1981), 84–85.

SOBOLEV, S. L.

[1] About the motion of a symmetrical top whose cavity is filled with a liquid, *J. Appl. Math. Theoretical Physics,* **3**(1960), 20–55.

SZ.-NAGY, B.; FOIAŞ, C.

[1] *Analyse harmonique des opérateurs de l'espace de Hilbert,* Akadémiai Kiadó, Budapest, 1967.

TRENOGIN, V. A.

[1] *Functional analysis* (Russian), Nauka, Moscow, 1980.

VAINBERG, M. M.

[1] *Variational method and monotone operator method in the theory of non-linear operators* (Russian), Nauka, Moscow, 1972.

VAINBERG, M. M.; AIZENGENDLER, P. G.

[1] Methods of investigation in the theory of the ramification of solutions, in *Itogy nauky, Mat. Anal.,* VINITI, Moscow, 1965, pp. 4–70.

VAINBERG, M. M.; TRENOGIN, V. A.

[1] *Ramification theory of the solution of non-linear equations,* (Russian), Nauka, Moscow, 1969.

VESENTINI, E.

[1] On the subharmonicity of the spectral radius, *Boll. Un. Math. Ital.*, **4**(1968), 427–429.

[2] Complex geodesics, *Comp. Math.*, **44**(1981), 375–394.

[3] Complex geodesics and holomorphic maps, *Sympos. Math.*, **26**(1982), 211–230.

VIGUÉ, J.-P.,

[1] Points fixes d'applications holomorphes dans un domaine borne convexe de $\mathbb{C}^n$, *Trans. Amer. Math. Soc.*, **289**(1985), 345–353.

WITTSTOCK, G.

[1] Über invariante Teilraume zu positiven Transformationen in Raumen mit indefiniter Metrik, *Math. Ann.*, **3**(1972), 167–175.

YOSIDA, K.

[1] Mean ergodic theorem in Banach spaces, *Proc. Imp. Acad. Tokio*, **14**(1938), 292–294.

YUZHAKOV, A. P.

[1] On the application of multiple logarithmic residue for the decomposition of implicit functions in power series (Russian), *Mat. S.*, **97:2**(1975), 177–192.

[2] On estimates of the convergence domain and the remainder of the multiple Lagrange series for systems of implicit functions (Russian), *Sib. Mat. J.*, **21:5**(1980), 176–179.

YUZHAKOV, A. P.; TSIKH, A. K.

[1] On the multiplicity of zeros of systems of holomorphic functions (Russian), Sib. Mat.J., **1:3**(1978), 693–697.

ZABREIKO, P. P.; KRASNOSELSKII, M. A.

[1] On a method for obtaining new fixed point principles (Russian), *Dokl. Akad. Nauk SSSR*, **176:6**(1967), 1233-1235.

ZAIDENBERG, M. G.; KREIN, M. G.; KUCMENT, P. A.; PANKOV, A. A.

[1] Banach bundles and linear operators (Russian), *Uspehi Mat. Nauk*, **30**(1975), 101–157.

# List of Symbols

| | |
|---|---|
| $\delta_{\mathrm{w}} F(x_0, h)$ | the first weak variation of the (nonlinear) operator $F$ at a point $x_0$ |
| $\delta F(x_0, h)$ | the first variation of the (nonlinear) operator $F$ at a point $x_0$ |
| $\mathrm{D}F(x_0, h)$ | the Gâteaux differential of the (nonlinear) operator $F$ at a point $x_0$ |
| $\mathrm{d}F(x_0, h)$ | the Fréchet differential of the (nonlinear) operator $F$ at a point $x_0$ |
| $V_{\mathrm{w}}(x_0, \mathfrak{Y})$ | the class of all (nonlinear) operators acting from a normed space $\mathfrak{X}$ into a normed space $\mathfrak{Y}$ and having first weak variation at a point $x_0$ |
| $V(x_0, \mathfrak{Y})$ | the class of all (nonlinear) operators acting from a normed space $\mathfrak{X}$ into a normed space $\mathfrak{Y}$ and having first variation at a point $x_0$ |
| $(\mathrm{P})\int$ | Pettis integral (weak integral) |
| $\mathcal{H}_\rho(\mathfrak{D}, \mathfrak{Y})$ | the class of all $\rho$-holomorphic (holomorphic with respect to $\rho$-topology) operators $F : \mathfrak{D} \to \mathfrak{Y}$, where $\mathfrak{D}$ is a $\rho$-open set in $\mathfrak{X}$ and $\mathfrak{X}$, $\mathfrak{Y}$ are normed spaces |
| $\mathcal{H}(\mathfrak{D}, \mathfrak{Y})$ | the class of all holomorphic operators $F : \mathfrak{D} \to \mathfrak{Y}$, where $\mathfrak{D}$ is an open set in $\mathfrak{X}$ and $\mathfrak{X}$, $\mathfrak{Y}$ are normed spaces |
| $\sigma(A)$ | the spectrum of a linear operator $A$ |
| $r(A)$ | the spectral radius of a linear operator $A$ |
| $J_\nu(x)$ | an indefinite metric on a Banach space |
| $\mathfrak{K}_+$ ($\mathfrak{K}_-$) | the set of all non-negative (non-positive) vectors in a Banach space with indefinite metric |
| $\mathfrak{K}_{++}$ ($\mathfrak{K}_{--}$) | the set of all positive (negative) vectors in a Banach space with in definite metric |
| $\mathfrak{M}_+$ ($\mathfrak{M}_-$) | the class of all maximal non-negative (non-positive) subspaces of a Banach space with indefinite metric |

# Subject Index

**A**

Aizenberg, L. 202
analytic manifold 233
analytic operator 53
Azizov, T. 261

**B**

Banach principle 134
Banach Steinhaus type theorem 100
bifurcation point 233
Bijection 4

**C**

Cauchy inequalities 87
completely continuous operator 21, 30, 55, 144
converse of Banach principle 136, 173

**D**

defect number 247

**E**

ergodic theorem 131

**F**

focusing operator 252
Fréche differential 36, 213
Fredholm operator 120, 208, 233, 234

**G**

Gâteaux differential 35, 53
*G*-metric 239

**H**

holomorphic function 83
holomorphic convex hull 218
Hausdorff space 12, 14
Hahn-Banach-Suchomlynoff theorem 27
Herve, M. 181, 184
Hölder condition 190

**I**

Implicit Function Theorem 160, 164, 189, 203
invertible operator 169, 197, 198
injection 4

**K**

Krasnoselskii, M. 57, 137, 157, 170, 196, 199
Krein space 240, 259

Krein, M. 240, 253, 259

**L**

Liapunov equations of ramification 227

local chart 212

**M**

mapping 3

maximum principle for spectral radius 111

metric space 13

metrizability 15, 21, 200

Montel property 18, 23, 99

**N**

negative vector 240

neutral vector 240

non-expansive mapping 136, 186, 226

normally solvable operator 118

**O**

operator

- proper 182
- $\mathcal{G}$-analytic 53
- $\delta$-analytic 53
- $\mathcal{F}$-analytic 54
- Hammerstein 42, 45, 149, 157
- Nemytski 42
- $\rho$-holomorphic 202
- Urysohn 41

**P**

positive vector 240

**R**

regular fixed point 182

relativity compactness 12

Rudin, W. 185, 233

**S**

Schmidt equation of ramification 228

Schmidt lemma 120

semi-definite lineal 241

Shabat, B. 200

Schauder principle 137

Schwarz lemma 93

small solution 201

smooth manifold 212

splitable operator 129, 226

Stein manifold 220, 234

strictly convex space 28

submanifold 216

Suffridge, T. 138

**T**

topological

- isomorphism 12
- product 10
- space 5

Trenogin, V. 137, 157, 170, 189

**V**

Vainberg, M. 137, 157

Vesentini, E. 110, 182

**U**

unbinding of an equation 196

**Y**

Yoshida, K. 131

Yuzhakov, A. 200

Titles previously published in the series

# OPERATOR THEORY: ADVANCES AND APPLICATIONS
BIRKHÄUSER VERLAG

1. **H. Bart, I. Gohberg, M.A. Kaashoek:** Minimal Factorization of Matrix and Operator Functions, 1979, (3-7643-1139-8)
2. **C. Apostol, R.G. Douglas, B.Sz.-Nagy, D. Voiculescu, Gr. Arsene** (Eds.): Topics in Modern Operator Theory, 1981, (3-7643-1244-0)
3. **K. Clancey, I. Gohberg:** Factorization of Matrix Functions and Singular Integral Operators, 1981, (3-7643-1297-1)
4. **I. Gohberg** (Ed.): Toeplitz Centennial, 1982, (3-7643-1333-1)
5. **H.G. Kaper, C.G. Lekkerkerker, J. Hejtmanek:** Spectral Methods in Linear Transport Theory, 1982, (3-7643-1372-2)
6. **C. Apostol, R.G. Douglas, B. Sz-Nagy, D. Voiculescu, Gr. Arsene** (Eds.): Invariant Subspaces and Other Topics, 1982, (3-7643-1360-9)
7. **M.G. Krein:** Topics in Differential and Integral Equations and Operator Theory, 1983, (3-7643-1517-2)
8. **I. Gohberg, P. Lancaster, L. Rodman:** Matrices and Indefinite Scalar Products, 1983, (3-7643-1527-X)
9. **H. Baumgärtel, M. Wollenberg:** Mathematical Scattering Theory, 1983, (3-7643-1519-9)
10. **D. Xia:** Spectral Theory of Hyponormal Operators, 1983, (3-7643-1541-5)
11. **C. Apostol, C.M. Pearcy, B. Sz.-Nagy, D. Voiculescu, Gr. Arsene** (Eds.): Dilation Theory, Toeplitz Operators and Other Topics, 1983, (3-7643-1516-4)
12. **H. Dym, I. Gohberg** (Eds.): Topics in Operator Theory Systems and Networks, 1984, (3-7643-1550-4)
13. **G. Heinig, K. Rost:** Algebraic Methods for Toeplitz-like Matrices and Operators, 1984, (3-7643-1643-8)
14. **H. Helson, B. Sz.-Nagy, F.-H. Vasilescu, D.Voiculescu, Gr. Arsene** (Eds.): Spectral Theory of Linear Operators and Related Topics, 1984, (3-7643-1642-X)
15. **H. Baumgärtel:** Analytic Perturbation Theory for Matrices and Operators, 1984, (3-7643-1664-0)
16. **H. König:** Eigenvalue Distribution of Compact Operators, 1986, (3-7643-1755-8)
17. **R.G. Douglas, C.M. Pearcy, B. Sz.-Nagy, F.-H. Vasilescu, D. Voiculescu, Gr. Arsene** (Eds.): Advances in Invariant Subspaces and Other Results of Operator Theory, 1986, (3-7643-1763-9)
18. **I. Gohberg** (Ed.): I. Schur Methods in Operator Theory and Signal Processing, 1986, (3-7643-1776-0)

19. **H. Bart, I. Gohberg, M.A. Kaashoek** (Eds.): Operator Theory and Systems, 1986, (3-7643-1783-3)
20. **D. Amir:** Isometric characterization of Inner Product Spaces, 1986, (3-7643-1774-4)
21. **I. Gohberg, M.A. Kaashoek** (Eds.): Constructive Methods of Wiener-Hopf Factorization, 1986, (3-7643-1826-0)
22. **V.A. Marchenko:** Sturm-Liouville Operators and Applications, 1986, (3-7643-1794-9)
23. **W. Greenberg, C. van der Mee, V. Protopopescu:** Boundary Value Problems in Abstract Kinetic Theory, 1987, (3-7643-1765-5)
24. **H. Helson, B. Sz.-Nagy, F.-H. Vasilescu, D. Voiculescu, Gr. Arsene** (Eds.): Operators in Indefinite Metric Spaces, Scattering Theory and Other Topics, 1987, (3-7643-1843-0)
25. **G.S. Litvinchuk, I.M. Spitkovskii:** Factorization of Measurable Matrix Functions, 1987, (3-7643-1883-X)
26. **N.Y. Krupnik:** Banach Algebras with Symbol and Singular Integral Operators, 1987, (3-7643-1836-8)
27. **A. Bultheel:** Laurent Series and their Pade Approximation, 1987, (3-7643-1940-2)
28. **H. Helson, C.M. Pearcy, F.-H. Vasilescu, D. Voiculescu, Gr. Arsene** (Eds.): Special Classes of Linear Operators and Other Topics, 1988, (3-7643-1970-4)
29. **I. Gohberg** (Ed.): Topics in Operator Theory and Interpolation, 1988, (3-7634-1960-7)
30. **Yu.I. Lyubich:** Introduction to the Theory of Banach Representations of Groups, 1988, (3-7643-2207-1)
31. **E.M. Polishchuk:** Continual Means and Boundary Value Problems in Function Spaces, 1988, (3-7643-2217-9)
32. **I. Gohberg** (Ed.): Topics in Operator Theory. Constantin Apostol Memorial Issue, 1988, (3-7643-2232-2)
33. **I. Gohberg** (Ed.): Topics in Interplation Theory of Rational Matrix-Valued Functions, 1988, (3-7643-2233-0)
34. **I. Gohberg** (Ed.): Orthogonal Matrix-Valued Polynomials and Applications, 1988, (3-7643-2242-X)
35. **I. Gohberg, J.W. Helton, L. Rodman** (Eds.): Contributions to Operator Theory and its Applications, 1988, (3-7643-2221-7)
36. **G.R. Belitskii, Yu.I. Lyubich:** Matrix Norms and their Applications, 1988, (3-7643-2220-9)
37. **K. Schmüdgen:** Unbounded Operator Algebras and Representation Theory, 1990, (3-7643-2321-3)
38. **L. Rodman:** An Introduction to Operator Polynomials, 1989, (3-7643-2324-8)
39. **M. Martin, M. Putinar:** Lectures on Hyponormal Operators, 1989, (3-7643-2329-9)
40. **H. Dym, S. Goldberg, P. Lancaster, M.A. Kaashoek** (Eds.): The Gohberg Anniversary Collection, Volume I, 1989, (3-7643-2307-8)
41. **H. Dym, S. Goldberg, P. Lancaster, M.A. Kaashoek** (Eds.): The Gohberg Anniversary Collection, Volume II, 1989, (3-7643-2308-6)

42. **N.K. Nikolskii** (Ed.): Toeplitz Operators and Spectral Function Theory, 1989, (3-7643-2344-2)
43. **H. Helson, B. Sz.-Nagy, F.-H. Vasilescu, Gr. Arsene** (Eds.): Linear Operators in Function Spaces, 1990, (3-7643-2343-4)
44. **C. Foias, A. Frazho:** The Commutant Lifting Approach to Interpolation Problems, 1990, (3-7643-2461-9)
45. **J.A. Ball, I. Gohberg, L. Rodman:** Interpolation of Rational Matrix Functions, 1990, (3-7643-2476-7)
46. **P. Exner, H. Neidhardt** (Eds.): Order, Disorder and Chaos in Quantum Systems, 1990, (3-7643-2492-9)
47. **I. Gohberg** (Ed.): Extension and Interpolation of Linear Operators and Matrix Functions, 1990, (3-7643-2530-5)
48. **L. de Branges, I. Gohberg, J. Rovnyak** (Eds.): Topics in Operator Theory. Ernst D. Hellinger Memorial Volume, 1990, (3-7643-2532-1)
49. **I. Gohberg, S. Goldberg, M.A. Kaashoek:** Classes of Linear Operators, Volume I, 1990, (3-7643-2531-3)
50. **H. Bart, I. Gohberg, M.A. Kaashoek** (Eds.): Topics in Matrix and Operator Theory, 1991, (3-7643-2570-4)
51. **W. Greenberg, J. Polewczak** (Eds.): Modern Mathematical Methods in Transport Theory, 1991, (3-7643-2571-2)
52. **S. Prössdorf, B. Silbermann:** Numerical Analysis for Integral and Related Operator Equations, 1991, (3-7643-2620-4)
53. **I. Gohberg, N. Krupnik:** One-Dimensional Linear Singular Integral Equations, Volume I, Introduction, 1992, (3-7643-2584-4)
54. **I. Gohberg, N. Krupnik:** One-Dimensional Linear Singular Integral Equations, Volume II, General Theory and Applications, 1992, (3-7643-2796-0)
55. **R.R. Akhmerov, M.I. Kamenskii, A.S. Potapov, A.E. Rodkina, B.N. Sadovskii:** Measures of Noncompactness and Condensing Operators, 1992, (3-7643-2716-2)
56. **I. Gohberg** (Ed.): Time-Variant Systems and Interpolation, 1992, (3-7643-2738-3)
57. **M. Demuth, B. Gramsch, B.W. Schulze** (Eds.): Operator Calculus and Spectral Theory, 1992, (3-7643-2792-8)
58. **I. Gohberg** (Ed.): Continuous and Discrete Fourier Transforms, Extension Problems and Wiener-Hopf Equations, 1992, (ISBN 3-7643-2809-6)
59. **T. Ando, I. Gohberg** (Eds.): Operator Theory and Complex Analysis, 1992, (3-7643-2824-X)
60. **P.A. Kuchment:** Floquet Theory for Partial Differential Equations, 1993, (3-7643-2901-7)
61. **A. Gheondea, D. Timotin, F.-H. Vasilescu** (Eds.): Operator Extensions, Interpolation of Functions and Related Topics, 1993, (3-7643-2902-5)

62. **T. Furuta, I. Gohberg, T. Nakazi** (Eds.): Contributions to Operator Theory and its Applications. The Tsuyoshi Ando Anniversary Volume, 1993, (3-7643-2928-9)
63. **I. Gohberg, S. Goldberg, M.A. Kaashoek:** Classes of Linear Operators, Volume 2, 1993, (3-7643-2944-0)
64. **I. Gohberg** (Ed.): New Aspects in Interpolation and Completion Theories, 1993, (ISBN 3-7643-2948-3)
65. **M.M. Djrbashian:** Harmonic Analysis and Boundary Value Problems in the Complex Domain, 1993, (3-7643-2855-X)
66. **V. Khatskevich, D. Shoiykhet:** Differentiable Operators and Nonlinear Equations, 1993, (3-7643-2929-7)

GPSR Compliance

*The European Union's (EU) General Product Safety Regulation (GPSR) is a set of rules that requires consumer products to be safe and our obligations to ensure this.*

*If you have any concerns about our products, you can contact us on ProductSafety@springernature.com*

In case Publisher is established outside the EU, the EU authorized representative is:

Springer Nature Customer Service Center GmbH
Europaplatz 3
69115 Heidelberg, Germany

**Batch number: 10577884**

Printed by Printforce, the Netherlands